ADVANCES IN MATERIALS SCIENCE RESEARCH

VOLUME 7

ADVANCES IN MATERIALS SCIENCE RESEARCH

Additional books in this series can be found on Nova's website
under the Series tab.

Additional E-books in this series can be found on Nova's website
under the E-book tab.

ADVANCES IN MATERIALS SCIENCE RESEARCH

ADVANCES IN MATERIALS SCIENCE RESEARCH

VOLUME 7

MARYANN C. WYTHERS
EDITOR

Nova Science Publishers, Inc.
New York

Copyright © 2012 by Nova Science Publishers, Inc.

All rights reserved. No part of this book may be reproduced, stored in a retrieval system or transmitted in any form or by any means: electronic, electrostatic, magnetic, tape, mechanical photocopying, recording or otherwise without the written permission of the Publisher.

For permission to use material from this book please contact us:
Telephone 631-231-7269; Fax 631-231-8175
Web Site: http://www.novapublishers.com

NOTICE TO THE READER

The Publisher has taken reasonable care in the preparation of this book, but makes no expressed or implied warranty of any kind and assumes no responsibility for any errors or omissions. No liability is assumed for incidental or consequential damages in connection with or arising out of information contained in this book. The Publisher shall not be liable for any special, consequential, or exemplary damages resulting, in whole or in part, from the readers' use of, or reliance upon, this material. Any parts of this book based on government reports are so indicated and copyright is claimed for those parts to the extent applicable to compilations of such works.

Independent verification should be sought for any data, advice or recommendations contained in this book. In addition, no responsibility is assumed by the publisher for any injury and/or damage to persons or property arising from any methods, products, instructions, ideas or otherwise contained in this publication.

This publication is designed to provide accurate and authoritative information with regard to the subject matter covered herein. It is sold with the clear understanding that the Publisher is not engaged in rendering legal or any other professional services. If legal or any other expert assistance is required, the services of a competent person should be sought. FROM A DECLARATION OF PARTICIPANTS JOINTLY ADOPTED BY A COMMITTEE OF THE AMERICAN BAR ASSOCIATION AND A COMMITTEE OF PUBLISHERS.

Additional color graphics may be available in the e-book version of this book.

Library of Congress Cataloging-in-Publication Data

ISSN: 2159-1997

ISBN: 978-1-61209-821-0

Published by Nova Science Publishers, Inc. † New York

CONTENTS

Preface		vii
Chapter 1	Thermal Conductivity of Ceramics - From Monolithic to Multiphase, from Dense to Porous, from Micro to Nano *Willi Pabst and Jan Hostaša*	1
Chapter 2	Recent Advanced in Tin Dioxide Materials:Developments in Thin Films,Nanowires and Nanorods *Z. W. Chen, Z. Jiao, M. H. Wu, C. H. Shek,C. M. L. Wu and J. K. L. Lai*	113
Chapter 3	Current Methods of Dental Ceramic and Metal Surface Treatment for Bonding *Boonlert Kukiattrakoon*	167
Chapter 4	Nitrocellulose in Propellants: Characteristics and Thermal Properties *Mª Ángeles Fernández de la Ossa, Mercedes Torre and Carmen García-Ruiz*	201
Chapter 5	Studies on Energy Properties, Thermal Behaviors,Combustion Properties and Burning Rate Prediction of Composite Modified Double Base Propellant Containing HNIW *Si-Yu Xu, Feng-Qi Zhao, Jian-Hua Yi, Hong-Xu Gao, and Rong-Zu Hu*	221
Chapter 6	Conjugated Polymers: Propriety and Application in the Development of Electrochromic, Electroluminescent and Photovoltaic Devices *Marcos Roberto de Abreu Alves, Hállen Daniel Rezende Calado, Claudio Luis Donicci and Tulio Matencio*	249
Chapter 7	Polycrystalline Silicon Thin Films on Glass for Photovoltaic Applications *Nicolás Budini, Javier A. Schmidt, Fermín M. Ochoa, Pablo A. Rinaldi, Roberto D. Arce and Román H. Buitrago*	269

Chapter 8	Catalytic Effect of Transition Metal Oxides Nanoparticles in Composite Propellants *José Luis de la Fuente*	**285**
Chapter 9	A Simple Model for Surface Relief in the Ocean Crust *A. L. Volynskii and S. L. Bazhenov*	**297**
Index		**311**

PREFACE

Materials science encompasses four classes of materials, the study of each of which may be considered a separate field: metals, ceramics, polymers and composites. This volume gathers important research from around the globe in this dynamic field including research on the thermal conductivity of ceramics; recent advances in tin dioxide materials; dental ceramic and metal surface treatment for bonding; characteristics and thermal properties of nitrocellulose in propellants; application of electrochromic, electroluminescent and photovoltaic devices and polycrystalline silicon thin films

Chapter 1- Thermal conductivity is the key property for thermal insulation applications. For many of these applications, ceramics are the materials of choice, especially in chemically aggressive environments and at high temperatures. The thermal conductivity of ceramics can be significantly reduced by porosity, whereas the influence of grain size is usually much weaker, except for truly nanocrystalline ceramics. This chapter deals with the thermal conductivity of ceramics and ceramic composites, its temperature dependence and its dependence on composition (in the case of composites), porosity and grain size. Following a brief introduction, it is shown how the effective thermal conductivities of dense polycrystalline materials can be calculated from the components of the conductivity tensor for single crystals. Subsequently, the dependence of thermal conductivity on porosity and composition (phase content) is discussed on the basis of micromechanical bounds, mainly one-point (Wiener) and two-point (Hashin-Shtrikman) bounds, as well as recently developed relations (exponential relation, sigmoidal average), and the use of phase mixture models for estimating the grain size influence on the thermal conductivity is briefly explained. The heat transfer mechanisms occurring in ceramics (conduction, convection, radiation) are concisely summarized, and a detailed survey of thermal conductivity values, based on classical and recent literature data, is given for different classes of ceramics, including the most important representatives of the classes of pure, doped and mixed oxides, silicates and non-oxides (carbides, nitrides, borides). Also the thermal conductivity of inorganic glasses is briefly discussed, and the thermal conductivity of ceramics is compared with other material classes (metals, polymers, liquids, gases). The temperature dependence of thermal conductivity is summarized in the form of empirical relations, and values are tabulated for the most important ceramic phases. Based on an extensive literature search, and reasonable porosity corrections, reliable thermal conductivity values are given for dense (pore-free) materials that can be used as reference values for property estimation and materials design.

Chapter 2- Tin oxide is a unique material of widespread technological applications, particularly in the field of gas sensors, dye-based solar cells, transparent conducting electrodes, and catalyst supports. New assessment strategies for tin dioxide (SnO_2) functional materials are of fundamental importance in the development of micro/nano-devices. However, the as-grown SnO_2 materials typically possess a high density of defects, which would degrade their properties. Therefore, the synthesis of defect-free SnO_2 materials is of great interest. In order to provide guidance for the search of better SnO_2 functional materials with suitable optical and electrical properties, it is necessary to investigate the temperature effects of SnO_2 thin film and nanostructured materials. It was found that the influence of annealing temperature on material properties is especially remarkable. However, some challenges are still remained in the motivation to clarify the intricate aspects of SnO_2 thin film and nanostructured materials as well as the applications. This chapter summarizes the recent new research in our group concerning the microstructures and properties of SnO_2 materials, including thin films, nanowires and nanorods. The present work mainly focuses on the synthesis, characterization and applications of SnO_2 thin films and nanowires by using pulsed laser deposition techniques as well as the SnO_2 nanorods by using micro-emulsion method. It is an interdisciplinary work that integrates the areas of physics, chemistry and materials science. The results may enable novel SnO_2 functional materials with appropriate microstructures to be tailor made for a large number of applications and provide new opportunities for future study of SnO_2 architectures with the goal of optimizing functional material properties for specific applications.

Chapter 3- This chapter focuses on the current methods of dental ceramic and metal surface treatment intraorally for the purpose of repairing or bonding to ceramic restorations. Fractured ceramics in ceramic restorations cause esthetic problems, functional problems, loss of time and economic cost to patients for fabricating a new one. Repairing the fractured restoration is an alternative choice for the patients. However, successful repair requires the bonding of repair materials to appropriately prepared surfaces of ceramics or metals. Ceramic surface treatments can be treated by both mechanical and chemical methods. Mechanical methods include grinding with coarse diamond burs, air abrasion, etching with hydrofluoric acid or acidulated phosphate fluoride gel, and laser irradiation, whilst chemical bonding involves silica coated aluminium oxide, and silane treatment. Similarly, metal surface preparations incorporate mechanical procedures such as grinding with carborundum discs or diamond burs, air abrasion, and chemical etching or electrolytic etching. Chemical bonding to metal has also been treated by tin plating, silica coated aluminium oxide, and primers or silane application. Furthermore, this chapter has also included the method of choice in clinical applications for current ceramic and metal surface treatment

Chapter 4- Nitrocellulose was discovered, by the German-Swiss chemist C.F. Schönbein, in the first half of the nineteenth century but remains to have a great interest today. Nitrocellulose has a similar aspect to cotton (white and fibrous texture). It is a nitrate cellulose ester polymer with β (1→4) bonds between monomers, produced from the nitration of cellulose. Its chemical formula is $[C_6H_7O_2(OH)_{3-x}(ONO_2)_x]_n$, where x indicates the hydroxyl groups exchanged by nitro groups. This macromolecule has different applications depending on their degree of nitration. Nitrocellulose with a low degree of nitration is applied in paints, lacquers, varnishes, inks, etc., while nitrocellulose with a high degree of nitration (>12.5%) is used in explosives. Within the nitrocellulose-containing explosives are included dynamites and propellants. Propellants containing nitrocellulose are smokeless gunpowders, which are widely used by the international military community for propelling projectiles. Depending on gunpowder's composition (active components), they can be classified as: i) single-base gunpowders, which contain mainly nitrocellulose, ii) double-base gunpowders, which contain nitrocellulose and other explosive substance (nitroglycerin, dinitroethylenglycol or dinitrotoluene), and iii) triple-base gunpowders, which are composed by nitrocellulose and two other explosive substances (nitroglycerin or dinitroethylenglycol and nitroguanidine).

Nitrocellulose characteristics, i.e., its high molar mass, complex structure, and unusual chemical behavior, make difficult to perform ordinary studies of this polymer. For these reasons, most works on the characterization of nitrocellulose in propellants have been based on measurements of physico-chemical properties such as molar mass distributions, viscosity, and specific refractive index of nitrocellulose by Size Exclusion Chromatography (SEC) with different types of detectors.

The thermal properties of nitrocellulose are important since it must be burned for its use as propellant. These properties have mainly been studied by determining, using spectrometric techniques (mainly infrared and mass spectrometry), the by-products released during the thermal degradation of nitrocellulose by means of pyrolysis or incineration processes. The most important kinetics parameters of thermal decomposition of nitrocellulose (activation energy, enthalpy, critical explosion and decomposition temperature, etc.) have been studied by typical thermal analytical techniques such as Differential Scanning Calorimetry (DSC) and ThermoGravimetry (TG) or Differential Thermal Analysis (DTA).

The aim of this book chapter is to provide an updated overview (from 1999 until nowadays) of the characteristics and thermal properties of nitrocellulose of high nitrogen content used in propellants.

Chapter 5- A series of composite modified double base propellants containing HNIW (HNIW-CMDB propellant) are prepared. Their properties including energy properties, thermal behaviors, non-isothermal decomposition reaction kinetics, combustion properties and burning rate prediction are investigated. The results show that the energy property parameters of HNIW-CMDB propellant, such as theory specific impulse, characteristic velocity, oxygen coefficient, combustion temperature, heat of explosion at constant volume, heat of combustion and the average relative molecular mass of combustion gas increase linearly with increasing HNIW content. The burning rate of HNIW-CMDB propellants with HNIW content less than 50% increases linearly with increasing pressure between 2MPa and 22MPa, and the effect of HNIW content on burning rate is negligible under the same pressure. The apparent activation energy (Ea), pre-exponential constant (A) and mechanism function of thermal decomposition reaction under both 4 MPa and 7MPa are obtained by non-isothermal method. There is obvious dark zone above the burning surface of HNIW-CMDB propellant when HNIW content is less than 28%. There are some luminous flame-lets from burning surface in the dark zone. There is no obvious dark zone above the burning surface of HNIW-CMDB propellant when HNIW content is more than 28%. The temperature distribution of combustion wave of HNIW-CMDB propellant is obtained under 6MPa. The model for predicting the burning rate of HNIW-CMDB propellant is established based on one-dimension gas phase reaction theory. The burning rates of HNIW-CMDB propellant are calculated and the results are in agreement with the tested data.

Chapter 6- In this chapter, the authors initially introduce the general topic of principle electrochemistry techniques applied in the synthesis and characterization of electrochromic materials. The authors then go on to discuss conjugated polymers and the electronic structure and electrochemical proprieties of these materials. Conjugated polymers have the advantage of being synthesized with a modifiable structure to provide the electronic and mechanical properties that they desire. In this context, they will approach synthesis processes in a third topic, showing how they can control the electronic properties of a conducting polymer by modifying its main structure and side chains. As a fourth and final topic to be discussed in this chapter, the authors describe the application of conjugated polymers in developing electrochromic devices (as well as the desired properties of these materials and usual characterization techniques for electrochromic devices), electroluminescent devices and photovoltaics.

Chapter 7- In this chapter, the authors briefly present the state of the art in some deposition and processing techniques that can be used to obtain thin polycrystalline silicon (poly-Si) films on glass substrates. In particular, they focus on the ultimate advances to date related to the technique of metal induced crystallization of amorphous silicon, which allows to produce large crystalline grains (sizes over 100 μm) at low temperatures (lower than 600°C) by means of thermally activated solid phase transformations. Specifically, the authors discuss in more detail the use of nickel as the metallic inductor of silicon crystallization, and its influence on grain nucleation and growth. They also analyze the relevant concept of the solid phase epitaxial crystallization of an amorphous silicon thin film, deposited onto a previously crystallized seed layer, to produce poly-Si. The effects of doping on crystallization are presented, together with the problems that it entails. Taking in account all of the issues presented, the authors propose a suitable process, scalable to industrial levels, which will allow to produce a competitive large-grained poly-Si solar cell with high efficiency and at low costs.

Chapter 8- The use of powder additives to modify the burning rate of solid propellant and other energetic materials has been the topic of much study for several decades. This chapter is a brief review to describe the effects on the performances of ammonium perchlorate (AP) based composite propellants by mean of employing nano-powder of transition metal oxides (TMOs), as for example iron (III) oxide (Fe_2O_3), cupric oxide (CuO) and copper chromite ($CuCr_2O_4$), as burning rate catalysts in comparison to traditional micro-fillers. The solid composite propellants with microparticles as catalyst seem to be less stable due to oversensitivity to pressure variations, but the nanostructured composite propellants yield high stable burning rates over a broad pressure range. In addition, the incorporation of these nanoparticles in the formulations of these energetic materials can also improve their combustion and thermal properties, according to the characterization obtained by thermal analysis, with techniques such as differential scanning calorimetry (DSC) and thermogravimetry analysis (TGA). These results indicate the excellent benefits found in using these nanoparticles as an additive for solid rocket propulsion applications.

Chapter 9- The method for modeling of the development of the surface relief near the mid-ocean ridges is developed. Structural and mechanical behavior of polymer films with a thin rigid coating is analyzed. The behavior of such systems under applied tensile and compressive stress is accompanied by the formation of a regular wavy surface relief and by regular fragmentation of the coating. The above phenomena are universal. Both phenomena (stress-induced development of a regular wavy surface relief and regular fragmentation of the coating) are provided by the specific features of mechanical stress transfer from a compliant soft support to a rigid thin coating. Structural and mechanical properties of the Earth are modeled by a soft polymer substrate coated with a thin rigid film. This method models the development of a system of folds and transform faults near the mid-ocean ridges was modeled. This approach allows estimate the effective thickness of the ocean crust.

In: Advances in Materials Science Research, Volume 7　　ISBN 978-1-61209-821-0
Editor: Maryann C. Wythers　　© 2012 Nova Science Publishers, Inc.

Chapter 1

THERMAL CONDUCTIVITY OF CERAMICS – FROM MONOLITHIC TO MULTIPHASE, FROM DENSE TO POROUS, FROM MICRO TO NANO

Willi Pabst and Jan Hostaša

Department of Glass and Ceramics
Institute of Chemical Technology, Prague (ICT Prague)
Technická 5, 166 28 Prague, Czech Republic

ABSTRACT

Thermal conductivity is the key property for thermal insulation applications. For many of these applications, ceramics are the materials of choice, especially in chemically aggressive environments and at high temperatures. The thermal conductivity of ceramics can be significantly reduced by porosity, whereas the influence of grain size is usually much weaker, except for truly nanocrystalline ceramics. This chapter deals with the thermal conductivity of ceramics and ceramic composites, its temperature dependence and its dependence on composition (in the case of composites), porosity and grain size. Following a brief introduction, it is shown how the effective thermal conductivities of dense polycrystalline materials can be calculated from the components of the conductivity tensor for single crystals. Subsequently, the dependence of thermal conductivity on porosity and composition (phase content) is discussed on the basis of micromechanical bounds, mainly one-point (Wiener) and two-point (Hashin-Shtrikman) bounds, as well as recently developed relations (exponential relation, sigmoidal average), and the use of phase mixture models for estimating the grain size influence on the thermal conductivity is briefly explained. The heat transfer mechanisms occurring in ceramics (conduction, convection, radiation) are concisely summarized, and a detailed survey of thermal conductivity values, based on classical and recent literature data, is given for different classes of ceramics, including the most important representatives of the classes of pure, doped and mixed oxides, silicates and non-oxides (carbides, nitrides, borides). Also the thermal conductivity of inorganic glasses is briefly discussed, and the thermal conductivity of ceramics is compared with other material classes (metals, polymers, liquids, gases). The temperature dependence of thermal conductivity is summarized in the form of empirical relations, and values are tabulated for the most important ceramic phases. Based on an extensive literature search, and reasonable porosity corrections, reliable thermal conductivity values are given for dense (pore-free) materials that can be used as reference values for property estimation and materials design.

1. INTRODUCTION

Thermal conductivity is the key property for thermally insulating materials (e.g. in the refractories industry) and for other materials designed to fulfil special tasks of thermal management beside other functions (e.g. in the electronics industry). It is also an important parameter for assessing such complex thermomechanical modes of behavior as thermal shock resistance, and a thorough knowledge of its temperature dependence is indispensible for the performance and construction of high- and low-temperature equipment. Many cutting edge technologies (e.g. the development of thermonuclear fusion reactors) are dependent on reliable information concerning the thermal conductivity of various materials. Unfortunately, the situation is rather involved, because information concerning the thermal conductivity of different materials is mostly incomplete (due to lacking or insufficient compositional and microstructural information) and often contradictory, not only in different authors' papers, but also in textbooks, reference handbooks and data bases. But the situation is even worse: current journal publications as well as the common textbook literature are not getting tired in repeating apparently general formulae for assessing the dependence of the effective thermal conductivity on volume fractions (of multiphase materials or composites) and porosity, which are applicable only in special situations under quite restrictive assumptions. Many of these formulae are more than 50 years old and the non-specialist in the field as well as the practicing materials scientist and engineer must get the impression that nothing new has been added to this traditional canon during the last five decades. On the other hand, the grain size influence on thermal conductivity is less thoroughly investigated, and its assessment usually relies on a few simplistic models, mostly based on the concept of interfacial resistances (Kapitza resistances), also more than half a century old.

Thermal conductivity values of different materials cover a range between approx. 0.001 W/mK (for micro- or mesoporous materials and aerogels at or around room temperature) to more than 10000 W/mK (for metals, some modifications of carbon and some ceramics at cryogenic temperatures), i.e. more than eight orders of magnitude. Even if we confine our interest to ceramics at room temperature and exclude the extreme cases of carbon (e.g. diamond and graphite) and micro- or mesoporous materials, the range of possible thermal conductivity values is still between 0.025 W/mK (for air filling the pores) and approx. 250-500 W/mK (for AlN, BeO and SiC), i.e. at least four orders of magnitude. Unfortunately, although the vast majority of solid phases making up common ceramic microstructures has thermal conductivities in the relatively narrow range 1-100 W/mK, it is not possible to reliably estimate the thermal conductivity of the ceramic material based on the knowledge of the composition in terms of solid phases (i.e. their volume fractions) alone, because porosity is the main factor influencing the thermal conductivity.

Indeed, the effect of porosity can override all other effects (composition and temperature), whereas grain size effects, which may be expected to play a role in nanocrystalline ceramics, are always of minor importance compared to porosity (at least as long as the interfaces are perfect, i.e. without micro- or nanocracks). Of course, in order to take the effect of porosity into due account, the porosity must be known for the sample in question. This is not always the case in journal papers and often ignored in textbooks, reference handbooks and databases. But even a precise knowledge of the porosity, albeit indispensable, usually does not solve the problem completely, because a universal model for the porosity dependence does not exist (the reason is that porosity is but one of many other microstructural descriptors of porous media).

Therefore it is not surprising that even for the experienced materials scientist and engineer it is sometimes not easy to extract reliable information on thermal conductivity from literature data. However, such information is urgently needed for design purposes and for predicting (estimating) the performance of materials in applications.

The present treatise is meant to serve as a practical guide to the thermal conductivity of ceramics, mainly for materials scientists and engineers. It consists of ten sections, which correspond roughly to four parts: first a general part (Sections 1-4), in which universally acknowledged as well as new concepts are introduced and widely valid results are presented in the form of micromechanical bounds or model relations, second a supplementary part on the grain size dependence of thermal conductivity (Section 5), third a part in which heat transfer is discussed from an atomistic point of view (according to the mainstream textbook tradition) and conduction is contrasted to convection and radiation (Section 6), and fourth a special part Sections 7-10), in which the thermal conductivity of ceramics and their constituent phases is presented, discussed and compared (mutually and with other material classes, such as glasses, metals, polymers, liquids and gases). Based on this guide the reader should be able, apart from realistically assessing the influence of convection or radiation contribution, to quantitatively estimate the thermal conductivity of polycrystalline multiphase ceramics, including porosity and grain size effects. It has been attempted to adhere to this basic division as strictly as is adequate for a more or less rigorous treatment, but not to slavery.

The treatment is principally – in the first two parts (Sections 1-5) strictly – based on continuum theory (rational thermomechanics) and micromechanics (theory of composites), but not without allowing for brief excurses to statistical physics (kinetic theory) and basic quantum theory, mainly in the third part (Section 6), wherever these were deemed useful to remind the reader of traditional (solid state physics and materials science) textbook conceptions and models – used mainly to explain certain features of individual material behavior (using the phonon concept) or the influence of radiation (photons). We emphasize, however, that while the first, general, part contains a rigorous, quantitative treatment (Sections 1-4), the explanations in the second, supplementary part are mostly of a qualitative *ad hoc* character, and their apparent plausibility (suggested by textbook tradition) may be deceptive in some cases. In the first two parts (Sections 1-5) all assumptions are explicitly stated, whereas in the third, supplementary part (Sections 6) only special results are presented and the fourth, special part (Sections 7-10) bears the character of a commented reference handbook for special materials.

The impetus for this treatise has been twofold. In the first place, it is based on the conviction that many rigorous models and relations – derived during the last 50 years or so (not to speak of the more recent ones) – are well known in micromechanics (theory of composites), but have still not found their way into textbooks of materials science and technology, as if the authors of these textbooks were reluctant to read micromechanics textbooks in fear of discovering that parts of their tradition are actually wrong. Second, it has grown out of the authors' needs and ambition to find reliable thermal conductivity data in the literature and their finding that there is an extremely high degree of confusion concerning the latter in today's journal papers, handbooks and data bases. Indeed, the degree of confusion is so high that it is possible to defend almost any measured conductivity value in a peer-reviewed paper by invoking complicated and often absurd arguments and by recourse to the influence of hidden (and of course accidentally uncharacterized) variables. In this sense this treatise is intended to provide the necessary tools for estimating the influence of composition and microstructure (porosity and grain size) on the effective thermal conductivity, when the thermal conductivity of the constituent phases is known. In order to achieve the latter, a critical assessment of literature data is part of this treatise. Of course, in contrast to the first part – which may be considered in a sense complete, ultimate and timeless – it has to be emphasized that mainly the fourth part of this treatise (Sections 7-10) reflects only the attempt of a fair shortcut through the current state of knowledge and available data. Experimental data collected from various literature sources are critically compared. However, the reliability of reported experimental data is generally dependent on the equipment available at the time and place of the measurement, the materials' purity, the availability of microstructural data for the material and, last but not least, the skill of the operator who has measured and the credibility of the author who has published the results. Therefore – and also with regard to the fact that a data collection of this kind is necessary incomplete – we admit that the final decision on recommended thermal conductivity values has without doubt a subjective bias, and we acknowlegde that the opinion of other authors might be different in certain respects. Nevertheless, we have done their best to arrive at unprejudiced conclusions that might be of some value for theoretical materials scientists as well as practicing engineers.

The content of this treatise is organized as follows: In Section 2 we introduce thermal conductivity as a tensorial material property from the viewpoint of rational thermomechanics (Fourier's law) and present expressions for the thermal conductivity of single crystals and dense polycrystalline materials, in Section 3 we give the micromechanical bounds for the effective thermal conductivity of multiphase materials (composites) in general (including the special case of two-phase materials and porous media), in Section 4 we discuss the most popular predictive model relations for the effective thermal conductivity of multiphase materials (with special emphasis on porous media), in Section 5 we present the so-called phase mixture model and its application to estimate the grain size dependence of thermal conductivity, and in Section 6 we summarize atomistic and quantum concepts of heat transfer (lattice conduction and phonons) and some traditional concepts for the description of heat transfer mechanisms beyond true conduction (in particular, radiative heat transfer). In the last four sections thermal conductivity data are presented for oxides (Section 7), silicates and glasses (Section 8) and non-oxides (Section 8), followed by a comparative section on other material classes, in particular metals and polymers (Section 10). SI units have been used throughout this treatise, but both the Celsius temperature (°C) and the (absolute, thermodynamic) Kelvin temperature (K), have been used, depending on which is more appropriate in a given context. All thermal conductivites are given in (W/mK).

2. THERMAL CONDUCTIVITY AS A MATERIAL PROPERTY OF SINGLE CRYSTALS AND POLYCRYSTALLINE MATERIALS

From the viewpoint of continuum theory (rational thermomechanics) material properties are coefficients of linear constitutive equations (Pabst 2005a). The constitutive equation for the heat flux vector **q** (Fourier's law) for anisotropic solids of arbitrary symmetry is

$$\mathbf{q} = -\mathbf{K}\,\mathrm{grad}\,T, \qquad (1)$$

where $\mathrm{grad}\,T$ is the temperature gradient and **K** is the heat conductivity or thermal conductivity tensor. It is a second-order tensor, usually assumed to be symmetric. Generally, with respect to second-order tensor properties not only amorphous materials but also cubic single crystals exhibit isotropic behavior (Nye 1985). Assuming **K** to be symmetric amounts to acknowledging the so-called Onsager relations of linear irreversible thermodynamics (Hellwege 1988). Unfortunately, the Onsager relations have turned out not to be valid in general (Truesdell 1984) and thus the assumed symmetry is indeed not guaranteed for all crystal classes (point groups of symmetry) (Pabst 2005a, Wang 1984). This result, however, is not very known in the scientific community and therefore it is common practice to assume symmetry in general and the consequences of its asymmetry in certain crystal classes are not well investigated so far. We will implicitly follow this practice in the present treatment, being aware of the fact that in a more rigorous treatment (e.g. when specializing more on single crystals) it might be useful to adopt the alternative standpoint. We note in passing that the whole theory of composites (micromechanics) assumes symmetry of the thermal conductivity tensor, cf. e.g. (Torquato 2002).

One of the reasons why the scientific community is so reluctant to acknowledge the asymmetry of the thermal conductivity tensor may be the fact that symmetric second-order tensors can be transformed to principal axes (with real eigenvalues and eigenvectors collinear with the crystallographic axes), while asymmetric ones cannot. This, of course, is a considerable simplification of further analysis, since in this case the maximum number of independent tensor components is three (for triclinic, monoclinic and orthorhombic single crystals and for orthotropic composites), i.e.

$$\mathbf{K} = \begin{pmatrix} k_a & 0 & 0 \\ 0 & k_b & 0 \\ 0 & 0 & k_c \end{pmatrix}, \qquad (2)$$

and for materials of higher symmetry the number of independent components is further reduced. Obviously all tabulated data, e.g. in handbooks and monographs on thermophysical properties (Grimvall 1999) and even crystal physics (Chojnacki 1979, Haussühl 1983, Nye 1985, Voigt 1928), implicitly assume the thermal conductivity tensor to be symmetric as well (otherwise the maximum number of thermal conductivity components would not be restricted to the three principal conductivities).

For all materials with an axis of rotational symmetry (i.e. trigonal, tetragonal and hexagonal monocrystals, also called uniaxial crystals, as well as transversely isotropic composites) there are two independent components (Grimvall 1999, Nye 1985), a parallel component in the direction of the rotational axis $k_\parallel$ and a perpendicular component $k_\perp$ (in all directions perpendicular to the axis), i.e.

$$\mathbf{K} = \begin{pmatrix} k_\perp & 0 & 0 \\ 0 & k_\perp & 0 \\ 0 & 0 & k_\parallel \end{pmatrix}, \tag{3}$$

while for cubic single crystals and isotropic materials, including macroscopically isotropic (also called quasi-isotropic (Grimvall 1999) or statistically isotropic (Torquato 2002)) composites, there is only one (k).

Thus, for isotropic materials, including macroscopically isotropic composites, the thermal conductivity tensor $\mathbf{K}$ (like all other second-order tensors) reduces to an isotropic tensor, determined by only one scalar component, i.e.

$$\mathbf{K} = k\,\mathbf{1}, \tag{4}$$

where $\mathbf{1}$ is the (second-order) unit tensor and k the thermal conductivity (heat conductivity), cf. (Pabst and Gregorová 2007). Obviously, the constitutive equation for the heat flux in isotropic materials (Fourier's law) is

$$\mathbf{q} = -k\,\mathrm{grad}\,T. \tag{5}$$

Fourier's law in this form is valid for amorphous materials (e.g. glasses), cubic crystals and, in an averaged sense (precisely defined in homogenization theory and micromechanics (Torquato 2002)), for polycrystalline or even multiphase materials (in other words, heterogenous materials) that are macroscopically (statistically) isotropic. In the latter case the averaged material property k is called effective thermal conductivity. In order to simplify notation, the symbol k – without subscript – is used in this treatise to denote both the thermal conductivity of glasses and cubic crystals and the effective thermal conductivity of statistically isotropic heterogeneous materials.

The effective thermal conductivity of dense, single-phase, polycrystalline materials can be calculated from those of the individual single crystals (in this context often called crystallites or "grains") by employing a homogenization procedure similar to the Voigt- and Reuss-type schemes applied for the elastic moduli (Voigt 1889, Reuss 1929, Hill 1952), cf. also (Pabst and Gregorová 2004a, 2006a, Pabst et al. 2004). This amounts to considering the single-phase polycrystalline material as statistically isotropic due to the assumed random orientation of crystallites. Only under this assumption is it possible to proceed in the standard micromechanical treatment of multiphase polycrystalline materials (composites), where the polycrystalline material itself is considered as a homogenized matrix or skeleton phase. It must be emphasized, that of course cases may occur where this assumption does not hold, e.g. when crystallites exhibit preferential orientation. Such cases (*viz.* anisotropic multiphase microstructures) have to be treated according to the theory of fiber or sandwich composites and are beyond the scope of the present treatise.

When the thermal conductivity components for a type of single crystals are k_a, k_b and k_c, and the thermal conductivity tensor is assumed to be symmetric as is usually done (see above), then the effective thermal conductivity of a dense (e.g. pore-free), statistically isotropic polycrystalline material made out of these crystals is bounded from below and above by the harmonic and the arithmetic mean of the single crystal tensor components (principal conductivites, i.e. eigenvalues of the conductivity tensor), respectively, i.e.

$$\left[\frac{1}{3}\left(\frac{1}{k_a} + \frac{1}{k_b} + \frac{1}{k_c}\right)\right]^{-1} \leq k \leq \frac{1}{3}(k_a + k_b + k_c) \tag{6}$$

(Molyneux bounds for polycrystalline materials) (Molyneux 1970), cf. (Grimvall 1999, Torquato 2002). The r.h.s. is the arithmetic mean, the l.h.s. the harmonic mean of the principal conductivities. Note that the Molyneux bounds for statistically isotropic polycrystalline materials are the conductivity counterparts of the Voigt and Reuss bounds for the elastic moduli of polycrystalline materials. Similar to the elastic case, the two bounds are often very close, the most prominent exception being graphite, see Appendix A. The Molyneux upper bound (arithmetic mean) is optimal in the sense that it is realizable or attainable by certain (laminated) microstructures (Grimvall 1999, Torquato 2002). However, no microstructure is known to realize the lower bound (harmonic mean). This is the reason why for thermal conductivity it is common practice to take the Molyneux upper bound itself – and not the (arithmetic) average of the upper and lower bound as in the case of elastic properties (Voigt-Reuss-Hill average (Green 1998)) – as the final value of the effective thermal conductivity of statistically isotropic polycrystalline materials, at least as long as the degree of anisotropy is low enough, see Appendix A.

Suggestions for an improved, i.e. tighter lower bound range from the geometric mean $(k_a k_b k_c)^{1/3}$ (Schulgasser 1976, 1977) to the solution of the cubic equation

$$k_-^3 + (k_a + k_b + k_c)k_-^2 - 4k_a k_b k_c = 0 \tag{7}$$

(Avellaneda et al. 1988, Nesi and Milton 1991), which is for materials consisting of uniaxial crystals with $k_a = k_b = k_\perp$ and $k_c = k_\parallel$

$$k_- = \frac{\left(k_\parallel^2 + 8 k_\perp k_\parallel\right)^{1/2} - k_\parallel}{2} \tag{8}$$

when $k_\parallel > k_\perp$, $k_\parallel$ and $k_\perp$ being interchanged when $k_\parallel < k_\perp$ (Grimvall 1999). The validity of both relations is questionable (see below). Note that the effective conductivity of (single-phase) polycrystalline materials in 2D (e.g. cross-sections of transversely isotropic materials) is

$$k = \sqrt{k_a k_b}, \qquad (9)$$

i.e. the geometric mean of the principal conductivities, irrespective of the details of microstructure, but according to (Torquato 2002) no corresponding equality is known in 3D.

Nevertheless, when – in addition to the assumed overall isotropy of the polycrystalline material – the grains (crystallites) are (assumed to be) isometric (equiaxed), more restrictive bounds are available (Hashin and Shtrikman 1963). In particular, for isotropic materials consisting of isometric crystallites with uniaxial crystal symmetry (i.e. trigonal, tetragonal or hexagonal) these bounds are

$$k_\perp \frac{4k_\perp + 5k_\parallel}{7k_\perp + 2k_\parallel} < k < k_\parallel \frac{7k_\perp + 2k_\parallel}{k_\perp + 8k_\parallel} \qquad (10)$$

(if $k_\parallel > k_\perp$; otherwise the bounds are reversed). These Hashin-Shtrikman bounds for polycrystalline materials are always within the aforementioned Molyneux bounds, cf. Appendix A, but can be in conflict with both the Schulgasser and the Avellaneda lower bound suggestions. Although in practice the difference is usually negligible (except for the case of graphite), this contradiction shows that the definitive form of the lower bound may still be an item of dispute.

As an alternative to the micromechanical bounds a model approach can be invoked. In particular, the effective medium approximation result for statistically isotropic polycrystalline materials made of (not necessarily isometric) crystallites with uniaxial crystal symmetry is

$$k = \frac{1}{4}\left[k_\perp + \left(k_\perp^2 + 8 k_\perp k_\parallel\right)^{1/2}\right] \qquad (11)$$

(Bruggeman 1935, Grimvall 1999), which is the solution of the general effective-medium equation ($i = a, b, c$)

$$\sum_i \frac{k - k_i}{2k + k_i} = 0 \qquad (12)$$

(Helsing and Helte 1991, Grimvall 1999). It is evident from the calculated values in Appendix A that the effective-medium result obeys the Hashin-Shtrikman bounds.

From an atomistic-structural point of view thermal conductivity is usually higher in the directions of strong bounds. This dependence of the thermal conductivity on the crystal structure is responsible for the anisotropy of this material property. In particular, for layered structures and chain structures, the thermal conductivity is higher within the planes and along the chains, respectively (Chojnacki 1979). For uniaxial crystals an anisotropy ratio can be defined as

$$\varepsilon = \frac{k_\parallel}{k_\perp} - 1. \qquad (13)$$

For a statistically isotropic polycrystalline material consisting of weakly anisotropic (i.e. $|\varepsilon| \ll 1$) crystallites one obtains the approximation

$$k \approx k_\perp \left(1 + \frac{\varepsilon}{3} - \frac{4\varepsilon^2}{27}\right) \qquad (14)$$

(Grimvall 1999). For very weak anisotropy, when only linear terms are retained, the linear approximation of this relation,

$$k \approx k_\perp \left(1 + \frac{\varepsilon}{3}\right) = \frac{k_\parallel + 2k_\perp}{3}, \qquad (14)$$

is identical to the Molyneux upper bound as well as the linear approximations of the Hashin-Shtrikman bounds and the effective-medium result (Grimvall 1999). The calculated example values in Appendix A show that – except for the case of graphite – the differences between the different approaches are small enough to be neglected for most practical purposes. That means that the linear approximation turns out to be sufficient in practice for all but the most extreme cases of anisotropy (e.g. graphite), where it will be very difficult to achieve a statistically isotropic microstructure anyway.

3. MICROMECHANICAL BOUNDS FOR THE EFFECTIVE THERMAL CONDUCTIVITY OF MULTIPHASE MATERIALS

Micromechanics (Markov 2000, Nemat-Nasser and Hori 1999, Torquato 2002), also called statistical continuum theory (Beran 1968), theory of composites (Milton 2002) or theory of heterogeneous media (Sahimi 2003), is the most general theory for describing the macroscopic behavior, in particular the effective properties, of polycrystalline materials and multiphase composites, including the important special cases of two-phase composites and porous media, in dependence on the microstructure. Therefore, here we give a brief summary of the rigorous bounds from the viewpoint of exact theory. We recall the one- and two-point bounds (Wiener bounds and Hashin-Shtrikman bounds) and give a detailed explanation of the three-point bounds (Beran bounds). Other bounds are briefly mentioned. All expressions in this section are given for uniform, statistically isotropic two-phase media in three-dimensional space, if not explicitly stated otherwise. In accordance with common practice it is implicitly assumed that all phases are isotropic, that means in this section a polycrystalline microstructure consisting of one type of crystallites with different (random) orientation is treated as a single phase to which a scalar thermal conductivity can be assigned (which is by itself an effective conductivity, of course, obtained by averaging according to the rules set out in Section 2).

Rigorous bounds on the effective properties generally incorporate microstructural information beyond volume fractions in the form of n-point correlation functions. They can be derived using variational principles (e.g. the Hashin-Shtrikman variational principle) or other methods. When the trial fields are based on contrast expansions, one obtains so-called contrast bounds (one-point, two-point, three-point, four-point), in other cases cluster bounds (multiple-scattering bounds) or unconventional bounds such as security-spheres bounds (Torquato 2002). Beside these thermal conductivity bounds there are so-called cross-property bounds (elementary and translational), which will be treated here only peripherally. For more details the reader is referred to (Torquato 2002) and (Markov 2000, Milton 2002, Sahimi 2003). Contrast bounds are the most popular and the most widely used bounds in practice.

Excellent approximations for the special case of porous materials (with phase 1 being the solid phase and phase 2 being the pore phase) is readily obtained by using the bounds for two-phase composites and setting $k_2 = 0$. Formal agreement with the rest of this treatise (concerning e.g. the special cases of matrix-inclusion microstructures or porous media) can be achieved by simply replacing the index 1 for the matrix properties by "0" and by renaming $\phi_2 \equiv \phi$ and $\phi_1 \equiv 1 - \phi$. In this section we closely follow the best currently available monograph in the field (Torquato 2002).

3.1. Contrast Bounds

When in a multiphase material (composite or porous) the properties of the individual phases (constituents) are known, it can be expected that certain bounds on the effective property of the multiphase material (mixture) can be given without further microstructural information apart from compositional information (phase volume fractions). For the thermal conductivity this is indeed the case, cf. (Beran 1968, Markov 2000, Milton 2002, Nemat-Nasser and Hori 1999, Sahimi 2003, Torquato 2002). It seems very plausible, for example, that in a multiphase material, in which the influence of phase boundaries is negligible, the resultant value of an effective property should neither be lower nor higher than the values of the phase properties. Therefore it can be concluded, for example, that the effective thermal conductivity of the multiphase material k must be some kind of average value of all the phase conductivities k_i. The most general type of average is the general power mean (weighted by the phase volume fractions ϕ_i) (Pabst and Gregorová 2004a). Interestingly, however, it is found that the effective thermal conductivity of a multiphase material always lies between two special averages, $viz.$ the (volume-weighted) arithmetic and harmonic mean.

3.1.1. One-Point Bounds

The one-point bounds (Wiener bounds) for the effective thermal conductivity of multiphase composites are

$$k_W^- = \langle k^{-1} \rangle^{-1} \leq k \leq \langle k \rangle = k_W^+ \tag{15}$$

(Wiener 1912), where the brackets denote the volume-weighted arithmetic average, i.e.

$$k_W^+ = \langle k \rangle = \sum \phi_i \, k_i \,, \tag{16}$$

$$\left(k_W^-\right)^{-1} = \langle k^{-1} \rangle = \sum \frac{\phi_i}{k_i}, \tag{17}$$

with k_i being the phase conductivities and ϕ_i being the volume fractions of the phases (summation symbols are understood to be over all i phases, i.e. over all values of i). In the micomechanical literature the Wiener bounds are called one-point bounds, as a consequence of the fact that only the one-point correlation functions (i.e. the volume fractions) are required as input information (Torquato 2002).

These bounds express the fact that the effective conductivity of a composite cannot be lower than the (volume-weighted) harmonic average and not higher than the (volume-weighted) arithmetic average of the phase conductivities. The bounds hold for multiphase materials (composites), including two-phase ones, with arbitrary symmetry and microstructure. In the case of porous materials the lower bound is zero, when the conductivity of the pore phase is neglected (which amounts to considering pores as vacuous voids), i.e. the effective thermal conductivity of porous materials is principally not bounded from below: that means, in principle microstructures are thinkable that make the effective conductivity zero at any volume fraction of pores. The same holds for all contrast bounds of higher order.

In the special case of two-phase composites these relations (and the Wiener bounds) reduce to

$$k_W^+ = \langle k \rangle = \phi_1 \, k_1 + \phi_2 \, k_2 \tag{18}$$

and

$$k_W^- = \langle k^{-1} \rangle^{-1} = \frac{k_1 k_2}{\phi_1 \, k_2 + \phi_2 \, k_1}. \tag{19}$$

Of course, since in the case of two-phase materials one of the two volume fractions is redundant ($\phi_1 \equiv 1-\phi$ and $\phi_2 \equiv \phi$), the Wiener bounds can also be written as

$$k_W^+ = (1-\phi)k_1 + \phi k_2 \qquad (20)$$

and

$$\frac{1}{k_W^-} = \frac{1-\phi}{k_1} + \frac{\phi}{k_2} \Rightarrow k_W^- = \frac{k_1 k_2}{(1-\phi)k_2 + \phi k_1}, \qquad (21)$$

respectively. When one of the phases are pores (with zero conductivity, i.e. $k_2 = 0$, in the case of vacuous pores) and the conductivity of the solid matrix phase is denoted as $k_1 \equiv k_0$, the lower Wiener bounds degenerate to zero identically and the upper Wiener bounds reduce to

$$k_W^+ = k_0 (1-\phi) \qquad (22)$$

(in this case ϕ denotes the volume fraction of pores and is called porosity).

The upper Wiener bound can be realized by a model microstructure with layers or fibers of different thermal conductivity arranged parallel to the heat flux direction. It corresponds to the so-called "mixture rule". On the other hand, the lower Wiener bound corresponds to a model material with series arrangement of the phases (layers) and heat flux in a direction perpendicular to the layers. That means, the two Wiener bounds (corresponding to the parallel and series models, respectively) can be realized by materials with anisotropic microstructure (e.g. fiber composites and sandwich structures). In the first case, the overall thermal conductivity of the composite structure is determined mainly by the highly conducting phases, in the second mainly by the low-conductivity (insulating) phases (Kingery 1960, Kingery et al. 1976).

As mentioned above, from a mathematical point of view, the two Wiener bounds can be viewed as special cases of the general power mean (general power average weighted by volume fractions)

$$k = \left(\sum \phi_i k_i^N\right)^{1/N}. \qquad (23)$$

This general power mean (Pabst and Gregorová 2004a) reduces to the upper Wiener bound (arithmetic average) if $N = 1$ and to the lower Wiener bound (harmonic average) if $N = -1$, but includes other possible averages (e.g. the geometric average) as well. Evidently, the physical statement behind this relation is that the effective thermal conductivity of a multiphase composite is in any case bounded by the thermal conductivities of the constituent phases. For real materials this statement is true as long as there is no miscibility of the phases (mutual solubility, e.g. due to reactions at high temperature) and as long as the effects of phase boundaries can be ignored, i.e. as long as the grains (crystallites) are sufficiently large for grain size effects to be negligible.

It is evident, that if in a sandwich structure (stacked layers) one phase (i.e. one layer) is a void phase (pore phase) with negligible thermal conductivity, this layer will represent an insulating barrier, i.e. the overall thermal conductivity in this direction is zero (infinite thermal resistance). Therefore the lower Wiener bound degenerates to zero in the case of porous materials, i.e. no rigorous lower bound can be given for porous materials with arbitrary microstructure.

Note that the Wiener bounds depend solely on the phase volume fractions and do not require any further information or assumption concerning the microstructure: they are valid for arbitrary microstructures, isotropic and anisotropic. Since the volume fractions correspond to one-point correlation functions (Markov 2000, Torquato 2002) the Wiener bounds are also called one-point bounds. Of course, depending on the phase contrast (ratio of the phase properties k_1 and k_2), the one-point bounds (Wiener bounds) can be relatively wide, often too wide to be of predictive value in practice. Using variational principles, Hashin and Shtrikman (Hashin and Shtrikman 1962) have shown that the Wiener bounds can be improved when more information is known on the microstructure (two-point correlation functions) and have derived the best possible bounds on the effective thermal conductivity of macroscopically (statistically) isotropic two-phase composites given just volume-fraction information.

3.1.2. Two-Point Bounds

The two-point bounds (Hashin-Shtrikman bounds, in the sequel abbreviated as "HS bounds") for the thermal conductivity of two-phase composites are

$$k_{HS}^- = k_L^{(2)} \leq k \leq k_U^{(2)} = k_{HS}^+, \tag{24}$$

$$k_{HS}^+ = k_U^{(2)} = \langle k \rangle - \frac{\phi_1 \phi_2 (k_2 - k_1)^2}{\langle \tilde{k} \rangle + 2k_1} = \phi_1 k_1 + \phi_2 k_2 - \frac{\phi_1 \phi_2 (k_1 - k_2)^2}{3k_1 - \phi_1 (k_1 - k_2)}, \tag{25}$$

$$k_{HS}^- = k_L^{(2)} = \langle k \rangle - \frac{\phi_1 \phi_2 (k_2 - k_1)^2}{\langle \tilde{k} \rangle + 2k_2} = \phi_1 k_1 + \phi_2 k_2 - \frac{\phi_1 \phi_2 (k_1 - k_2)^2}{3k_2 + \phi_2 (k_1 - k_2)}, \tag{26}$$

$$\langle \tilde{k} \rangle = \phi_1 k_2 + \phi_2 k_1 \tag{27}$$

for $k_1 \geq k_2$ (Hashin and Shtrikman 1962). Note that, although actually two-point bounds, i.e. incorporating the two-point probability function $S_2^{(i)}$, in the special case of statistically isotropic materials, this two-point correlation function need not be explicitly known (Markov 2000, Torquato 2002). Since for isotropic media the key integral involving $S_2^{(i)}$ depends only on the end points of the function $S_2^{(i)}$, i.e. on ϕ_i and ϕ_i^2 (Torquato 2002), the HS bounds for statistically isotropic materials explicitly involve only the volume fractions (i.e. for porous materials only the porosity). In other words, for statistically isotropic materials only the volume fractions must be known (in the special case of two-phase materials only one of the volume fractions) in order calculate the HS bounds. We would like to emphasize that the HS upper bound is identical to the Maxwell or Maxwell-Eucken relation (Maxwell 1873, Eucken 1932, 1933, Francl and Kingery 1954), which has been derived for spherical inclusions dispersed in a matrix, see Section 4, and can be written in the form (Okaz et al. 1986)

$$k = k_1 \cdot \frac{k_2 + 2k_1 + 2\phi_2(k_2 - k_1)}{k_2 + 2k_1 + \phi_2(k_2 - k_1)} \tag{28}$$

(dispersed phase with index 2, continuous phase with index 1). This relation has been rederived several times in different contexts and several other equivalent formulations are known (Christensen 1979, Grimvall 1999, Hashin and Shtrikman 1962, Kerner 1956, Landauer 1978, Markov 2000, Salmang and Scholze 1982, Torquato 2002).

In the case of porous media – considered as a special case of two-phase composites, in which the thermal conductivity of the pore phase can be neglected compared to that of the solid phase (i.e. $k_2 \approx 0$ and denoting $k_1 \equiv k_0$ and $\phi_2 \equiv \phi$) – the HS upper bound for the effective thermal conductivity is

$$k_{HS}^+ = k_0 \left(\frac{1-\phi}{1+\phi/2} \right). \tag{29}$$

As in the case of one-point (Wiener) bounds, also the two-point (HS) lower bounds degenerate to zero identically in the case of porous materials.

Similar to the one-point bounds (Wiener bounds), also the two-point bounds (HS bounds) are optimal in the sense that they correspond to realizable microstructures. These Hashin-Shtrikman microstructures (also called "Hashin assemblages" (Markov 2000), "composite spheres model" (Christensen 1979) or "coated-spheres model" (Torquato 2002)) are composed of spheres which consist of a core with conductivity k_2 and radius a, surrounded by a concentric shell of conductivity k_1 and radius b. The two-dimensional (2D) analog of this idealized model microstructure is called "composite cylinders model" or "coated-cylinders model" (Christensen 1979). These composite spheres (or circular cylinder cross-sections in 2D) have an infinitely broad size distribution and are in a space-filling arrangement (or area-filling arrangement in 2D), resulting in an overall inclusion volume fraction of $\phi_2 = (a/b)^3$ (or $\phi_2 = (a/b)^2$ in 2D). The effective conductivity of this model microstructure can be generally written as

$$k = \langle k \rangle - \frac{(k_2 - k_1)^2 \phi_1 \phi_2}{\langle \widetilde{k} \rangle + (d-1)k_1}, \tag{30}$$

where d is the space dimension (2 or 3). If $k_2 \geq k_1$ this relation corresponds to the HS lower bound and its counterpart with the phase labels reversed corresponds to the HS upper bound (if $k_2 \geq k_1$; otherwise *vice versa*). Note that for this microstructure (geometrical model) phase 2 is always disconnected or "well separated" (i.e. does not percolate) except in the trivial case $\phi_2 \equiv \phi = 1$. In practice, the 2D case applies to the transverse conductivity of transversely isotropic composites with aligned fibers. Alternative formulations (given in (Christensen 1979)) for the effective thermal conductivity of the Hashin-Shtrikman model are for isotropic two-phase composites (coated spheres model in 3D)

$$k = k_1 \left[1 + \frac{\phi}{k_1/(k_2 - k_1) + (1-\phi)/3} \right], \tag{31}$$

whereas for transversely isotropic two-phase composites (coated cylinders model or 2D case) it is

$$k_T = k_m \left[1 + \frac{\phi}{k_m/(k_f - k_m) + (1-\phi)/2} \right] \tag{32}$$

in the transverse direction and

$$k_L = \langle k \rangle = (1-\phi)k_m + \phi k_f \tag{33}$$

in the longitudinal (axial) direction (Christensen 1979). In these equations the indices "1" and "*m*" refer to the matrix phase, while indices "2" and "*f*" refer to the inclusion phase. Note that the index-free volume fraction ϕ always refers to the inclusion phase (wherever applicable). We note in passing that the index "*f*" (denoting primarily "fibers") should not be read to preclude the possibility that also platelets may be preferentially oriented and thus cause transverse isotropy.

Note that, in contrast to the relatively wide Wiener bounds, which are not realizable for isotropic materials, the much tighter Hashin-Shtrikman bounds are optimal, i.e. realizable for certain isotropic materials in the sense that microstructures can be devised for which these bounds can be attained exactly (Torquato 2002). Without microstructural information of higher order (for real materials) or more special assumptions (for model materials) these bounds cannot be improved, cf. (Beran 1968, Markov 2000, Milton 2002, Nemat-Nasser and Hori 1999, Sahimi 2003, Torquato 2002). The microstructure corresponding to the Hashin-Shtrikman bounds is the so-called Hashin assemblage ((Hashin 1962, Markov 2000) also called coated-spheres model or composite spheres model (Christensen 1979, Torquato 2002)) consisting of polydisperse composite spheres containing concentric spherical inclusions. In the case of macroscopically isotropic composites (or porous materials) in 3D the Hashin assemblage can be imagined as a composite (or porous material) consisting of spheres (hollow in the case of porous materials) with a certain wall thickness and an infinitely wide size distribution that enables space filling with a fractal microstructure. For transversely isotropic composites the analogous model in 2D is the coated-cylinders model (composite-cylinders model (Christensen 1979)).

3.1.3. Three-Point Bounds

Three-point bounds for the effective thermal conductivity of two-phase composites are

$$k_L^{(3)} \leq k \leq k_U^{(3)}, \tag{34}$$

$$k_L^{(3)} = \langle k \rangle - \frac{\phi_1 \phi_2 (k_2 - k_1)^2}{\langle \tilde{k} \rangle + 2 \langle k^{-1} \rangle_\xi^{-1}}, \tag{35}$$

$$k_U^{(3)} = \langle k \rangle - \frac{\phi_1 \phi_2 (k_2 - k_1)^2}{\langle \tilde{k} \rangle + 2 \langle k \rangle_\xi}, \tag{36}$$

$$\langle k \rangle_\xi = \xi_1 k_1 + \xi_2 k_2, \tag{37}$$

(Beran bounds) (Beran 1965), where the three-point microstructural parameter ξ_i lies in the range $[0,1]$ and has the property $\xi_1 + \xi_2 = 1$. It is known today, however, that the lower Beran bound is not the best possible three-point lower bound for the thermal conductivity. An improved three-point lower bound is the Milton-Torquato bound

$$\frac{k_L^{(3)}}{k_1} = \frac{1 + (1 + 2\phi_2)\beta_{21} - 2(\phi_1 \xi_2 - \phi_2)\beta_{21}^2}{1 + \phi_1 \beta_{21} - (2\phi_1 \xi_2 + \phi_2)\beta_{21}^2}, \tag{38}$$

(Milton 1984, Torquato 1985a), where $\beta_{21} = (k_2 - k_1)/(k_2 + 2k_1)$, see Appendix B. This optimal lower bound is exactly realized for space-filling doubly-coated composite spheres (Torquato 2002). Similar to the aforementioned one- and two-point bounds mentioned above and to the lower Beran bound, it is also zero in the case of porous materials. The same is true for the four-point Milton bounds (Milton 1981), which contain an additional microstructural parameter, see (Torquato 2002). Five-point bounds for heterogeneous materials, and the measurement of the corresponding higher-order correlation functions for real microstructures, are beyond presently available technology (Torquato 2002).

In order to apply (evaluate) the three-point bounds for composites (or porous materials) in practice, either the three-point probability function has to be determined experimentally by image analysis (Corson 1974) (or tomography), or a model microstructure is assumed (e.g. realistic random models constructed by computer simulation) for which the three-point microstructural parameter can be calculated in dependence of the volume fraction (or porosity). A detailed treatment of three-point bounds is beyond the scope of this treatise (for more information see (Torquato 2002)), but the following points should be noted: In the case of so-called symmetric-cell materials (SCM, i.e. materials with phase-inversion symmetry, for which $\xi_2 = 1/2$ at $\phi_1 = \phi_2 = 1/2$), ξ_2 can be obtained for arbitrary volume fractions ϕ_2. For isotropic symmetric-cell materials made up of space-filling spheroidal cells (pores), which must be randomly oriented and possess a size distribution down to the infinitesimally small, one obtains (Miller 1969, Milton 1982, Torquato 2002):

Spherical cells:

$$\xi_2 = \phi_2, \tag{39}$$

Needle-like cells:

$$\xi_2 = \phi_1/4 + 3\phi_2/4 = \frac{1 + 2\phi_2}{4}, \tag{40}$$

Disk-like cells:

$$\xi_2 = \phi_1 = 1 - \phi_2. \tag{41}$$

Similarly, for cubic cells (corresponding to a 3D random checkerboard configuration) we have $\xi_2 = 0.11882\phi_1 + 0.88118\phi_2$ (Helsing 1994), and analogous relations have been found in the 2D case for triangular, square and hexagonal cells (Torquato 2002). In the case of dilute dispersions (or in the low-porosity limit) of randomly oriented spheroidal inclusions (pores) the three-point parameters can be approximated by 0 for spherical cells, 1/4 for long and thin needles (fibers), 1 for long and thin disks (platelets) and 0.11882 for randomly stacked cubes. As an example of an SCM consider a mixture of two ceramic powders A and B that have been densely sintered and exhibit a polycrystalline ("grainy") microstructure. Depending on the volume fraction of A and B, crystallites A or crystallites B may form a connected matrix phase (and B or A, respectively, the corresponding inclusions dispersed in it), or the microstructure can be bicontinuous.

For dense random dispersions other than SCMs, no closed-form expressions are available, but the microstructural parameters have been calculated for several model microstructures:

- Identical (i.e. monodisperse) overlapping (i.e. fully penetrable) spheres, e.g. sintered particles or pores that are allowed to be interconnected with other pores (Berryman 1985, Torquato and Stell 1983, 1985),
- Identical (i.e. monodisperse) "hard" (non-overlapping, i.e. impenetrable) spheres, e.g. particles with point contacts or closed, isolated pores) (Lado and Torquato 1986, Miller and Torquato 1990, Torquato et al. 1987),

- Polydisperse "hard" (non-overlapping, i.e. impenetrable) spheres, e.g. particles or pores with a size distribution down to the infinitesimally small (Miller and Torquato 1990, Thorvert et al. 1990).

Values of the three-point parameters for these microstructures can be found in (Torquato 2002). For monodisperse overlapping spheres we have (Torquato 1985b, Torquato et al. 1987)

$$\xi_2 \approx 0.5615\phi_2 \tag{42}$$

as a reasonable approximation for almost the whole range of volume fractions, while in the case of monodisperse non-overlapping spheres we have (Beasley and Torquato 1986, Torquato et al. 1987),

$$\xi_2 = 0.21068\,\phi_2 - 0.04693\,\phi_2^{\,2} + 0.00247\,\phi_2^{\,3} + \ldots \tag{43}$$

as a good approximation up to $\phi_2 \approx 0.60$, i.e. about 94 % of the maximally random jammed density value for spheres, which is approx. 0.64, see (Torquato 2002).

In the case of binary mixtures with two different and widely separated particle sizes (or porous materials with a bidisperse pore size distribution) one obtains (Thorvert et al. 1990)

$$\xi_2 \approx 0.35534\phi_2, \tag{44}$$

whereas for a system with infinitely many different and widely separated particle sizes (or porous material with polydisperse pores) the corresponding result is (Thorvert et al. 1990)

$$\xi_2 \approx 0.5\,\phi_2 \tag{45}$$

(Note that the bidisperse results lie midway between the monodisperse and the polydisperse results; thus the principal effect of polydispersivity in size is to increase the three-point parameter ξ_2, see (Miller and Torquato 1990)).

From these results it is evident that for a broad class of microstructures with spherical particles or pores commonly encountered in practice the microstructural parameter ξ_2 is bounded from above by the value of the symmetric-cell material (SCM) with spherical cells, i.e. $\xi_2 \leq \phi_2$. Of course, this result cannot be valid in general, since the Hashin-Shtrikman coated-spheres geometry, for which $\xi_2 \to 1$, is a counterexample. Nevertheless, for those media, for which it holds, ξ_2 can be eliminated in favor of the volume fraction ϕ_2 in the three-point upper bound to get the following improved upper bound on the effective conductivity (Torquato 2002):

$$k \le \langle k \rangle - \frac{\phi_1 \phi_2 (k_2 - k_1)^2}{\langle \widetilde{k} \rangle + 2\langle k \rangle}. \tag{46}$$

Three-point bounds can provide relatively sharp estimates of the effective conductivity for finite phase contrasts (i.e. phase conductivity ratios k_1/k_2 or k_2/k_1), and even for infinite (or zero) phase contrasts (as in the case of porous materials) one of the these bounds will often provide a good approximation to reality. For example, experimental data for porous sandstone (Sugawara and Yoshizawa 1962) are reported to be close to the three-point upper bound, since the conducting quartz phase (modeled by overlapping spheres) is above the percolation threshold of $\phi_{2c} \approx 0.29$, where the microstructure is bicontinuous and both phases (i.e. pores and quartz grains) are connected (note that in 3D the overlapping-spheres model is connected, i.e. forms a bicontinuous microstructure, in the range $0.29 \le \phi_2 \le 0.97$ (Torquato 2002)). Simulation results by (Kim and Torquato 1992) have shown that the three-point upper bound provides a good estimate for 3D systems of overlapping, perfectly insulating spherical inclusions (i.e. pores with vanishing thermal conductivity) in a solid matrix. However, somewhere in the bicontinuous range ($0.29 \le \phi_2 \le 0.97$) a transition (percolation) to the lower bound (zero in this case) must occur.

3.2. Other Bounds

3.2.1. Cluster Bounds

When one of the phases is present in the form of well-defined inclusions (e.g. closed pores), this specific qualitative statement (microstructural information) can be incorporated to derive so-called cluster bounds (Torquato 1986), in different context also called "multiple scattering bounds" (Torquato 2002). First-order (i.e. three-point) cluster bounds for the effective conductivity were derived by (Weissberg 1963) for a system of perfectly insulating, fully penetrable spheres of equal radius and generalized to arbitrary phase contrast by (DeVera and Strieder 1977) and to arbitrary degrees of interpenetrability by (Torquato 1986) (Torquato cluster bounds). They have been generalized by (Markov 1998) to the case of spheroidal inclusions (including their extremely oblate case, *viz.* cracks). These bounds coincide with the so-called "noninteracting-cracks approximation", thus rigorously proving that crack interactions always reduce the conductivity of a microcracked solid.

The cluster bounds for impenetrable spheres are identical to the three-point Beran bounds (Beasley and Torquato 1986), whereas for overlapping spheres the cluster bounds are not as sharp as the best three-point contrast bounds, but they are simple and analytical (DeVera and Strieder 1977, Torquato 1986). For the special case of overlapping insulating spheres, the cluster upper bounds are even simpler and very similar to the three-point bounds:

$$\frac{k}{k_1} \le \frac{\phi_1}{1 - \ln \phi_1 / 2} \tag{47}$$

(Torquato 2002). Similar to the contrast bounds, also the cluster lower bounds are zero for porous materials.

3.2.2. Security-Spheres Bounds

Both contrast bounds and cluster bounds are conventional bounds in the sense that they diverge from one another in the limit of infinite phase contrast. For example, when phase 2 is superconducting relative to phase 1 the finite-order upper bounds diverge to infinity, i.e. they treat phase 2 as a percolating phase even if it may actually not be percolating (Torquato 2002). On the other hand, when phase 2 is perfectly insulating relative to phase 1, the lower bounds degenerate to zero, i.e. the percolation threshold of phase 2 predicted by the lower bound is zero (e.g. a porosity of 0 %). This is due to the fact that the limited microstructural information that these bounds incorporate is insufficient to distinguish phase connectivity. In certain cases the remaining bound yields a useful estimate of the effective conductivity (Torquato 1985b), but this is of course not a general rule.

With regard to this situation, unconventional bounds have been derived that account for connectedness information in such a way that the bounds do not diverge in the limit of infinite phase contrast. In the so-called "security-spheres" concept (Bruno 1991, Torquato and Rubinstein 1991), an inclusion sphere (phase 2) is surrounded by a larger "security sphere" in such a way that the concentric-shell region between the actual sphere and the surface of the security sphere contains only matrix phase (phase 1). The security-spheres lower bound for conductivity (Torquato-Rubinstein lower bound) in the special cases of perfectly insulating ($k_2 = 0$) spherical inclusions (i.e. pores) is given by an integral expression (Torquato and Rubinstein 1991), where the pore diameter, the (dimensionless) pore distance, the pore volume fraction (e.g. porosity) and the nearest-neighbor probability density function (or the minimum interparticle distance (Bruno 1991)) are required as input information, see (Torquato 2002). Unlike conventional lower bounds, the security-spheres lower bounds remain nonzero in the case of perfectly insulating cavities (provided the integral is convergent), see (Bruno 1991, Torquato and Rubinstein 1991, Torquato 2002). For periodic lattices (e.g. cubic lattice of spherical pore cavities, i.e. perfectly insulating void inclusions with $k_2/k_1 = 0$) the security-spheres bound can be expressed in non-integral form (Bruno bound) via the gamma function (Bruno 1991), see also (Torquato 2002).

3.2.3. Cross-Property Bounds

Cross-property relations in general are useful if one property is more easily measured than another. Since the effective properties reflect certain microstructural information about the medium, one might expect that one could extract useful information about one effective property given an accurate (experimental or theoretical) determination of another effective property (Torquato 2002). Rigorous, i.e. generally valid, cross-property relations are mostly inequalities (i.e. in the form of bounds). Only in special cases, e.g. for properties with formally the same constitutive equation or in exceptional cases, e.g. the Levin relation, they take on the form of equalities. Cross-property bounds for conductivity and elastic moduli are due to (Berryman and Milton 1988, Cherkaev and Gibiansky 1992, 1993, Milton 1981, 1984, Prager 1969, Torquato 1992).

For isotropic two-phase media of arbitrary topology having nonnegative phase-Poisson ratios and $K_2/K_1 \leq k_2/k_1$ the relative bulk modulus and the relative conductivity are related via the elementary bound (Milton 1984, Torquato 1992)

$$\frac{K}{K_1} \leq \frac{k}{k_1} \tag{48}$$

This cross-property bound has interesting implications. In particular, for a porous material with phase 2 being the void phase (pore phase), the effective conductivity and elastic moduli have the same percolation thresholds and near the threshold ($\phi_1 \rightarrow \phi_{1c}^+$) they obey the scaling laws,

$$k \propto (\phi_1 - \phi_{1c})^t, \tag{49}$$

$$K \propto (\phi_1 - \phi_{1c})^f, \tag{50}$$

where t and f are critical exponents. As a consequence of the above cross-property bounds one obtains the inequalities (Torquato 1992)

$$f \geq t. \tag{51}$$

The same result is valid with respect to the shear modulus. Thus, the critical exponents for the elastic moduli are generally greater than the critical exponent for the conductivity, in agreement with theoretical and experimental studies of continuum percolation (Feng et al. 1987, Smith and Lobb 1979, Benguigui 1986). There are other types of cross-property bounds (e.g. so-called translation bounds), which are far beyond the scope of the present treatise, see (Berryman and Milton 1988, Markov 2000, Torquato 2002) and the references cited there.

4. MODEL RELATIONS FOR THE EFFECTIVE THERMAL CONDUCTIVITY OF MULTIPHASE MATERIALS

4.1. Effective-Medium Approximations

The most important effective-medium approximations (EMAs) are

- Maxwell-type approximations,
- Self-consistent approximations,
- Differential approximations.

All of these EMAs rely on the knowledge of the effective property in the infinitely dilute limit and therefore on the solutions of the single-inclusion problem, cf. Appendix B. The oldest EMA is that due to Maxwell (Maxwell 1873).

4.1.1. Maxwell Approximation

The Maxwell approximation (Maxwell 1873), in conductivity context also called Maxwell-Eucken relation (Eucken 1932), is based on the model of a composite sphere (composed of spherical inclusions of phase 2) embedded (dispersed) in an infinite matrix of phase 1. Maxwell-type EMAs give good estimates of the effective conductivity at nondilute concentration, provided that the inclusions are well separated.

- Spherical inclusions:
  - Maxwell approximation, an explicit scheme for two-phase composites of arbitrary phase conductivities (also called Maxwell-Eucken or Maxwell-Garnett approximation (Landauer 1978), intimately related to the Clausius-Mossotti formula for the dielectric constant (Markov 2000)):

  $$\frac{k-k_1}{k+2k_1} = \phi\left[\frac{k_2-k_1}{k_2+2k_1}\right]. \tag{52}$$

  This is identical to the coated-spheres model (HS upper bound if $k_2 \leq k_1$).

  - Special case – perfectly insulating spheres ($k_2/k_1 = 0$):

  $$\frac{k}{k_1} = \frac{2(1-\phi)}{3-\phi}. \tag{53}$$

  - Special case – superconducting spheres ($k_2/k_1 = \infty$):

  $$\frac{k}{k_1} = \frac{1+2\phi}{1-\phi}. \tag{54}$$

  - Multiphase composites ((Benveniste 1986); this relation can also be derived via the so-called Mori-Tanaka method (Mori and Tanaka 1973)):

  $$\frac{k-k_1}{k+2k_1} = \sum \phi_i\left[\frac{k_i-k_1}{k_i+2k_1}\right]. \tag{55}$$

- Ellipsoidal inclusions:
  - Transversally isotropic composites (with aligned ellipsoidal inclusions):

$$\sum \phi_i (\mathbf{K} - \mathbf{K}_i) \cdot \mathbf{R}^{(i1)} = 0, \qquad (56)$$

where $\mathbf{R}^{(i1)}$ is the field intensity (or temperature gradient) concentration tensor

$$\mathbf{R}^{(i1)} = \left(1 + \mathbf{A} \frac{k_i - k_1}{k_1}\right)^{-1} \qquad (57)$$

and **A** the corresponding depolarization tensor, cf. Appendix B. This formula corresponds to the so-called Willis bounds (Willis 1977).

o Isotropic composites (with randomly orientated ellipsoidal inclusions):

$$\sum \phi_i (k - k_i) \cdot R^{(i1)} = 0, \qquad (58)$$

where the isotropic invariant $R^{(i1)} = \operatorname{tr} \mathbf{R}^{(i1)}/3$ of the field intensity concentration tensor $\mathbf{R}^{(i1)}$, following from the solution of the single-inclusion problem is given in Appendix B (Torquato 2002). Although for two-phase composites, e.g. porous media, this formula always lies between the isotropic HS bounds, it can violate these bounds for more than two phases (Norris 1989), implying that the formula does not correspond to realizable microstructures in this case (a typical drawback of Mori-Tanaka schemes (Mori and Tanaka 1973, Benveniste 1986)).

4.1.2. Self-Consistent Approximation

Self-consistent (SC) approximations, which have been proposed in (Bruggeman 1935) and elaborated in (Landauer 1952, Landauer 1978), are based on the model assumption that the material outside an inclusion can be considered as a homogeneous matrix medium whose effective conductivity is the unknown to be calculated. One then requires that the perturbations to a uniform field be zero on average (self-consistency requirement), see (Christensen 1979, Torquato 2002). For multiphase composites the results are:

• Spherical inclusions (so-called symmetric SC approximation):

$$\sum \phi_i \left[\frac{k_i - k}{k_i + 2k}\right] = 0 \qquad (59)$$

This implicit scheme is a quadratic equation for the effective conductivity, which in the two-phase case has the solution

$$k = \frac{\lambda + \sqrt{\lambda^2 + 8k_1 k_2}}{4},\qquad(60)$$

with

$$\lambda = k_1(3\phi_1 - 1) + k_2(3\phi_2 - 1).\qquad(61)$$

The deficiencies of SC approximations are discussed in detail in (Christensen 1979, Torquato 2002). Maybe the most well-known failure of SC approximations is the prediction of non-trivial but spurious percolation thresholds, e.g. in the case of perfectly insulating spherical inclusions.

o   Perfectly insulating spherical inclusions ($k_2/k_1 = 0$):

$$\frac{k}{k_1} = 1 - \frac{3}{2}\phi,\qquad(62)$$

i.e. the inclusion phase is predicted to percolate at $\phi_C = 2/3$.

o   Superconducting spherical inclusions ($k_2/k_1 = \infty$):

$$\frac{k}{k_1} = \frac{1}{1 - 3\phi},\qquad(63)$$

i.e. the inclusion phase is predicted to percolate at $\phi_C = 1/3$, independent of microstructural details (this does not correspond to reality, since the percolation threshold is known to be very sensitive to the microstructure).

Among other deficiencies, the SC approximations will generally violate bounds that improve upon the HS bounds for realistic models of dispersions (Torquato 2002). There are, however, certain microstructures, that realize the SC approximations: the so-called Milton structures, fractal-like hierarchical structures of granular media in which spherical grains of two phases of comparable size are well-separated with self-similarity on many length scales (Milton 1984, 1985), which are structures with phase-inversion symmetry, see (Torquato 2002). Revisiting the percolation predictions, the above SC formula for the limiting case of perfectly insulating spherical inclusions predicts that the porous medium no longer conducts at $\phi_C = 2/3$ (where the matrix phase becomes disconnected). For inclusion volume fractions in the range $1/3 \leq \phi \leq 2/3$ the microstructure of SC materials is bicontinuous. For the case of ellipsoidal inclusions the result is formally similar to that of the Maxwell-type approximation, but the matrix conductivity is replaced by an effective conductivity, see (Torquato 2002).

4.1.3. Differential Scheme Approximation

Unlike the symmetric SC approximation the differential scheme (DS) approximation, also introduced in (Bruggeman 1935), does not treat each phase symmetrically. When for a two-phase composite (matrix phase 1, inclusion phase 2) the effective conductivity $k(\phi)$ at one selected value of ϕ is known, then $k(\phi)$ is considered to be the effective host conductivity and $k(\phi + \Delta\phi)$ is the effective conductivity after replacing a small fraction of the composite host "phase" by inclusion phase. On average, a fraction $\Delta\phi/(1-\phi)$ of the composite host "phase" must be replaced in order to change the overall fraction of the inclusion phase to $\phi + \Delta\phi$. In the case of spherical inclusions, we can use the dilute-limit result for the effective conductivity to obtain

$$k(\phi + \Delta\phi) - k(\phi) = k(\phi)\left[\frac{k_2 - k(\phi)}{k_2 + 2k(\phi)}\right] \cdot \frac{3\Delta\phi}{1-\phi} \tag{64}$$

In the limit $\Delta\phi \to 0$, the following differential equation

$$(1-\phi)\frac{dk}{d\phi} = 3k(\phi)\left[\frac{k_2 - k(\phi)}{k_2 + 2k(\phi)}\right], \tag{65}$$

can be set up, with the "boundary" condition $k(\phi = 0) = k_1$. This differential equation can be integrated analytically to yield

$$\left(\frac{k_2 - k}{k_2 - k_1}\right)\left(\frac{k_1}{k}\right)^{1/3} = 1 - \phi \tag{66}$$

The embedding process in the DS approximation ensures that the initial matrix material remains connected in the resulting composite. Because of this feature, in the special case of composites with insulating inclusions (i.e. porous materials) the DS approximation corresponds to Archie's empirical law,

$$\frac{k}{k_1} = (1-\phi)^m, \tag{67}$$

where the exponent m varies from approx. 1.5 to 4 (Archie 1942, Sen et al. 1981). The DS approximation for this case ($k_2/k_1 = 0$) is

$$\frac{k}{k_1} = (1-\phi)^{3/2}. \tag{68}$$

This result may be compared is to the estimates given by the Maxwell-type and SC approximations, which are

$$\frac{k}{k_1} = \frac{2(1-\phi)}{2+\phi} = \frac{1-\phi}{1+\frac{\phi}{2}} \tag{69}$$

and

$$\frac{k}{k_1} = \frac{3(1-\phi)-1}{2} = 1 - \frac{3}{2}\phi, \tag{70}$$

respectively, in poor agreement with "Archie's law" (Markov 2000). Similar to the SC approximation also the DS approximation is realizable by hierarchical models and hence must lie between the HS bounds (Milton 1985). Both the Maxwell-type and the DS approximations correspond to well-separated inclusions. Therefore, they are not good estimates of the effective conductivities of dispersions in which the inclusions form large clusters. The DS approximation precludes the occurrence of a percolation threshold and predicts power law behavior in the limiting extreme cases (superconducting and perfectly inclusions, respectively). It can therefore be expected that it will provide a good description of non-percolating systems, e.g. certain porous materials like foams. For the case of ellipsoidal inclusions (aligned or randomly oriented) see (Torquato 2002).

4.2. Cluster Expansions

Maxwell's classic dilute-limit formula (Maxwell 1873) assumes that the (spherical) inclusions do not interact. Cluster expansions, on the other hand, attempt to extend the applicability of these results to higher volume fractions (non-dilute systems) by taking interactions of inclusions into account (Torquato 2002).

4.2.1. Dilute Systems
The effective conductivity of an isotropic dilute dispersion of spherical inclusions is (Maxwell 1873)

$$k = k_1 + 3k_1 \beta_{21} \phi + ..., \tag{71}$$

independent of the size distribution (even for non-spherical inclusions polydispersity does not come into play), with the so-called polarization coefficient ("polarizability") β_{21} defined as

$$\beta_{21} = \frac{k_2 - k_1}{k_2 + 2k_1}. \qquad (72)$$

This expression coincides with the first-order expansion of one of the HS bounds, i.e. the polarization coefficient β_{21} for spheres provides bounds on the corresponding coefficient for isotropic dispersions of (randomly oriented) inclusions of arbitrary shape. Specifically, if we write $k = k_1 + f k_1 \beta_{21} \phi + ...$, then we have $f \geq 3$. This is valid even for ellipsoidal and polyhedral inclusions (Hetherington and Thorpe 1992).

For the special case of superconducting spheres ($k_2/k_1 = \infty$) we have (Markov 2000)

$$\frac{k}{k_1} = 1 + 3\phi + ..., \qquad (73)$$

whereas in the special case of perfectly insulating spheres ($k_2 = 0$) one obtains

$$\frac{k}{k_1} = 1 - \frac{3}{2}\phi + ... \qquad (74)$$

The corresponding result for the case of aligned ellipsoids is (Torquato 2002)

$$\mathbf{K} = \mathbf{K}_1 + \mathbf{M}\phi + ..., \qquad (75)$$

where $\mathbf{M}$ is the polarization concentration tensor for the ellipsoid, given by

$$\mathbf{M} = (k_2 - k_1)\left[\mathbf{1} + \mathbf{A}\frac{(k_2 - k_1)}{k_1}\right]^{-1}, \qquad (76)$$

where $\mathbf{A}$ is the depolarization tensor, see Appendix B. In the case of aligned spheroids the effective conductivity tensor of the transversally isotropic composite, related to the principal directions, is given by

$$\mathbf{K} = \begin{pmatrix} K_{11} & 0 & 0 \\ 0 & K_{22} & 0 \\ 0 & 0 & K_{33} \end{pmatrix} \qquad (77)$$

where

$$K_{11} = K_{22} = k_1 + \frac{(k_2 - k_1)\phi}{1 + Q(k_2 + k_1)/k_1} + ..., \qquad (78)$$

$$K_{33} = k_1 + \frac{(k_2 - k_1)\phi}{1 + (1 - 2Q)(k_2 + k_1)/k_1} + ..., \qquad (79)$$

with Q being the shape factor for prolate and oblate spheroids, respectively and the volume fraction is related to number density ρ via the relation $\phi = \rho 4\pi a^2 c/3$. Note that for needle-shaped inclusions ($Q = 1/2$) K_{33} reduces to the arithmetic average and for disk-shaped inclusions ($Q = 0$) to the harmonic average $K_{33} = k_1[1 + (k_2 - k_1)\phi] + ...$, approximately. Tensorial polarization coefficients for other inclusion shapes are given in (Douglas and Garboczi 1995).

In the case of randomly oriented spheroidal inclusions a statistically isotropic composite results, whose effective thermal conductivity is

$$k = k_1 + M\phi + ..., \qquad (80)$$

where M is just the scalar polarization coefficient

$$M = \frac{k_2 - k_1}{3} \sum \frac{1}{1 + A_k (k_2 - k_1)/k_1}, \qquad (81)$$

with the A_k being the depolarization factors given in Appendix B (Polder and Van Santen 1946, Douglas and Garboczi 1995). For the special cases of needles and disks, respectively, we have the general expressions,

$$\frac{k}{k_1} = 1 + \frac{(k_2 + 5k_1)(k_2 - k_1)}{3k_1(k_1 + k_2)} \cdot \phi + ..., \qquad (82)$$

$$\frac{k}{k_1} = 1 + \frac{(2k_2 + k_1)(k_2 - k_1)}{3k_1 k_2} \cdot \phi + ... \qquad (83)$$

the first of which reduces to

$$\frac{k}{k_1} = 1 - \frac{5}{3}\phi + ... \qquad (84)$$

for the case of randomly oriented needle-like perfectly insulating pores, while the second remains undefined in the limiting case of infinitely thin disk-like pores (microcracks). This is the reason why in the theory of cracked solids the volume fraction concept (porosity) is replaced by the concept of crack density, see (Markov 2000). Note, however, that for finite thickness the coefficient can be calculated using the depolarization factors in Appendix B.

4.2.2. Non-Dilute Systems

The extension of the cluster expansion concept to non-dilute systems (by taking into account a successively larger number of interacting inclusions (or pores) goes back to (Rayleigh 1892) and has been elaborated later in the second half of the 20th century (Brown 1955, Jeffrey 1973, Torquato 1984). In general, the effective conductivity composites with (possibly overlapping) spheres can be written as

$$\frac{k}{k_1} = 1 + d\,\beta_{21}\,\phi + B\phi^2 + ..., \tag{85}$$

where B is a coefficient dependent on the zero-density limit of the radial distribution function (sum of the zero-density limits of the pair-blocking and pair-connectedness functions) and a functional of "cluster operators", see (Torquato 1984, Torquato 2002).

In the case of impenetrable spheres with the "most random" distribution (i.e. spherical inclusions avoiding overlap but not contact), the following extreme-contrast limits have been obtained (Jeffrey 1973):

- Superconducting inclusions ($k_2/k_1 = \infty$):

$$\frac{k}{k_1} = 1 + 3\phi + 4.51\phi^2 + ..., \tag{86}$$

- Perfectly insulating inclusions ($k_2/k_1 = 0$):

$$\frac{k}{k_1} = 1 - \frac{3}{2}\phi + 0.588\phi^2 + ... \tag{87}$$

On the other hand, in the case of impenetrable spheres with a "well-separated" distribution (i.e. spherical inclusions avoiding overlap and contact), the extreme-contrast limits are (Torquato 2002):

- Superconducting inclusions ($k_2/k_1 = \infty$):

$$\frac{k}{k_1} = 1 + 3\phi + 3\phi^2 + ..., \tag{88}$$

- Perfectly insulating inclusions ($k_2/k_1 = 0$):

$$\frac{k}{k_1} = 1 - \frac{3}{2}\phi + \frac{3}{4}\phi^2 + \dots. \tag{89}$$

Interestingly, the latter formulae can also be obtained by expanding the Maxwell approximation or the HS lower (upper) bound through second order in ϕ. For the case of arbitrary phase contrast (ratio of phase properties) see (Choy et al. 1998).

In the case of fully penetrable (overlapping) spheres (a limiting case of Torquato's cherry-pit model, for which the impenetrability parameter is equal to zero) the effective conductivity in the extreme-contrast limit is (Torquato 2002):

- Superconducting inclusions ($k_2/k_1 = \infty$):

$$\frac{k}{k_1} = 1 + 3\phi + 7.56\phi^2 + \dots, \tag{90}$$

- Perfectly insulating inclusions ($k_2/k_1 = 0$):

$$\frac{k}{k_1} = 1 - \frac{3}{2}\phi + 0.345\phi^2 + \dots. \tag{91}$$

For the case of arbitrary phase contrast (ratio of phase properties) see (Torquato 1985b). Interestingly, the Coble-Kingery type relation for porous media (Pabst 2005b) is intermediate between Eqs. (87) and Eq. (91), see Eq. (104) below. To summarize, through second order in ϕ, the effect of clustering is to increase k when the inclusions are superconducting and to decrease k when the inclusions are perfectly insulating. All cluster-expansions mentioned above are based on the solution of relatively simple two-body problems. Already three-body problems are very difficult to solve and higher-order expansions are not available (Torquato 2002). It should be mentioned that apart from cluster expansions there are so-called contrast expansions ("strong" or "weak"). These incorporate higher-order microstructural information via three-point parameters (which are related to three-point correlation functions, see Section 3) and predict microstructure-dependent percolation thresholds. The interested reader may consult (Torquato 2002) for this rather advanced topic.

4.3. Convenient Relations for the Prediction or Description of the Thermal Conductivity of Isotropic Two-Phase Composites and Porous Media

In this subsection we summarize the most convenient (i.e. easiest-to-use) relations for the parameter-free prediction or description (comparative fitting of measured data) of the effective thermal conductivity of (statistically isotropic) two-phase composites and porous materials (the latter as a special case of general composites). All of these relations, some of them being relatively new, have a strong foundation in, or connection with, the rigorous micromechanical results given in the preceding sections. Some of them only restate micromechanical key results in an alternative or more explicit form or from another point of view.

4.3.1. Halpin-Tsai Relation and Sigmoidal Average

Semi-empirical relations are generally not based on a specific model or theory. In many cases they have been found heuristically and lack a rigorous derivation. Nevertheless they can provide useful estimates of average values between bounds. Probably the most popular semi-empirical relation for the effective properties of composites is the Halpin-Tsai relation (Halpin and Kardos 1976). For thermal conductivity it can be written as

$$k = k_1 \cdot \left(\frac{1 + \zeta \chi \phi_2}{1 - \chi \phi_2} \right), \tag{92}$$

where the auxiliary parameter χ is related to another parameter ζ via the definition

$$\chi = \frac{k_2 - k_1}{k_2 + \zeta k_1}, \tag{93}$$

and ζ is in the range from 0 to ∞ (typically, phase "2" are the inclusions, e.g. fibers, and phase "1" the matrix). It can readily be shown that for $\zeta \to 0$ the Halpin-Tsai relation approaches the volume-weighted harmonic mean or the lower one-point bound (lower Wiener bound),

$$k_W^- = \frac{k_1 k_2}{\phi_1 k_2 + \phi_2 k_1}, \tag{94}$$

whereas for $\zeta \to \infty$ it approaches the volume-weighted arithmetic mean or the upper one-point bound (upper Wiener bound),

$$k_W^+ = \phi_1 k_1 + \phi_2 k_2. \tag{95}$$

Recently a new mixed average relation has been proposed, the "sigmoidal average", which is able to adequately describe S-shaped curves:

$$k = (1 - \phi) k_W^+ + \phi k_W^- \tag{96}$$

(Pabst and Gregorová 2010, Pabst and Hostaša 2010, Pabst et al. 2009). This relation ensures that the effective property remains close to the prevailing phase (if this applies) and thus does not violate the bounds. A definite advantage of our sigmoidal average relation over the Halpin-Tsai relation is the fact that the bounds occur explicitly in this relation and may readily be replaced by higher-order bounds, e.g. the two-point Hashin-Shtrikman bounds for isotropic microstructures, if desired. Thus a sigmoidal average may be also constructed using the upper and lower two-point (HS) bounds, i.e. according to the relation

$$k = (1-\phi)k_{HS}^+ + \phi k_{HS}^- . \tag{97}$$

Evidently, the sigmoidal average is an appropriate choice for describing the volume-fraction dependence of the effective properties of dense polycrystalline ("grainy") microstructures, corresponding to the aforementioned "symmetric-cell materials" (SCMs) of micromechanics.

4.3.2. Porous Materials as a Special Case of Two-Phase Composites

In the case of porous materials, with porosity ϕ ($=\phi_2$) and negligible thermal conductivity of the void phase, i.e. $k_2 \ll k_1$, it is convenient to define a relative thermal conductivity as

$$k_r \equiv \frac{k}{k_0}, \tag{98}$$

where k is the effective thermal conductivity of the porous material as a whole (as above) and k_0 ($= k_1$) the thermal conductivity of the solid phase alone (i.e. the matrix phase in the case of porous materials of the matrix-inclusion type = porous materials with essentially closed pores, or the skeleton phase in the case of bicontinuous porous materials = open-pore partially sintered or cellular solids; for convenience we will refer to k_0 as the solid phase thermal conductivity in the sequel, irrespective of the underlying concept of pore topology). In this case all the above relations are considerably simplified.

In particular, the upper Wiener bound, which describes a linearly decreasing line ("mixture rule") with a slope of -1, can be written as

$$k_r = \frac{k_W^+}{k_0} = 1 - \phi , \tag{99}$$

and the Hashin-Shtrikman upper bound (Maxwell-Eucken relation, describing a nonlinearly decreasing curve) simplifies to the relation

$$k_r = \frac{k_{HS}^+}{k_0} = \frac{1-\phi}{1+\phi/2} , \tag{100}$$

(Salmang and Scholze 1982, Santos 2003), which is at the same time a special case of a model relation proposed by Nielsen (Nielsen 1973, 1984, Pezzotti et al. 2000a,b).

Note that the Wiener bound relation is contained as a special case (when radiation effects are neglected) in the Loeb relation (Francl and Kingery 1954, Loeb 1954), see Section 6. It will therefore be useful for structures with cylindrical pore channels parallel to the temperature gradient (and heat flux) (Pouchon et al. 1998). Although for isotropic materials with isometric pores this simple relation will usually be wrong, it has been used earlier for porosity correction when a better alternative was lacking, e.g. (Kingery 1960, Kingery et al. 1954). We would like to remind the reader of the fact that both the Wiener lower bound and the HS lower bound are zero in the case of porous materials, see Section 3.

For very low porosities ($\phi \to 0$), where mutual interactions between the pores can be neglected (so-called dilute approximation) it is justified to assume a linear dependence of the relative thermal conductivity on the porosity. In the case of spherical pores this linear approximation is

$$k_r = 1 - \frac{3}{2}\phi \tag{101}$$

This linear relation, which may be called the dilute approximation (also called non-interaction approximation or one-particle approximation (Markov 2000)), is the solution of the single-inclusion problem of a spherical void in an infinite medium (see Appendix B). For other pore shapes the value of the coefficient will in general be different from 3/2.

4.3.3. Coble-Kingery-Type Relation for Porous Materials

Experience shows that usually the porosity dependence of the effective thermal conductivity is not linear. Thus, since it is clear that the linear relation claims validity and practical significance only in the case of very small porosities (dilute approximation), there have been numerous attempts to extend it to higher porosities by allowing for a nonlinear dependence. The simplest way to do so is the Coble-Kingery approach, which has originally been proposed in elasticity context (Coble and Kingery 1956), but can be adopted equally well for thermal conductivity (Pabst 2005b). This approach is based on the following argumentation: In order to get a non-linear relation, take the linear relation (dilute approximation), and add a quadratic term in ϕ in order to obtain a second-order polynomial, i.e.

$$k_r = 1 - [k]\phi + \alpha \phi^2, \tag{102}$$

where the value of the first-order coefficient $[k]$ is a function of pore shape (e.g. the aspect ratio in the case of spheroids), see Appendix B, and the value of the second-order coefficient α is determined via the necessary condition that $k_r = 0$ at least for $\phi = 1$ (which is necessary in order not to violate the upper Wiener bound). In general one obtains (Pabst and Gregorová 2004a, 2007)

$$k_r = 1 - [k]\phi + ([k]-1)\phi^2. \tag{103}$$

In the special case of porous materials with spherical pores this reduces to the Coble-Kingery-type relation

$$k_r = 1 - \frac{3}{2}\phi + \frac{1}{2}\phi^2 \qquad (104)$$

This relation yields values that are for all porosities ($0 < \phi < 1$) below the Hashin-Shtrikman upper bound (Maxwell-Eucken relation) and reduces to the dilute approximation (self-consistent relation) in the dilute limit ($\phi \to 0$), as required. Moreover, the second-order coefficient (0.5) is between the second-order coefficient of the cluster expansion for impenetrable insulating spheres (Jeffrey 1973) and the second-order coefficient of the cluster expansion for fully penetrable insulating spheres (Torquato 1985b) (0.588 and 0.345, respectively), cf. (Pabst 2005b).

4.3.4. Effective Medium Approximations for Porous Materials

As mentioned before, apart from the Maxwell-Eucken relation, which is based on an effective field approach, there are two popular relations based on effective medium approaches, the self-consistent approximation

$$\phi_1 \frac{k_1 - k}{k_1 + 2k} + \phi_2 \frac{k_2 - k}{k_2 + 2k} = 0 \qquad (105)$$

and the differential approximation

$$\frac{dk}{d\phi_2} = \frac{3k(k_2/k_1 - k/k_1)}{(1 - \phi_2)(k_2/k_1 + 2k/k_1)} \qquad (106)$$

(Bruggeman 1935, Cernuschi et al. 2004, Hasselman et al. 1987, Landauer 1978, Markov 2000, McLachlan 1985, 1986, McLachlan et al. 1990, 2001, Phan-Thien and Pham 2000, Torquato 2002). Both equations are implicit in the effective thermal conductivity k, i.e. for general multiphase composites they have to be solved iteratively. Similar to the Maxwell-Eucken relation they have been derived for material models with spherical inclusions (phase 2) in a continuous matrix (phase 1).

For porous materials the self-consistent approximation and the differential approximation are given by simple explicit relations

$$k_r = 1 - \frac{3}{2}\phi \qquad (107)$$

and

$$k_r = (1 - \phi)^{3/2}, \qquad (108)$$

respectively (Bruggeman 1935, Cernuschi et al. 2004, Hu et al. 1998, Lu et al. 2001, Markov 2000, Phan-Thien and Pham 2000, Torquato 2002). Evidently, the self-consistent approximation is in this case equivalent to the aforementioned dilute approximation. The same expression follows from the Maxwell-Eucken relation for small porosities (Cernuschi et al. 2004) and can be regained from all subsequent relations in the limiting case of dilute ($\phi \rightarrow 0$) systems of (non-interacting) spherical pores. Although both relations, the self-consistent relation and the differential relation, have been derived for a model material with isolated (closed) spherical pores, they can in practice be used up to 67 % and 100 % porosity, respectively, because they do not violate the HS upper bound. The differential approximation can be conveniently extended to non-spherical (spheroidal) pores, possibly with preferential orientation (Lu et al. 2001, Ondracek and Schultz 1973, Ondracek 1987, Schultz 1981, Schultz and Wedemeyer 1986).

Another useful effective medium approximation is the exponential relation (Pabst and Gregorová 2008)

$$k_r = \exp\left(\frac{-\frac{3}{2}\phi}{1-\phi}\right). \tag{109}$$

This exponential relation, which is closely related to the so-called Mooney relation in suspension rheology (Mooney 1951) and the corresponding relation in elasticity context (Pabst and Gregorová 2004b), can be derived via a functional equation approach, see also (Tichá et al. 2005), and, similar to the Maxwell-Eucken relation, the differential relation and the Coble-Kingery-type relation above, reduces to the dilute approximation (in this case equivalent with the self-consistent relation) in the case of very small porosity, as required. It has been sucessfully used to predict the effective thermal conductivity of porous alumina ceramics at room temperature and at elevated temperature (Pabst and Gregorová 2007, Živcová et al. 2009).

4.3.5. Fit Relations for Porous Materials

Sometimes the purely empirical relation

$$k_r = \frac{1-\phi}{1+b\phi} \tag{110}$$

(with fit parameter b) is called Maxwell-Eucken relation (Elbel and Vollath 1988, Pouchon et al. 1998) and, similarly, the empirical relation

$$k_r = 1 - c\phi \tag{111}$$

(with fit parameter c) is sometimes called modified Loeb relation (Pouchon et al. 1998, Santos 2003). A value of $c = 4/3$ is sometimes propagated (Klemens 1991, Nieto et al. 2004, Schlichting et al. 2001), but its justification is doubtful. Moreover, empirical power-law relations of the type

$$k_r = (1-\phi)^N \qquad (112)$$

are frequently used for fitting purposes (Archie 1942, Raghavan et al. 2001, Schulz 1981). It is highly confusing, however, to set $c = 2.5$ or $N = 2.5$, as is sometimes done (Clarke 2003, Pouchon et al. 1998), since this value (although it may incidentally be in relatively good agreement with measured data) gives the wrong impression of being derived from a model similar to the Einstein model for the effective viscosity of suspensions (Markov 2000, Torquato 2002). Interestingly, both finite-element calculations (Bakker et al. 1995) and experiments with highly porous aerogels (Fricke 1988, Fricke and Tillotson 1997, Lu et al. 1995) have lead to the conclusion that $N \approx 1.7 \pm 0.2$, which is rather close to the value of 3/2 predicted for spherical pores. A more general power-law relation of the form

$$k_r = \left(1 - \frac{\phi}{\phi_C}\right)^N \qquad (113)$$

has been used by McLachlan, who interpreted the critical porosity ϕ_C and the critical exponent N in terms of percolation theory (McLachlan 1985, 1986, McLachlan et al. 1990, 2001), cf. (Sahimi 1994, 2003, Stauffer and Aharony 1985, Zallen 1983).

As an alternative to McLachlan's relation, which can be viewed as a two-parameter fit relation, a one-parameter fit relation has been proposed recently (Pabst and Gregorová 2006b),

$$k_r = \left(1 - \frac{\phi}{2}\right) \cdot \left(1 - \frac{\phi}{\phi_C}\right). \qquad (114)$$

In both cases the parameter ϕ_C can be interpreted as a critical porosity (percolation threshold), at which the thermal conductivity becomes practically zero, similar to the situation in elasticity context (Pabst and Gregorová 2004c). The latter relation, however, has the advantage that only one fit parameter is present and that in the absence of a critical porosity, i.e. in the case $\phi_C = 1$, the Coble-Kingery-type relation is regained. This guarantees that the HS upper bound will never be violated.

4.3.6. Ondracek Bounds for Porous Materials

A general model equation has been derived by Ondracek and coworkers (Nikolopoulos and Ondracek 1983, Ondracek 1986) for the thermal conductivity of porous materials with a matrix-inclusion type microstructure containing spheroidal pores. Apart from porosity and the aspect ratio R of the spheroid, this equation contains a shape factor F, ranging from 0 to 0.5 (1/3 for spherical pores), and an orientation factor $\cos^2 \alpha$, ranging from 0 to 1 (1/3 for random orientation). For porous materials it can be written in the form

$$k_r = (1-\phi)^{\frac{1-\cos^2\alpha}{1-F}+\frac{\cos^2\alpha}{2F}}.\tag{115}$$

In the case of spherical pores this equation reduces to the power-law relation

$$k_r = (1-\phi)^{3/2},\tag{116}$$

and in the case of randomly oriented infinitely prolate pores (cylindrical pore channels) one obtains

$$k_r = (1-\phi)^{5/3}.\tag{117}$$

Based on his general equation and the empirical finding (partly supported by plausibility and surface energy arguments), that the aspect ratio of oblate (lenticular) pores developed during sintering cannot be smaller than 0.1, Ondracek (Ondracek 1982) derived upper as well as lower (!) "bounds" for the effective conductivity of sintered porous materials. Although these bounds have been termed "first-order", "second-order" and "third-order bounds" they should not be confused with the rigorous micromechanical bounds (one-, two- and three-point bounds). Instead, they correspond to different pore orientations and pore shape factors as follows:

- "First-order bounds":

  - Upper bound ($F = 0$ and $\cos^2\alpha = 0$, corresponding to cracks with infinitely small aspect ratio oriented parallel to the field direction, i.e. with their normals perpendicular to it, or $F = 0.5$ and $\cos^2\alpha = 1$, corresponding to pore channels oriented parallel to the field direction):

  $$k_r^{I+} = 1-\phi \tag{118}$$

  - Lower bound ($F = 0.0696$ and $\cos^2\alpha = 1$, corresponding to oblate / lenticular pores with the minimum aspect ratio of 0.1, oriented parallel to the field direction, i.e. with their normals perpendicular to it):

  $$k_r^{I-} = (1-\phi)^7 \tag{119}$$

- "Second-order bounds":
  - Upper bound ($F = 0$ and $\cos^2\alpha = 1/3$, corresponding to randomly oriented cracks):

  $$k_r^{II+} = \frac{1-\phi}{1+\phi/2} \tag{120}$$

- Lower bound ($F = 0.0696$ and $\cos^2 \alpha = 1/3$, corresponding to randomly oriented oblate / lenticular pores with the minimum aspect ratio of 0.1):

$$k_r^{II-} = (1-\phi)^3 \qquad (121)$$

- "Third-order bounds":

  - Upper bound ($F = 1/3$ and $\cos^2 \alpha = 1/3$, corresponding to spherical pores or randomly oriented closed, unconnected isometric pores):

  $$k_r^{III+} = (1-\phi)^{3/2} \qquad (122)$$

  - Lower bound ($F = 0.0696$ and $\cos^2 \alpha = 1/3$, corresponding to randomly oriented closed, unconnected oblate / lenticular pores with the minimum aspect ratio of 0.1):

  $$k_r^{III-} = (1-\phi)^3 \qquad (123)$$

Note that the "first, second and third-order" upper bounds correspond to the Wiener bound for arbitrary microstructures, the Hashin-Shtrikman bound for isotropic microstructures and the power-law prediction for isotropic microstructures with isometric pores, respectively. Of course, the "third-order upper bound" cannot be a rigorous bound, since the isotropic Hashin-Shtrikman microstructure / coated-spheres model (which consists of closed, unconnected pores) would exceed it. Nevertheless, the so-called "third-order bounds" are reported to give excellent agreement with numerous experimental data published in the literature (Nikolopoulos and Ondracek 1983).

5. PHASE MIXTURE MODELS FOR THE GRAIN SIZE DEPENDENCE OF THE THERMAL CONDUCTIVITY

5.1. Motivation

The traditional micromechanical treatment of effective properties does not take into account interface effects. In other words, grain boundaries are treated as sharp interfaces (of zero thickness or volume) without properties. If not explicitly stated differently, it is tacitly assumed in micromechanics that the interface is perfect in the sense that it makes no contribution e.g. to the thermal resistance of a body (or the thermal resistivity of a material). Therefore it is not possible, within the traditional micromechanical framework, to assess the grain size dependence of the effective thermal conductivity, which may be expected to play a significant role mainly in nanocrystalline materials. In order to estimate grain size effects on properties (and to assess the technological potential of the nano-approach in quantitative terms) it is necessary to dispose of certain modeling tools that allow an explicit calculation of the grain size dependence and predict the properties of nanocrystalline materials even in cases where current processing technologies do not yet allow the preparation of defect- and pore-free nanomaterials in bodies of sufficient size for direct experimental verification. In this section one category of these tools is presented: phase mixture models. Although phase mixture models are being applied since the 1990es for the purpose of predicting the mechanical properties of nanomaterials (Aifantis 1993, 1994, Aust et al. 2004, Carsley et al. 1995, Chaim 1997, Chaim and Hefetz 2004, Ehre and Chaim 2008, Estrin et al. 2004, Jiang and Weng 2004, Khadar et al. 2003, Kim 1998, Kim and Bush 1999, Kim et al. 2000a, 2000b, 2001, Makhlouf et al. 2009, Nair and Khadar 2008, Palumbo et al. 1990, Sakai et al. 1999, Wang et al. 1993, 1995, 1997, Yeheskel et al. 2005), their application to thermal conductivity is of a quite recent date (Pabst 2009, Pabst and Gregorová 2010, Pabst et al. 2009). In fact, in this field it was much more popular for a long time to model the grain boundary as a sharp interface and to assign an interfacial resistance to this grain boundary (Gasch et al. 2008, Gordon et al. 1994, Hasselman and Johnson 1987, Reeves et al. 1991, Smith et al. 2003a, 2003b, Tessier-Doyen et al. 2007, Yang et al. 2002). The phase mixture approach, in which the grain boundary is considered as a region of small but finite thickness, is an alternative to the latter approach. Although the interfacial-resistance approach has been used for quite a long time, interfacial-resistance models tend to fail for nanocrystalline materials with very small grain sizes, because they often predict values far below the conductivity value of the corresponding defective or glassy material (amorphous limit prediction), which is at least 1.1 W/mK (Cahill et al. 1992, Clarke 2003, Carke and Phillpot 2005, Schelling and Phillpot 2001, Sun et al. 2009, Walker and Anderson 1984). The phase mixture approach, on the other hand, has the advantage that it will never predict conductivities below this amorphous limit (if experimental data seem to indicate this, it can be conjectured that porosity effects are dominating over grain size effects). Moreover, mechanical and thermal properties can be treated in a common framework (Pabst and Gregorová 2010).

The present section is divided into several subsections. Subsection 5.2 introduces the general concept of phase mixture modeling and the "equivalent porosity" concept. In Subsection 5.3 the cubic core-shell unit-cell geometry is defined. This model geometry is the simplest to calculate and seems to be a reasonable model for isotropic microstructures. Concerning other unit cell shapes and anisotropic microstructures the reader should refer to (Pabst and Gregorová 2010). In Subsection 5.4 the grain size dependence of thermal conductivity is discussed in general for ceramics with different conductivity levels.

5.2. Phase Mixture Models and Micromechanical Bounds

In phase mixture models single-phase polycrystalline materials are considered as two-phase composites with a crystalline core phase (interior of the crystallites, with properties corresponding to the dense bulk phase) and a highly disordered grain boundary (GB) phase of finite thickness and lower density (and much higher defect-concentration), covering the crystalline cores. The properties of the GB phase are different from that of the bulk phase in a specific way, depending on the property considered and the material in question. Since the GB region is strongly disordered (glass-like or gas-like (Gleiter 1989)) it seems reasonable to assume, that the grain boundary properties are close to those of an amorphous glass phase corresponding in chemical composition to the crystalline core (bulk) phase. The grain boundary thickness is usually estimated to be around 0.5–2 nm (Adams et al. 1989, Chaim 1997, Ehre and Chaim 2008, Epperson et al. 1989, Jiang and Weng 2004, Kirchheim et al. 1988, Mütschele and Kirchheim 1987, Phillpot et al. 1995, Ping et al. 1995, Sakai et al. 1999, Thomas et al. 1990, Wang et al. 1996, Wen and Yan 1995). Therefore, in cases where specific information is lacking (e.g. from TEM), a value of 1 nm appears to be a reasonable estimate for the GB thickness.

For two-phase composites, when the volume fraction of one phase is the only microstructural information available (in a two-phase composite the volume fraction of the other phase is then redundant), it is well known that the effective thermal conductivity k of a material is bounded from above by the volume-weighted arithmetic mean (upper Wiener bound),

$$k_W^+ = \phi_1 k_1 + \phi_2 k_2 \tag{124}$$

and from below by the volume-weighted harmonic mean (lower Wiener bound),

$$k_W^- = \frac{k_1 k_2}{\phi_1 k_2 + \phi_2 k_1}, \tag{125}$$

see Section 4, (Wiener 1912) and (Markov 2000, Milton 2002, Nemat-Nasser and Hori 1999, Torquato 2002). When isotropy can be assumed, the two-point bounds (upper and lower Hashin-Shtrikman (HS) bounds (Hashin and Shtrikman 1962)) adopt an especially simple form (Torquato 2002) for the thermal conductivity, *viz.*

$$k_{HS}^+ = \phi_1 k_1 + \phi_2 k_2 - \frac{\phi_1 \phi_2 (k_1 - k_2)^2}{3k_1 - \phi_1 (k_1 - k_2)}, \tag{126}$$

(upper HS bound) and

$$k_{HS}^- = \phi_1 k_1 + \phi_2 k_2 - \frac{\phi_1 \phi_2 (k_1 - k_2)^2}{3k_2 + \phi_2 (k_1 - k_2)} \tag{127}$$

(lower HS bound) when $k_1 > k_2$ (otherwise the bounds are reversed).

Without further, very specific, microstructural information the HS bounds cannot be improved, see Section 4. Therefore, it would be useful to have a guideline concerning which of these bounds should be used for specific microstructures or else, what kind of average between the two bounds might be used when more specific information is lacking. When the phase contrast is small, e.g. a factor of 2, there is no point (one can use any type of average, since the difference between them is negligible with respect to the precision expected for the prediction), but the question is crucial for composites with phase contrast far away from unity (more than one order of magnitude). In particular, when the phase contrast approaches infinity, the lower bounds become zero (i.e. useless in practice) and the simple arithmetic average leads to wrong conclusions in the case small second phase volume fractions. The question of averaging (homogenization) with relation to phase mixture modeling has been treated in detail in (Pabst and Gregorová 2010).

In fact, there are certain microstructures for which one of the HS bounds are correct: the so-called HS microstructures, also called coated-spheres models (Christensen 1979, Torquato 2002) or Hashin assemblages (Hashin 1962, Markov 2000). These are microstructures in which one phase is always connected, while the other remains disconnected for all volume fractions (except for 0 and 1), and the aforementioned core-shell model used in the phase mixture approach is very similar to them. Composites with this type of microstructure remain of matrix-inclusion type for all volume fractions (except for 0 and 1), i.e. do not exhibit percolation thresholds (Torquato 2002). One phase (the core or bulk phase) is always embedded in (or at least covered with a thin film of) the second phase (the shell or GB phase), so that the overall behavior will always be dominated by the shell phase, i.e. close to one of the bounds. In particular, when the material is isotropic and the shell phase is the low-property-value phase, the effective property will be close to the lower HS bound.

A second, practically important, type of microstructures is represented by so-called "symmetric-cell materials" (Miller 1969, Torquato 2002), which can be imagined as a mixture of two types of grains in a 3D random checkerboard arrangement. In this case percolative behavior is expected, i.e. as long as one phase is dominating the behavior is that of a matrix-inclusion composite, with the effective property determined by the matrix phase (i.e. being close to one of the bounds). At a certain volume fraction of the second phase the effective property shifts to approach the opposite bound, in extreme cases in the form of a percolation threshold. The simplest average relation that is able to describe this shift from the upper bound to the lower bound is the "sigmoidal average" mentioned in Section 4, see also (Pabst and Gregorová 2010, Pabst et al. 2009):

$$k_\sigma = \phi_1 k^+ + \phi_2 k^-, \tag{128}$$

where k^+ and k^- are the upper and lower bounds (Wiener or HS). We emphasize that the sigmoidal average remains admissible even if the lower bound degenerates to zero (i.e. even in the case of one phase property value being zero). Therefore it may principally be used to estimate the porosity dependence of the effective thermal conductivity. Although we will use this relation in the present section only in connection with porosity, it should be mentioned that it also opens the door for modeling dense composite ceramics, e.g. alumina-zirconia composites, or extending the phase mixture approach to composites (in contrast to the single-phase ceramics treated here).

Taking into account the upper Wiener bounds and upper Hashin-Shtrikman bounds for the thermal conductivity of porous materials (Pabst and Gregorová 2006b, 2007, Pabst et al. 2007), i.e.

$$k_r = 1 - \phi, \tag{129}$$

$$k_r = \frac{1-\phi}{1+\phi/2}, \tag{130}$$

respectively, the corresponding sigmoidal average relations are

$$k_r = (1-\phi)^2 , \tag{131}$$

$$k_r = \frac{(1-\phi)^2}{1+\phi/2} . \tag{132}$$

These straightforward relations are different from other relations derived by micromechanical approaches (Pabst and Gregorová 2010) and both relatively close to the exponential relation

$$k_r = \exp\left(\frac{-\tfrac{3}{2}\phi}{1-\phi}\right), \tag{133}$$

which is well known to provide realistic predictions for the porosity dependence in (isotropic) materials with (closed or open) convex pores (Pabst and Gregorová 2006b, 2007, 2008, Tichá et al. 2005, Živcová et al. 2009). The advantage of the sigmoidal averages is that they can be generalized to composites with arbitrary phase contrasts, whereas no corresponding generalization of the exponential relation is known. Apart from the well-known power-law relation,

$$k_r = (1-\phi)^{3/2}, \tag{134}$$

which may be useful for highly porous cellular materials (Gibson and Ashby 1997, Pabst and Gregorová 2006b, 2007, 2008), the sigmoidal averages and the exponential relation seem to be the most useful predictive relations (estimates) for the porosity dependence of the effective thermal conductivity. Figure 1 compares the different relations available for estimating this porosity dependence. A look into the literature reveals that many experimental data are close to the exponential relation or the sigmoidal average relations, cf. (Pabst and Gregorová 2007, 2008, Živcová et al. 2009) and the papers cited therein.

It is often very difficult to distinguish between grain size and porosity effects. Although both tend to reduce the thermal conductivity in general, the grain size effect is very small compared to the porosity effect. That means, very small – possibly undetected or undetectable – porosities may have an effect corresponding to significant grain size reduction (Pabst and Gregorová 2010). Sometimes, especially in nanocrystalline materials, nanosized pores may be present that represent *de facto* regions of reduced density (and presumably increased structural disorder, similar to the GB regions). In this case it may be difficult even conceptually to distinguish grain size and porosity effects. In any case, to each (calculated) grain size effect (thermal conductivity reduction), a certain porosity value can be assigned (via one of the aforementioned predictive relations) that would have the same effect. We call this the "equivalent porosity concept".

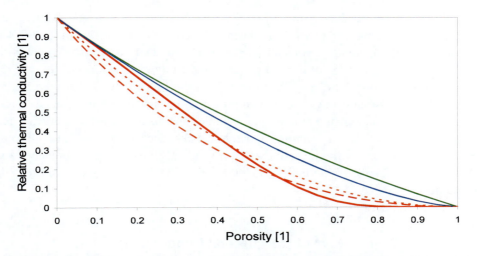

Figure 1. Relative thermal conductivity of porous materials according to different predictive relations (estimates for statistically isotropic materials); upper Hashin-Shtrikman bound (green), power-law relation (blue), exponential relation (red full), sigmoidal average of the Wiener bounds (red dotted), sigmoidal average of the Hashin-Shtrikman bounds (red dashed).

It has to be emphasized, that current technology does usually not allow a direct local experimental measurement of the GB phase properties. Therefore, at this point one must make assumptions based on estimates in the literature, partly based on integral (macroscale) measurements, partly on simulation results. An estimate of the GB thermal conductivity can be made using the amorphous limit formula (Clarke 2003, Clarke and Phillpot 2005)

$$k_{min} \approx 0.87 k_B \rho^{1/6} E^{1/2} \left(\frac{n \cdot N_A}{M} \right)^{2/3}, \qquad (135)$$

where k_B is the Boltzmann constant, N_A the Avogadro constant, ρ the density, E the Young's modulus, n the number of ions in the unit cell and M the mean atomic mass of the ions in the unit cell. When using this formula, the amorphous limit is e.g. 2.9 W/mK for alumina (corundum, α-Al_2O_3), 2.6 W/mK for magnesia (MgO), 2.3 W/mK for magnesia-alumina spinel ($MgAl_2O_4$), 2.1 W/mK for titania (rutile, TiO_2) and 1.1-2.0 W/mK for the silicates considered there, viz. 1.1 W/mK for Y_2SiO_5, 1.4 W/mK for γ-$Y_2Si_2O_7$, 1.7 W/mK for mullite ($Al_6Si_2O_{13}$) and 2.0 for forsterite (Mg_2SiO_4) (Sun et al. 2009). The amorphous limit for zirconia has been estimated to be 1.1 W/mK (Cahill et al. 1992, Schelling and Phillpot 2001). These values can be recommended as estimates of the grain boundary thermal conductivity. Of course, in systems where such information is lacking, it is reasonable to calculate an extremal estimate, based on the value of 1.1 W/mK, which is a reasonable estimate for the amorphous limit (minimum thermal conductivity) in the absence of a better alternative (Pabst 2009, Pabst and Gregorová 2010). Note that the value of 1.1 W/mK is a good estimate also for the thermal conductivity of silicate glasses (range 0.6-1.4 W/mK, see Section 8) and is sometimes recognized as a more or less universal value (Kingery 1960, Kingery et al. 1976). For Si_3N_4 a strong grain size dependence of the thermal conductivity has been reported, which can be modeled using a thermal conductivity value of 0.6 W/mK for the grain boundary phase (Kushan et al. 2007). Also for SiC ceramics a significant grain size dependence of the thermal conductivity has been reported in the range 2-8 μm (60-90 W/mK) (Jang and Sakka 2008), but a GB conductivity has not been reported. It has to be emphasized, that – although the exact value of the GB phase conductivity may be disputable – the results of the phase mixture model are relatively insensitive to this parameter. Other possible errors, intrinsic to this kind of modeling approach, are much more significant (e.g. the GB thickness estimate), and the choice of the appropriate average relation is much more crucial than an error in the estimate of the GB phase conductivity (Pabst and Gregorová 2010). In order to make the results of this section invariant with respect to the GB conductivity value (and the thermal conductivity level of the bulk phase) the grain size dependence is given for relative conductivities here.

5.3. Grain Size Dependence of the Thermal Conductivity

Grains in polycrystalline, including nanocrystalline, materials are usually modeled as polyhedra with flat faces. From the viewpoint of simplicity and computational efficiency it is advantageous to model the material as monodisperse, i.e. as composed of identical polyhedra of the same size. Only a few of these polyhedra, of course, are able to fill space completely (without voids), and of those which are, the most popular one is the cube. Seemingly closer to reality is the tetrakaidecahedron or truncated octahedron (also called "Kelvin cell"), a polyhedron consisting of six square and eight hexagonal faces. Except for these two, which are ideally suited as models for isometric grains, there are three prismatic shapes (based on with regular triangles, squares, or regular hexagons as cross-sections) which are also space-filling. The latter three can be used to model anisometric grain shapes, i.e. prolate (elongated) and oblate (flattened) grains, see (Pabst and Gregorová 2010). The simplest space-filling arrangement of cubes is a simple cubic arrangement and that of Kelvin cells body-centered cubic whereas the spatial arrangements of prismatic shapes may be nematic or smectic (Pabst and Gregorová 2010). It has been shown in (Pabst and Gregorová 2010) that the predictions based on different isometric shapes are very close, as long as the volume-equivalent sphere diameter is chosen as a common size measure. Therefore, here we confine ourselves to cubic unit cells, which may be imagined as being in simple cubic arrangement (although the arrangement is in fact irrelevant here).

The volume-equivalent sphere diameter D_C is related to the cube edge length a_C via the relation

$$D_C = a_C \sqrt[3]{\frac{6}{\pi}}. \qquad (136)$$

Similar relations are available for the volume-equivalent sphere diameter of a tetrakaidecahedron and for anisometric grain shape models (e.g. tetragonal, trigonal or hexagonal prisms), where additionally a shape measure (aspect ratio) has to be taken into account, see (Pabst and Gregorová 2010). When a_C denotes the edge length of the cube and t the GB thickness, the following simple formulae are needed for calculating the grain size dependence of the effective thermal conductivity of model materials with (two-phase) cubic unit cells:

- Volume of the whole unit cell: $V = a_C^3$
- Volume of the crystalline core phase in the unit cell: $V_{core} = (a_C - t)^3$
- Volume fraction of the crystalline core phase in the unit cell: $\phi_{core} = \dfrac{(a_C - t)^3}{a_C^3}$
- Volume fraction of the GB phase in the unit cell: $\phi_{GB} = 1 - \dfrac{(a_C - t)^3}{a_C^3}$

The corresponding expressions for tetrakaidecahedra and anisometric unit cells (trigonal, tetragonal and hexagonal prisms) can be found in (Pabst and Gregorová 2010). For isometric shape models (cubes and tetrakaidecahedra, as well as the prismatic shapes with aspect ratio 1) the only value of interest is an orientationally averaged effective property. Since the calculation made for isometric shapes uses only the volume fractions of one unit cell as input information, the cubic symmetry of the spatial cell arrangement is in fact irrelevant and an overall average value is automatically obtained.

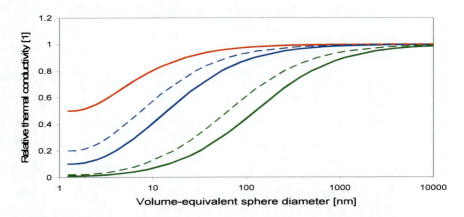

Figure 2. Grain size dependence of the relative thermal conductivity for dense (pore-free) isotropic polycrystalline ceramics, calculated via the lower HS bound for different phase contrasts between the highly conducting crystalline core phase and the badly conducting GB phase (red full: 2, blue dashed: 5, blue full: 10, green dashed: 50, green full: 100).

Figure 2 shows the grain size dependence of the relative thermal conductivity for dense (pore-free) isotropic polycrystalline ceramics, calculated via the lower HS bound for different phase contrasts between the highly conducting crystalline core phase and the badly conducting GB phase. Phase contrasts in the range 2-5 correspond e.g. to zirconia and many silicates, phase contrasts between 10 and 50 to alumina and many other oxide ceramics and phase contrasts of order 100 to beryllia and some non-oxide ceramics, see Sections 7-9. As expected, the grain size effect is generally more significant for ceramics with a high phase contrast between the crystalline cores and the GB phase. These phase mixture modeling results indicate that – in contrast e.g. to the Young's modulus (Pabst and Gregorová 2010) –

significant thermal conductivity reductions can be obtained by reducing the grain size. This has to be taken into account especially in the case of nanocrystalline ceramics. With anisometric grains, the potential is even larger in principle, but aspect ratios very far from unity have to occur in order for the shape effect to be significant. Shifts in the grain size dependence by more than one order of magnitude can hardly be expected, not even for the most extreme aspect ratios (Pabst and Gregorová 2010).

Tables 1 and 2 show the maximum grain sizes (volume-equivalent sphere diameters) D_i (in (μm)) corresponding to 10, 20, 50, 80, 90, 95, 98 and 99 % of the thermal conductivity of the crystalline core phase (coarse-grained dense ceramic) and the corresponding equivalent porosities (in (%)), calculated according to different porosity models (see above). The porosities listed would have the same effect in reducing the effective thermal conductivity as a grain size reduction to D_i.

Table 1. Maximum grain sizes (volume-equivalent sphere diameters) in (μm) corresponding to 10, 20, 50, 80, 90, 95, 98 and 99 % of the thermal conductivity of the crystalline core phase for polycrystalline ceramics of different phase contrasts between crystalline core and amorphous GB phase, calculated via the phase mixture model using cubic unit cells; phase contrasts between 2 and 5 are typical for zirconia and silicate ceramics, phase contrasts between 10 and 50 for many oxide ceramics (e.g. alumina) and phase contrasts of order 100 for beryllia and some non-oxide ceramics (e.g. silicon carbide)

Critical grain size label (quantile)	Phase contrast (ratio of the thermal conductivities of the crystalline core phase and the amorphous GB phase)					
	2	5	10	30	50	100
D_{10}	-	-	1	5	8	15
D_{20}	-	1	4	11	17	32
D_{50}	1	6	14	40	64	126
D_{80}	10	29	55	155	254	502
D_{90}	22	63	122	347	570	1128
D_{95}	47	133	256	731	1203	2382
D_{98}	122	341	658	1882	3099	6140
D_{99}	246	689	1328	3800	6266	12405

The values indicate that for a material with a phase contrast value of 2 between core phase and GB phase (e.g. zirconia and some silicates) a porosity of only 1 % (which may be undetectable) is equivalent (with respect to its effect on the thermal conductivity) to a grain size of 122 nm. Conductivity reductions by 5, 10 and 20 % require for this material either grain sizes of 47, 22 and 10 nm or porosities of 2-3, 4-7 and 8-14 % (depending on the model chosen for the porosity dependence), respectively. That means in this case grain size effects can only be expected for truly nanocrystalline samples with a grain size significantly below 100 nm. On the other hand, for materials with phase contrast values between 10 and 50 (typical for oxides, e.g. alumina) a porosity of 1 % is equivalent to a grain size of 0.7-3.1 μm, and conductivity reductions by 5, 10 and 20 % require for this material grain sizes of 256-1203 nm, 122-570 nm and 55-254 nm or porosities of 2-3, 4-7 and 8-14 % (as above), respectively. In the other extreme case, namely for highly conducting materials (e.g. beryllia and some non-oxides) the grain size effect is much larger and may be easily observed even for ceramics with relative coarse (micron-sized) grains: a porosity of 1 % is equivalent to a grain size of 6.1 μm, and conductivity reductions by 5, 10 and 20 % require grain sizes of 2.4 μm, 1.1 μm and 0.5 μm (or porosities of 2-3, 4-7 and 8-14 % as above), respectively.

Table 2. Equivalent porosities in (%) for the thermal conductivity corresponding to different grain sizes, calculated according to different porosity models; the porosities listed reduce the effective thermal conductivity to 10, 20, 50, 80, 90, 95, 98 and 99 % of the thermal conductivity of the coarse-grained dense ceramic and would have the same effect as a grain size reduction corresponding to the critical grain size label (see Table 1)

Critical grain size label (quantile)	Equivalent porosity (%) according to the			
	Exponential relation	Power-law	Sigmoidal average of the HS bounds	Sigmoidal average of the Wiener bounds
D_{10}	60.6	78.5	63.0	68.4
D_{20}	51.8	65.8	49.1	55.3
D_{50}	31.6	37.0	24.1	29.3
D_{80}	12.9	13.8	8.2	10.6
D_{90}	6.6	6.8	3.9	5.1
D_{95}	3.3	3.4	1.9	2.5
D_{98}	1.3	1.3	0.8	1.0
D_{99}	0.7	0.7	0.4	0.5

Thus, since the GB conductivity is of order 1 W/mK (ranging from approx. 0.5 to 3.0 W/mK), the grain size effect in reducing the thermal conductivity is largest for highly conducting ceramics. In practice that means that not much is gained by grain size reduction in materials with an intrinsically low conducting bulk phase (crystalline core phase), which are usually chosen for insulation purposes (e.g. zirconia). In this case, grain sizes of order 10-100 nm would be required to yield measurable decreases in thermal conductivity. Such small grain sizes can hardly ever be obtained in dense materials (e.g. by sintering). However, as shown by the equivalent porosity concept, when the material is not completely dense, porosity effects will usually dominate over grain size effects or even make them negligible.

Concluding this section it may be said that, apart from porosity, which must be very well characterized in the forefront (not a trivial task for nanocrystalline materials), the most influential parameter determining the calculated grain size dependence in phase mixture modeling is the GB thickness, for which we have used the estimate 1 nm here. This value refers to ceramics which have been densified without assistance of a liquid phase (melt). As soon as the ceramic composition itself is multiphase and prone to glass formation, the GB thickness may of course attain arbitrary values. In this case experimental investigations by electron microscopy (preferentially high-resolution-TEM combined with image analysis) are indispensable and should precede the application of the phase mixture model.

6. HEAT TRANSFER MECHANISMS

In the sense of statistical physics (kinetic theory) heat – basically a phenomenological concept, *viz.* a measurable form of energy – corresponds on the atomistic level to the motion of atoms or larger atomic complexes, e.g. molecules. Heat transfer in gases occurs by the translational motion of gas atoms or molecules. The thermal conductivity of (macroscopically stagnant) gases is determined by collisions of gas molecules (for flowing gases convection contributes to heat transfer as well). A well-known result of the kinetic theory of gases is the equation for the (true) thermal conductivity of gases,

$$k_{gas} = \frac{1}{3}\rho c_V v L, \qquad (137)$$

where c_V is the specific heat at constant volume, ρ the density, v the (average) translational velocity of the gas atoms (or molecules) and L their mean free path length, see e.g. (Grimvall 1999, Hellwege 1988, Kittel 1988, Kopitzki 1989). In crystalline solids the atoms (or larger molecular complexes) are not free to move, but heat is accumulated in the vibrational states of the atoms in the lattice. These vibrational states correspond to elastic waves, which propagate through the solid with the velocity of sound (determined by the elastic constants and the bulk density) and are damped by all kinds of lattice defects, e.g. vacancies, interstitials, impurities substituting the regular lattice atoms (foreign atoms as well as different isotopes of the same element), grain boundaries and all types of structural disorder. In amorphous (glassy) solids, where the conception of a "lattice" is not useful any more and may at best be replaced by that of a "network", heat transfer by vibration is greatly impeded, similar to liquids. Therefore the thermal conductivity of glasses is low (0.5-1.4 W/mK), similar to that of common liquids (0.1-0.6 W/mK) (Gebhart 1993, Lide 2010, Touloukian et al. 1970). Now the wave-particle-dualism (De Broglie's principle) of quantum mechanics can be invoked to consider lattice vibrations (elastic waves) in crystalline solids as quasi-particles ("quanta") called phonons. Thus, the true conductivity in solids can be considered as mediated by phonons alone, while the apparent conductivity including the radiative contribution to heat flux (which is the quantity usually measured at high temperatures unless corrections are made), can be viewed as mediated by phonons and photons (i.e. the quasi-particles or quanta of electromagnetic waves or radiation). Of course, in contrast to electrons, both phonons and photons are bosons, i.e. they do not have a rest mass and can be created and annihilated.

When the phonon picture is adopted, an equation completely analogous to the previous one for gases can be used to describe the thermal conductivity of crystalline solids:

$$k_{lattice} = \frac{1}{3} \rho c_V v_{phonon} L_{phonon} \qquad (138)$$

In this equation ρ is the bulk density of the solid, L_{phonon} the mean free path length of the phonons (i.e. the average length of the path, after which the phonon population decreases to 1/e of the original value) and v_{phonon} is the group velocity (Grimvall 1999). Of course, for engineering purposes (e.g. for thermal conductivity prediction) this equation is useless, unless independently measured L_{phonon} values are available. This is usually not the case, since in most cases L_{phonon} is the quantity that is estimated on the basis of measured thermal conductivities. Nevertheless, the equation explains – at least in a qualitative manner – the temperature dependence of the thermal conductivity: at 0 K the specific heat and thus also the thermal conductivity is zero, then both increase with T^3, until the specific heat (at least in the classical harmonic approximation) becomes more or less constant (according to the Dulong-Petit rule); in this region another effect comes into play – the phonon density increases with increasing temperature and thus the mean free path of the phonons is reduced by inelastic phonon-phonon scattering (so-called Umklapp processes). The decrease in thermal conductivity of crystalline solids in this high-temperature region is

$$k_{lattice} \propto \frac{1}{T^x}, \qquad (139)$$

where x is theoretically between 1 (the theoretical result for three-phonon processes, e.g. "collisions" between two phonons, creating one phonon moving in another direction) and 2 (for four-phonon processes) (Ashcroft and Mermin 1976, Grimvall 1999). Since four-phonon processes are usually negligible (i.e. their probability is low), a simple 1/T-dependence of the thermal conductivity is usually assumed for crystalline solids in the high-temperature region. Of course, in between the power-law increase and the hyperbolic decrease (which can be preceded by an exponential decrease (Kopitzki 1989)) there must be a maximum of $k_{lattice}$ (typically at temperatures below 100 K, see Sections 7-9). Above room temperature the thermal conductivity of many crystalline ceramics is in the decreasing region with hyperbolic temperature dependence, and therefore it is possible in many cases to fit this decrease using a three-parameter relation of the type

$$k = a \cdot (T+b)^c \qquad (140)$$

(shifted power law). While pure crystalline solids (metals, semiconductors and non-metallic inorganic solids) have a distinct maximum in the thermal conductivity much below room temperature, with increasing lattice disorder the height of the maximum decreases and the maximum is shifted to higher temperatures. In very impure or disordered materials the maximum can be absent (Grimvall 1999). For glasses and liquids, both organic and inorganic, the thermal conductivity increases initially as T^2, and attains usually a more or less constant value (or exhibits a slightly increasing tendency) above room temperature, similar to highly disordered solids. At room temperature the mean free path length for phonons is approx. 5 nm (Kingery 1962); with increasing temperature (and thus increasing phonon density) the mean free path decreases and at approx. 1200 °C it attains the lattice dimensions (Salmang and Scholze 1982). From this point onwards the lattice conductivity (phonon conductivity or true thermal conductivity) attains a constant value, because the mean free phonon path cannot be smaller than the lattice dimensions (however, the apparent thermal conductivity, which is usually measured, may increase with increasing temperature, due to the radiative contribution, see below). As mentioned above, below room temperature the thermal conductivity increases with decreasing temperature, which is tantamount with an increase in the mean free path length of the phonons, because with decreasing temperature the phonon density decreases. Thus at very low (cryogenic) temperatures the mean free path length (i.e. the average distance travelled by the phonon from one collision to another) can exceed the sample dimensions. In this case the measured thermal conductivity becomes dependent on the sample size and ceases to be a material property in the strict sense of the word (Hellwege 1988). The reduction of L_{phonon} at crystal defects, impurities and grain boundaries in polycrystalline solids is essentially independent of temperature. Their influence can be very large at low temperatures, but can be negligibly small at higher temperatures (Salmang and Scholze 1982). High defect concentrations (e.g. in doped ZrO_2) and lacking long-range order (e.g. in glasses) are responsible for small mean free path lengths of the photons (in silica glass only approx. 0.5 nm (Salmang and Scholze 1982)) and thus low thermal conductivites, which are in this case only determined by the temperature dependence of the specific heat (e.g. the slight, almost linear, increase with temperature sometimes observed for glasses and for zirconia may be interpreted as a result of anharmonic contributions to the specific heat (Grimvall 1999)).

The significant increase of thermal conductivity at higher temperatures (in crystalline and non-crystalline solids, e.g. above 300 °C for clear colorless silicate glasses (Schill 1993), above 500 °C for clear silica glass (Salmang and Scholze 1982), and above 1300 °C for polycrystalline Al_2O_3 and MgO (Kingery et al. 1954)) has to be attributed to radiative heat transfer (Salmang and Scholze 1982). In other words that means that the conductivity value measured must be interpreted as an apparent thermal conductivity, consisting of a true lattice conductivity (phonon contribution) and a radiative contribution (photon contribution) $k_{radiation}$. It is common practice to assume additivity of the two, i.e. the total or apparent thermal conductivity is given by

$$k_{apparent} = k_{lattice} + k_{radiation}. \tag{141}$$

Radiative heat transfer through a solid phase is essentially determined by the the optical properties of the solid (more precisely, the real part n of the refractive index and the mean free path length of the photons L_{photon}, which is by itself a function of the imaginary part of the refractive index, i.e. the absorption coefficient) in the near-infrared region. The radiative contribution to thermal conductivity can be estimated via the relation

$$k_{radiation} = \frac{16}{3}\sigma n^2 T^3 L_{photon}, \tag{142}$$

where σ is the Stefan-Boltzmann constant ($5.67 \cdot 10^{-8}$ Wm^{-2}K^{-4}) (Modest 1993, Salmang and Scholze 1982, Schlegel 1999, Schulle 1990). Due to the strong T^3 dependence the radiative contribution to the apparent thermal conductivity increases rapidly with temperature. In practice even stronger temperature dependences are sometimes found (corresponding to exponents between 3.5 and 5 (Schulle 1990)).

Apart from for the solid phase(s) heat transfer occurs through pores if present. Pores (also called cells in the context of highly porous materials, so-called cellular materials) can be of different shape (convex or concave, oblate, prolate or isometric), open (interconnected) or closed (isolated), empty (vacuous voids) or filled with fluids. The pore-filling fluids can be liquids or gases. When the pores (e.g. cracks) are open, forced convection may occur as soon as the size (equivalent diameter) of the pore openings (throats, channels, cells, windows) is large enough and the pressure drop across the material high enough. This can be critical e.g. in wall and roof linings made of refractory insulation materials in high-temperature aggregates like furnaces and melters (Kutzendörfer and Máša 1991, Schlegel 1999). Heat transfer in this case is not a problem of the solid skeleton alone, but a combined hydrodynamic problem, which is beyond the scope of the present treatise. The interested reader should consult (Kaviany 1995, Öchsner et al. 2008). When the pores are closed, free convection due to density differences of gases in temperature gradients might principally occur in large pores. The question whether or not convection plays a role can be answered by estimating the Grashof number (Gibson and Ashby 1997, Kuneš 1989). However, due to the fact that gases form laminar, highly viscous boundary layers in the vicinity of solid surfaces (as long as the continuum approximation is valid, i.e. for sufficiently low Knudsen numbers), with increasing viscosity at higher temperatures, free convection can be completely excluded in practice for pores smaller than a few mm (Schlegel 1999).

The true thermal conductivity of vacuous pores or voids (vacuum) is of course zero. Gases at room temperature have thermal conductivities between 0.006 W/mK (Xe) and 0.182 W/mK (H$_2$), but most common gases and vapors have values in the range 0.013-0.031 W/mK, see Table C-1 (in Appendix C). The thermal conductivity of air is 0.026 W/mK at room temperature and 0.076 W/mK at 1000 °C. Water vapor has a slightly lower thermal conductivity at RT (0.017 W/mK) but a higher one (0.147 W/mK) at 1000 °C, see Table C-1.

The dependence of thermal conductivity on gas pressure has been discussed in (Mogro-Campero 1997). When the pore size is below the mean free path length (30-180 nm) the true thermal conductivity of the gas phase drop dramatically. This explains the fact that micro- or mesoporous materials with high porosity can have conductivities below that of stagnant air (mean free path length 60-65 nm). For other aspects of micro- and nanoscale heat transfer the interested reader may consult (Zhang 2007).

We note in passing that liquids exhibit a rather broad range of thermal conductivities. In particular, of course, metal melts can have very high conductivities (e.g. for aluminum 104 W/mK at 700 °C, for mercury 8.3 W/mK at room temperature), while most common pure liquids at or around room temperature have thermal conductivity values between 0.1 and 0.6 W/mK (Gebhart 1993, Lide 2010, Touloukian et al. 1970). Note, in particular, that water has an unusually high thermal conductivity for a pure liquid (0.57 W/mK close to 0 °C, 0.61 W/mK at room temperature and 0.68 W/mK at 100 °C). In glass melts the apparent thermal conductivity is very high (because of radiative heat transfer), although the true thermal conductivity is small (not very different from the RT value, i.e. of order 1 W/mK). Although not of central interest in this contribution, the thermal conductivity of all these liquids can play a role during processing, e.g. in water-saturated ceramic green bodies or the infiltration of ceramic preforms by metal or glass melts.

Vacuous or gas-filled pores generally reduce the true thermal conductivity of porous solids, while some liquids may in exceptional cases increase it (e.g. mercury with $k = 8.3$ W/mK). Even in the former case, however, the presence of pores may under certain circumstances lead to an increase of the apparent (total) thermal conductivity measured for a material. This is possible because radiative heat transfer, being of purely electromagnetic nature, does not require matter to proceed: all bodies constantly emit and absorb energy mutually, even if separated by a vacuum. This energy exchange cools down the warmer bodies and heats up the cooler ones (note that, in contrast to Fourier's law, there is no temperature gradient involved, only a temperature difference, i.e. a jump, over a distance). The radiative heat flux is governed by the Stefan-Boltzmann law

$$q_{radiation} = \sigma \varepsilon T^4, \tag{143}$$

where ε is the emissivity (i.e. the ratio of the energy emitted from a surface to the energy emitted by a black surface at the same temperature), σ the Stefan-Boltzmann constant and T the absolute temperature (Modest 1993, Schlegel 1999).

Radiative heat transfer is not affected very much by gases. In particular, mono- and biatomic gases (Ar, N_2, O_2, CO etc.) and air do not absorb, emit or reflect infrared radiation to any substantial degree (therefore they are called diathermic gases (Kutzendörfer and Máša 1991, Modest 1993, Schlegel 1999)). Thus, when pores are filled with these gases, the radiative contribution to heat transfer will be similar as for vacuous pores. Polyatomic gases (CO_2, H_2O, NH_3, SO_2 etc.), on the other hand, do have some absorptivity with respect to thermal radiation and will thus slightly reduce the radiative contribution. The radiative heat transfer between a body with a convex surface A_1 with temperature T_1 and emissivity ε_1 surrounded by a second body with a concave surface A_2 (i.e. $A_1 < A_2$) with temperature T_2 and emissivity ε_2 is calculated via the relation

$$q_{radiation} = \sigma \varepsilon_{combination} \left(T_1^4 - T_2^4 \right), \tag{144}$$

where $\varepsilon_{combination}$ is the combined emissivity of the two surfaces given by

$$\varepsilon_{combination} = \left[\frac{1}{\varepsilon_1} + \left(\frac{1}{\varepsilon_2} - 1 \right) \frac{A_1}{A_2} \right]^{-1} \tag{145}$$

(Kutzendörfer and Máša 1991, Modest 1993). In the special case of heat transfer between two flat surfaces or between the opposite walls inside a closed cavity (e.g. the walls of pores), $A_1 = A_2$. The emissivities vary between 0 and 1 (the latter denoting a black surface) and depend on the materials, the wavelength (or frequency) of the radiation (Modest 1993), and on temperature. Moreover, the emissivities depend on surface orientation, i.e. on the direction between the surface normal and the observer. Therefore, in practice (i.e. for the purpose of engineering calculations) the emissivities are often replaced by values averaged over all directions (so-called hemispherical emissivities) and wavelengths (so-called total hemispherical emissvities) (Modest 1993). The directional dependence of emissivity, however, is responsible for the fact that emissivities – far from being material properties in the strict sense of the word – are dependent on the state of the surface (smooth or rough, coated or uncoated) and thus indirectly e.g. on the grain size. Dielectrics (nonmetals) tend to have high emissivities, while metals tend to have lower emissivities, in the infrared (Modest 1993).

Table D-1 (in Appendix D) lists orientational values for the total (i.e. spectrally averaged) normal emissivities (i.e. in direction normal to the surface) of selected materials. Note, however, that the corresponding total hemispherical emissivities are higher (by 3-5 % for dielectrics and by up to 25 % for metals) (Modest 1993). As a rule it can be said, that smooth, polished pure metal surfaces have the lowest emissivities (approx. 0.02-0.10) while alloying, rough surfaces and oxide layers all enhance the emissivity. Nonmetals (dieletrics) have usually larger emissivities (0.20-0.99), often with a significant temperature dependence (at lower temperature the emissivity is usually higher). The grain size dependence of emissivities reported for some materials arises as a consequence of the surface roughness.

The spectrum (i.e. the wavelength dependence) of thermal radiation has a maximum which shifts to smaller wavelengths with increasing temperature. For example, at 200 °C the maximum is at 8 µm, while at 1700 °C it is at 2 µm (Schlegel 1999). Therefore thermal radiation, which is invisible to the human eye at room temperatature, becomes visible at higher temperatures, with a color that appear red at 800 °C, yellow at 1200 °C and white at 1600 °C (Schlegel 1999). According to Kirchhoff's law highly absorbing bodies are highly emissive, but the emissivity of highly reflective bodies is low (Schlegel 1999). Most polycrystalline ceramics are more or less opaque for thermal radiation because of the presence of grain boundaries, pores, microcracks and other defects that scatter radiation. Glasses and single crystals, however, can exhibit a high transmissivity for thermal radiation (with the absorption exhibiting an exponential decrease according to the Lambert-Beer law). Also thin layers, e.g. pore walls (cell walls) in highly porous (cellular) ceramic materials can be highly transmissive (Schlegel 1999). This – together with the aforementioned fact that the voids do not show any thermal resistance (even if gas-filled) – is the reason for the great influence of radiative heat transfer in highly porous materials, including cellular materials, insulating fiber mats and powder beds. As a consequence it may happen in highly porous materials that the apparent thermal conductivity increases with increasing porosity. While at room temperature this effect is completely negligible, except maybe for aerogels (the conducitivity minimum is at approx. 98 % porosity), it may play a role at higher temperatures. However, even at 1000 °C the conductivity minimum is for most materials still at 80 % porosity (Kutzendörfer and Máša 1991, Schlegel 1999, Schulle 1990), so that indeed very high porosities are required to exploit this effect in practice. Moreover, it can be expected that at such high porosities also convection will play a role, so that it will be extremely difficult to separate the three basic heat transfer mechanisms in such cases.

The radiative contribution due to pores can be estimated on the basis of the Loeb model (Loeb 1954). According to this simple model (which assumes idealized pore shapes and unique orientation in the case of non-sphericity) the radiative contribution due to pores can be written as

$$k_{Loeb} = 4\gamma D\varepsilon\sigma T^3, \qquad (146)$$

where D is the pore size (more precisely, the largest dimension of the gap in the direction of heat flow) and γ is a geometrical factor of order unity (2/3 for spherical pores, $\pi/4$ for cylindrical pores with axes perpendicular to the heat flow and 1 for laminar and cylindrical pores with axes parallel to the heat flow). As a result, the effective apparent thermal conductivity of a porous material due to this model is

$$\frac{k}{k_{lattice}} = 1 - \frac{P_A\left(1 - \bar{k}_{Loeb}/k_{lattice}\right)}{1 + \dfrac{\bar{k}_{Loeb}}{k_{lattice}} \cdot \dfrac{1-P_L}{P_L}}, \qquad (147)$$

where P_A is the areal fraction of pores in the cross-section perpendicular to heat flow, P_L the lineal fraction of pores parallel to heat flow and $\bar{k}_{Loeb}$ Loeb's expression for the radiative contribution to heat transfer with T^3 replaced by the mean of the cube of the absolute temperature of the body (Loeb 1954).

Evidently, radiation through porous materials is negligible compared to true conduction of the solid matrix or skeleton when the dimensionless ratio

$$\frac{\bar{k}_{Loeb}}{k_{lattice}}, \qquad (148)$$

is much smaller than unity. Assuming spherical pores and a pore surface emissivity of 0.38 ± 0.10 (corresponding for example to alumina, α-Al$_2$O$_3$, according to (Francl and Kingery 1954)) one obtains for a pore size (diameter) < 100 μm at a 500 °C a radiation contribution of $2.66 \cdot 10^{-3}$ W/mK, which is clearly negligible in comparison to the thermal conductivity of dense alumina at this temperature which is approx. 12 W/mK (Živcová et al. 2009).

In the case of uniform isotropic materials with randomly distributed pores the areal fraction equals the lineal fraction and both equal the volume fraction of pores, i.e. $P_A = P_L = \phi$, due to the Delesse-Rosiwal law (ϕ = porosity). Denoting the true (lattice, phonon) conductivity of the solid phase as k_0, Loeb's equation for isotropic materials would be

$$\frac{k}{k_0} = 1 - \frac{\phi\left(1 - \bar{k}_{Loeb}/k_0\right)}{1 + \dfrac{\bar{k}_{Loeb}}{k_0} \cdot \dfrac{1-\phi}{\phi}} = 1 - \frac{\phi^2\left(k_0 - \bar{k}_{Loeb}\right)}{k_0\phi + \bar{k}_{Loeb}(1-\phi)} \qquad (149)$$

In the special case when radiation is absent (i.e. $\bar{k}_{Loeb} = 0$) this reduces to the expression

$$\frac{k}{k_0} = 1 - \phi, \qquad (150)$$

which is the volume-weighted arithmetic mean (upper Wiener bound). As a consequence it is clear that the original Loeb model is actually not very appropriate for isotropic materials. Other, more sophisticated models, which are beyond the scope of the present treatise, are discussed in (Kaviany 1995, Schlegel 1999, Schulle 1990). Irrespective of the special model it is a fact that pores scatter photons very effectively, unless their size is not significantly (at least one order of magnitude, say) different from the wavelength of the radiation. Therefore, while a photon mean free path of approx. 10 cm has been calculated for α-Al$_2$O$_3$ single crystals at 750 °C (Lee and Kingery 1960), it can be as low as 400 μm when a porosity of only 0.25 % is present (Salmang and Scholze 1982).

7. THERMAL CONDUCTIVITY OF OXIDE CERAMICS

Oxides are usually electrical insulators (dielectrics), where heat transfer is essentially mediated by phonons, and therefore their thermal conductivity is commonly expected to be lower than that of metals and semiconductors, where also electrons act as heat carriers. However, at low temperatures oxide ceramics can have remarkably high thermal conductivities. For example, crystalline silica (SiO$_2$ in the form of low-quartz) can have a thermal conductivity of more than 1000 W/mK at 10-20 K (Berman et al. 1950, Childs et al. 1973, Hellwege 1988, Ibach and Lüth 1988), and completely defect-free alumina (α-Al$_2$O$_3$, corundum) has an estimated conductivity of 20000-25000 W/mK at 30-40 K (Berman 1953, Hellwege 1988, Kittel 1988), a value that comes close to the peak conductivities of highly conducting metals (aluminum, copper and silver, see below).

The first and so far most comprehensive systematic investigation of the thermal conductivity of oxide ceramics has been undertaken by Kingery and his coworkers at the MIT in the 1950s and has been published in a series of famous papers in the Journal of the American Ceramic Society, including (Francl and Kingery 1954, Kingery and McQuarrie 1954, Kingery et al. 1954). Since these papers are milestones of engineering science and may be considered as the beginning of modern ceramic science, we will recall the results published there in some detail in this section. It will become evident, that – although the appropriate formulae for correcting the porosity influence were not available at that time – these carefully measured data are still among the most reliable and best documented up to the present day. In this section we treat single oxides (pure oxides, possibly doped) as well as mixed oxides, but exclude silicates and glasses, which will be treated below (Section 8). For selected oxides, after recalling and reanalyzing the data of (Kingery et al. 1954), we first treat the thermal conductivity in the high-temperature region, i.e. from room temperature upwards to refractory temperatures, and then the low-temperature (cryogenic) region, wherever data are available. Since crystalline materials (single or polycrystalline) typically obey a $1/T$-temperature dependence (see Section 6), the temperature dependence of the thermal conductivity in the high-temperature region can often be fitted using a shifted power law of the type

$$k(T) = a \cdot (T+b)^c,\qquad(151)$$

where a, b and c are three fit parameters. Of course, this is only the case as long as there are no phase transitions in the temperature range of interest. Phase transitions usually lead to a discontinuity in the temperature dependence of the thermal conductivity. For example, in the thermal conductivity of cobalt oxide (CoO) the discontinuity at approx. 300 K (Watanabe 1993) is characteristic of a second-order phase transition (also called "lambda-transition" because of the corresponding singularity or lambda-shaped peak in the temperature dependence of the specific heat) (Kleber 1990) from the paramagnetic high-temperature phase to the ferromagnetic low-temperature phase at the Curie point of 298 K (Greenwood and Earnshaw 1993).

Figure 3 shows the temperature dependence of thermal conductivity of single-oxide ceramics (including doped zirconia, a defective structure, and one mixed-oxide ceramic, viz. magnesia-alumina spinel) according to the key paper of Kingery and coworkers (Kingery et al. 1954). The primary results for the data given in this paper and shown in Figure 3 have been obtained by very careful measurements and, in contrast to many more recent papers, the materials were well defined especially with respect to porosity. Therefore it is possible to make a porosity correction and extrapolate to full density. Each data point in this graph is denoted by a double symbol, because the way how to perform the porosity correction is by no means obvious (especially when the porosity is high, in those data up to 34 %). In Kingery's 1954 paper the simple mixture rule (i.e. the volume-weighted arithmetic mean or, in other words, the upper Wiener bound relation), i.e. the relation

$$k_0 = \frac{k}{1-\phi},\qquad(152)$$

has been used to correct for porosity ϕ (i.e. to obtain the solid phase conductivities k_0 from the measured effective conductivities k), obviously in the absence of a better alternative at that time, see (Francl and Kingery 1954). These are the bottom parts of the double symbols in Figure 3. It is well known, however, that in practice the influence of porosity is always stronger that predicted by the mixture rule (at least as far as isotropic materials are concerned), and the exponential relation, i.e.

$$k_0 = \frac{k}{\exp\left(\dfrac{-2\phi}{1-\phi}\right)},\qquad(153)$$

is a useful relation to estimate this influence much more realistically (Pabst and Gregorová 2007, 2008, Živcová et al. 2009). The top parts of the double symbols in Figure 3 correspond to this porosity correction (of course, when the porosity is low the two parts of the symbols are more or less coincident).

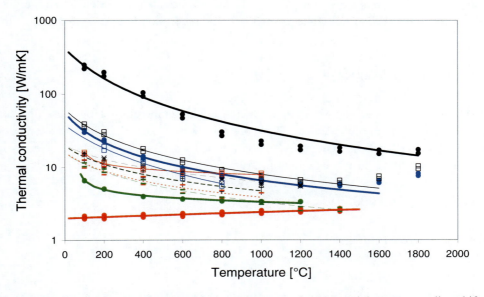

Figure 3. Thermal conductivity of oxide ceramics (Kingery et al. 1954) and the corresponding shifted power fits; all data denoted by identical double symbols (bottom symbol: Kingery's values corrected for porosity via the simple mixture rule, top symbol: Kingery's values corrected for porosity via the exponential relation) and average fit curves in between – Al_2O_3 (blue circles and thick blue curve), BeO (black circles and thick black curve), CaO (red squares and full thin red curve), MgO (black squares and full thin black curve), NiO (black plusses and dashed thin black curve), $MgAl_2O_4$ spinel (black crosses and dotted black curve), ThO_2 (horizontal green bars and dotted thin green curve), TiO_2 (green circles and thick green curve), UO_2 (horizontal red bars and dotted thin red curve), ZnO (blue squares and full thin blue curve) and ZrO_2 (red circles and thick red line).

We would like to emphasize that the influence of porosity cannot be weaker than predicted by the mixture rule, see Section 3, and will usually not be much stronger than predicted by the exponential relation, unless partial sintering and concave pore shape comes into play (Pabst and Gregorová 2008). Fit curves (and the linear regression line for zirconia) were calculated from the average values obtained as the arithmetic mean of the two porosity corrections. Figure 3 shows that the thermal conductivity of single-oxide ceramics from room temperature up to very high temperature is usually between those of zirconia and beryllia. Magnesia and alumina are among the relatively highly conducting ceramics, while titania is rather badly conducting, similar to most silicate ceramics, cf. Figure 14. Zirconia is the single-oxide ceramic with the lowest thermal conductivity. Note, however, that the zirconia ceramics shown here and in most other papers is not pure, but fully or partially stabilized. The dopant ions act as lattice defects that impede the propagation of lattice waves (in other words, reduce the mean free path of the phonons by scattering). Therefore, doped zirconia types (i.e. fully stabilized cubic zirconia, partially stabilized zirconia – PSZ – or tetragonal zirconia polycrystals – TZP) must be considered as materials with a defective structure (depending on the dopant content even a highly disordered one), with a thermal conductivity approaching the amorphous limit of 1.1 W/mK (Cahill et al. 1992). This explains the glass-like behavior of doped zirconia and the fact that the temperature dependence of the thermal conductivity of doped zirconia does not follow the *1/T*-relationship typical for crystalline materials. Therefore, it is not surprising that the temperature dependence of the thermal conductivity of zirconia cannot be satisfactorily fitted by a shifted power relation. Based on Kingery's data (Kingery et al. 1954) in Figure 3 the following linear regression line fit is obtained (with *T* in (°C)):

$$k(T) = 2.00 + 0.00041 \cdot T. \tag{154}$$

This fit seems to indicate a very small increase with temperature. However, this increase is completely negligible in practice. Therefore this finding is still in reasonable agreement with the fits in (Pabst and Gregorová 2007), based on more recent data, where slightly decreasing trends have been found, ranging from

$$k(T) = 2.93 - 0.00050 \cdot T \tag{155}$$

for tetragonal zirconia to

$$k(T) = 1.96 - 0.00028 \cdot T \tag{156}$$

for cubic zirconia (the so-called *t'* phase being in between these two). With respect to these findings, and also with respect to Figure 12, it can be concluded that in doped zirconia types, whether tetragonal or cubic, the thermal conductivity is between 2 and 3 W/mK at room temperature and there is essentially no temperature dependence (and no indication of radiat6ive heat transfer) up to at least 1400 °C. Since there is a general tendency for the thermal conductivity to decrease (from approx. 3 W/mK to approx. 2 W/mK) with dopant content (Pabst and Gregorová 2007), it can be surmised that Kingery's zirconia was a highly-doped one, i.e. fully stabilized zirconia (with a high content of cubic phase). Note that pure monoclinic zirconia without dopants (Mévrel et al. 2004, Raghavan et al 1998, 2001) exhibits the typical hyperbolic decrease with temperature and can be fitted using a shifted power relation (from RT to approx. 1200 °C, with *T* in (°C))

$$k(T) = 215 \, (T + 236)^{-0.612} \tag{157}$$

(Pabst and Gregorová 2007). The thermal conductivity of pure monoclinic zirconia and its temperature dependence is very similar to that of titania and most silicate phases. Empirical expressions (shifted power fits) for the other oxide ceramics in Figure 3 are given in Table 3. The quality of the fits is evident from Figure 3. It is evident that the shifted power fits describe the temperature dependence of the thermal conductivity satisfactorily (except for beryllia) as long as the radiation contribution can be neglected. For polycrystalline oxide ceramics this is the case up to temperatures of at least 1300 °C: in Figure 3 an increase of the measured (apparent) thermal conductivity is visible at temperatures higher than 1400 °C (for alumina and magnesia).

Table 3. Empirical fits of the temperature dependence of the thermal conductivity of oxide ceramics (for temperatures in (°C)); data from (Kingery et al. 1954), corrected for porosity by the mixture rule and the exponential relation (average fits)

	Average fit of Kingery's data, corrected for porosity
Al_2O_3	$k(T) = 5872 \cdot (T+121)^{-0.968}$
BeO	$k(T) = 2920188 \cdot (T+250)^{-1.603}$
CaO	$k(T) = 26.8 \cdot (T-80)^{-0.180}$
MgO	$k(T) = 15393 \cdot (T+169)^{-1.071}$
NiO	$k(T) = 777.7 \cdot (T+158)^{-0.724}$
spinel ($MgAl_2O_4$)	$k(T) = 752 \cdot (T+217)^{-0.673}$
ThO_2	$k(T) = 1674.5 \cdot (T+202)^{-0.878}$
TiO_2	$k(T) = 13.2 \cdot (T-69)^{-0.202}$
UO_2	$k(T) = 559 \cdot (T+150)^{-0.707}$
ZnO	$k(T) = 135816 \cdot (T+299)^{-1.434}$
ZrO_2	$k(T) = 2.00 + 0.00041 \cdot T$

Figure 4 shows more recent literature data of the thermal conductivity of alumina (α-Al_2O_3) from room temperature up to the high-temperature region, in comparison to the fit curves from Table 3 (based on Kingery's values alone), and two other fit curves, based on a larger amount of more recent data, viz. the fit curve according to (Munro 1997a),

$$k(T) = 5.85 + \frac{15360 \cdot \exp(-0.002T)}{T+516} \qquad (158)$$

and the curve according to (Pabst and Gregorová 2007),

$$k(T) = 3645 \cdot (T+194)^{-0.874} \qquad (159)$$

(both for temperatures in (°C)). It is evident, that the fit curve based on the Kingery's values alone, while in reasonable agreement at high temperatures, tends to overestimate the thermal conductivity of alumina at or near room temperature. The fit in (Pabst and Gregorová 2007), which is of the shifted-power type (i.e. a three-parameter fit), is quite close to the pure empirical four-parameter fit according to Munro (Munro 1997a), which provides probably the most precise representation of current thermal conductivity data on alumina (of course without any connection to the theoretically founded hyperbolic temperature dependence of crystalline solids).

We would like to emphasize that all data presented here and in the sequel are for micrometer-sized grains, if not explicitly stated differently. Concerning the grain size influence on thermal conductivity, a classical paper from the Kingery group (Charvat and Kingery 1957) did not find any grain size dependence for alumina in the size range 9-17 µm. Recent modeling results obtained via the phase mixture approach lead to the conclusion that grain size effects may become significant in alumina only for submicron or nanometric grain sizes (Pabst 2009, Pabst and Gregorová 2010). For example, in the case of isometric grain shape, a grain size below approx. 500 ± 250 nm is needed to achieve a thermal conductivity decrease by more than least 5 %, cf. Section 5 and (Pabst and Gregorová 2010) (for ceramics with lower conductivity the requirements are even stronger, e.g. to a achieve an only 5 % conductivity decrease in doped zirconia the grain size must be below 50 nm). That means, in our opinion, apparent grain size effects reported in the literature are in most cases due to (undetected or insufficiently characterized) porosity effects.

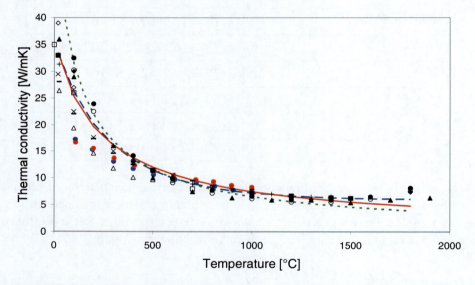

Figure 4. Thermal conductivity of alumina in the high-temperature region; full squares (Munro 1997a), empty circles (Kingery et al. 1954), full black circles: Kingery's data corrected for porosity by the exponential relation, empty triangles (Liu et al. 1995a), horizontal bars (Barea et al. 2003), crosses (Nieto et al. 2004), full triangles (Shackelford and Doremus 2008), empty diamonds (Miyayama et al. 1991), full diamonds (Lide 2010), plusses (Hasselman 1985), green circles (Santos et al. 1998) (empty circles: sintered at 1400 °C, full circles: sintered at 1500 °C; both corrected for porosity using the exponential relation), dotted curve: fit according to Table 3, dashed curve: fit according to (Munro 1997a), full curve: fit according to (Pabst and Gregorová 2007).

Figure 5 shows the thermal conductivity of alumina in the low-temperature region, together with the Kingery values (Kingery et al. 1954) and the fits according to (Munro 1997a) and (Pabst and Gregorová 2007). Obviously all measurements converge in the high-temperature region, but at low temperatures there are significant (order-of-magnitude) differences concerning the peak values of the conductivity and the corresponding temperatures. Of course, the reason is that at cryogenic temperatures the thermal conductivity is strongly influenced by defects in the solid structure (vacancies, impurities) and by the state of the microstructure (single crystals versus polycrystalline materials), including grain size.

As indicated by Figure 5, the data can be roughly subdivided into two groups, corresponding to single crystals and polycrystalline materials, respectively. This is confirmed by the 26 low-temperature curves for alumina in (Childs et al. 1973). Defect-free alumina (corundum) single crystals can attain peak conductivities of 20000-25000 W/mK at 30-40 K (Berman 1953, Childs et al. 1973, Hellwege 1988, Kittel 1988). In fact, electrically insulating materials with high thermal conductivity are key materials in low-temperature physics (cryophysical experiments and apparatuses). On the other hand, polycrystalline alumina ceramics exhibit peak conductivities of only 100-500 W/mK at 60-80 K (Childs et al. 1973, Lide 2010, Nemoto et al. 1985).

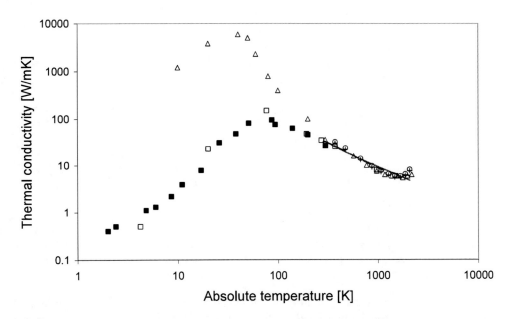

Figure 5. Thermal conductivity of alumina in the low-temperature region (full squares (Nemoto et al. 1985), empty squares (Lide 2010), empty triangles (Shackelford and Doremus 2008)) and high-temperature data for comparison (plusses (Kingery et al. 1954), empty circles: Kingery's data corrected for porosity by the exponential relation, dashed curve: fit according to (Munro 1997a), full curve: fit according to (Pabst and Gregorová 2007)).

Figure 6 shows the thermal conductivity of beryllia (BeO) from room temperature up to the high-temperature region, together with the shifted power fit (valid from room temperature up to approx. 1600 °C, with T in °C):

$$k(T) = 912870\,(T + 227)^{-1.428}. \tag{160}$$

The porosity dependence of the (room temperature) thermal conductivity of BeO shown in (Richerson 2006) indicates a value of 270 W/mK for a completely dense material (density 3.0 g/cm^3), 125 W/mK at density 2.4 g/cm^3 (porosity 20 %) and 62 W/mK at 1.8 g/cm^3 (porosity 40 %). Depending on purity, however, the value can be from approx. 280 W/mK for 99.5 % pure BeO down to approx. 200 W/mK for 97 % pure BeO (Richerson 2006). Maximum literature values (coming from low-temperature measurement equipment) are around 400 W/mK at room temperature (Slack 1973). The thermal conductivity of BeO has been investigated from 350 °C to 2200 °C in (Taylor 1962), and it is reported that up to 1700 °C a $1/T$ behavior has been found (Salmang and Scholze 1982). Nevertheless, Figures 3 and 4, as well the values in Table 5 below, show that the shifted power fit is too high, especially at temperatures around 1000 °C. In any case there are no indications of a radiation contribution to the thermal conductivity of beryllia up to approx. 1700 °C.

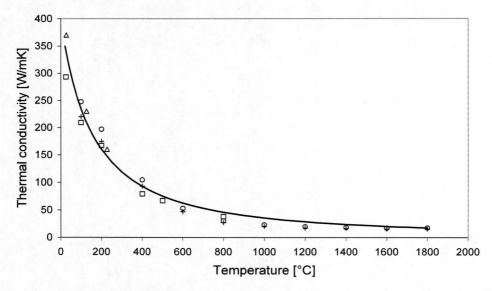

Figure 6. Thermal conductivity of beryllia (BeO) in the high-temperature region; empty triangles (Slack and Austerman 1971), empty squares (NIST 2003, Staff 1989), plusses: Kingery's values corrected for porosity via the mixture rule (Kingery et al. 1954), empty circles: Kingery's values corrected for porosity via the exponential relation.

Figure 7 shows the thermal conductivity of beryllia single crystals in the low-temperature region, together with the Kingery values for polycrystalline BeO ceramics (Kingery et al. 1954) and the other high-temperature values from the preceding Figure 6. Again, all measurements converge in the high-temperature region. For BeO single crystals, peak values of around 13000-14000 W/mK are achieved at approx. 40-50 K, see Figure 7, while peak values for polycrystalline BeO ceramics (not shown here) are about one order of magnitude lower, *viz.* 600-1700 W/mK at 80-150 K (Childs et al. 1973).

Literature data for sintered calcia (CaO) are rather scarce and contradictory. However, most room temperature values are in the range 20 ± 10 W/mK, while values at 1000 °C are approx. 8 ± 1 W/mK (Kingery et al. 1954, Martienssen and Warlimont 2005, Salmang and Scholze 1982, Staroň and Tomšů 2000). Of course, due to its easy hydroxylation, CaO is not a typical phase in ceramic microstructures.

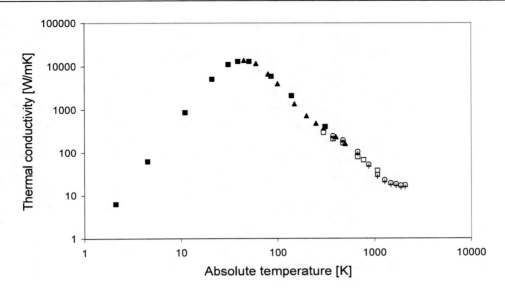

Figure 7. Thermal conductivity of beryllia (BeO); full squares (Slack 1973), full triangles (Slack and Austerman 1971), empty squares (NIST 2003, Staff 1989), plusses: Kingery's values corrected for porosity via the mixture rule (Kingery et al. 1954), empty circles: Kingery's values corrected for porosity via the exponential relation.

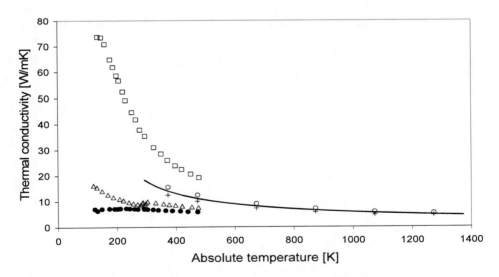

Figure 8. Thermal conductivity of cobalt, manganese and nickel oxide (CoO, MnO and NiO); empty triangles: CoO single crystal (99.99 % pure (Watanabe 1993)), full circles: MnO single crystal (99.99 % pure (Watanabe 1993)), empty squares: NiO single crystal (99.99 % pure (Watanabe 1993)), plusses: NiO ceramics (values corrected for porosity via the mixture rule (Kingery et al. 1954)), empty circles: NiO ceramics (Kingery's values corrected for porosity via the exponential relation), curve: average fit of Kingery's data.

Figure 8 shows the thermal conductivity of cobalt oxide (CoO), manganese oxide (MnO) and nickel oxide (NiO) according to (Watanabe 1993) together with the average fit of Kingery's data for NiO. The step-like discontinuity for CoO close to 300 K is characteristic of a second-order phase transition (lambda-transition) (Kleber 1990) from the paramagnetic high-temperature phase to the ferromagnetic low-temperature phase at the Curie point of 298 K (Greenwood and Earnshaw 1993). The fact that the Kingery data (Kingery et al. 1954) for NiO are lower than the more recent NiO data from (Watanabe 1993) can be attributed to the fact that in the latter case highly pure single crystals were measured, in contrast to the polycrystalline ceramic samples in the former case. The low-temperature peak conductivities indicated in Figure 8 are, however, approximately one order of magnitude lower than those in (Childs et al. 1973), where peak values of 60 W/mK at 35 K are presented for MnO and 450 W/mK at 45 K for NiO. The room temperature conductivities reported there are 11 W/mK and 45 W/mK for MnO and NiO, respectively (Childs et al. 1973).

Figure 9 shows the thermal conductivity of magnesia (MgO) from room temperature up to the high-temperature region, together with the shifted power fit (valid from room temperature up to approx. 1400 °C, with T in °C):

$$k(T) = 14386\,(T + 208)^{-1.045}. \qquad (161)$$

As mentioned before, the increase above 1400 °C must be attributed to the radiative contribution to heat transfer. Figure 10 shows the thermal conductivity of magnesia in the low-temperature region (Miyayama et al. 1991), together with the Kingery values (Kingery et al. 1954) and the other high-temperature values from the preceding Figure 9. The peak value in the graph is 290 W/mK at 27 K. However, in this case the low-temperature measurements do not converge to the high-temperature results here, raising some doubt on the reliability of these measurements. Nevertheless, the peak conductvities reported for MgO exhibit a wide range: from 65 W/mK to 3400 W/mK at temperatures between 20 and 40 K, while room temperature values are 32-65 W/mK (Childs et al. 1973).

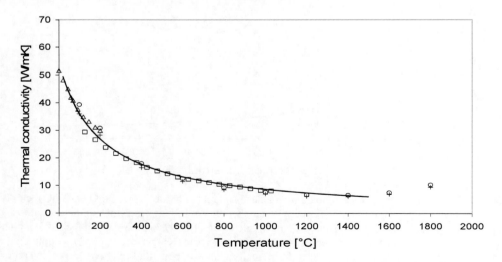

Figure 9. Thermal conductivity of magnesia (MgO) in the high-temperature region and fit curve from room temperature to 1400 °C; plusses: Kingery's values corrected for porosity via the mixture rule (Kingery et al. 1954), empty circles: Kingery's values corrected for porosity via the exponential relation, empty squares (Slifka et al. 1998), empty triangles (Watanabe 1993), the increase above 1500 °C must be attributed to radiative heat transfer.

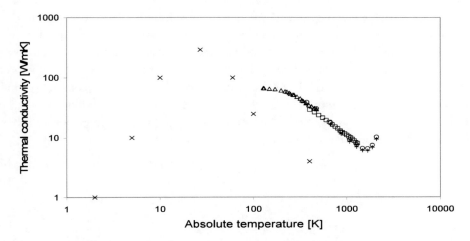

Figure 10. Thermal conductivity of magnesia (MgO) in the low-temperature region; plusses: Kingery's values corrected for porosity via the mixture rule (Kingery et al. 1954), empty circles: Kingery's values corrected for porosity via the exponential relation, empty squares (Slifka et al. 1998), empty triangles (Watanabe 1993), crosses (Miyayama et al. 1991).

Figure 11 shows the thermal conductivity of magnesia-alumina spinel (MgAl$_2$O$_4$) from room temperature up to the high-temperature region, together with the shifted power fit (valid from room temperature up to approx. 1200 °C, with T in °C):

$$k(T) = 391\,(T+128)^{-0.587}\,. \qquad (162)$$

The low-temperature peak conductivity of MgAl$_2$O$_4$ is approx. 170 W/mK at 60 K (Childs et al. 1973), while the room temperature value reported in (Childs et al. 1973) is 26 W/mK.

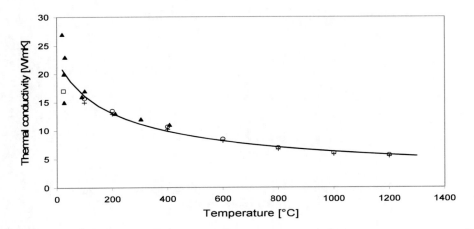

Figure 11. Thermal conductivity of magnesia-alumina spinel (MgAl$_2$O$_4$) in the high-temperature region and fit curve from room temperature to 1200 °C; plusses: Kingery's values corrected for porosity via the mixture rule (Kingery et al. 1954), empty circles: Kingery's values corrected for porosity via the exponential relation, full triangles (Burghartz and Schultz 1994), empty square (Shackelford and Doremus 2008).

Figures 12 and 13 show the thermal conductivity of zirconia (ZrO$_2$) ceramics in the high-temperature region. It is evident, that for doped zirconia some data in Figure 12 indicate a slight decrease, some a slight increase of conductivity with temperature (see Figure 3 and the discussion above). With respect to the small absolute values these trends can be considered as negligible, that means the thermal conductivity of doped zirconia (in contrast to undoped, monoclinic, zirconia, see Figure 13) can be considered as constant, within a range of 2-3 W/mK (tending to the lower end of this range for highly doped, e.g. dominantly cubic, zirconia, see (Pabst and Gregorová 2007)).

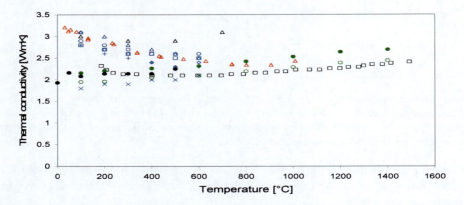

Figure 12. Thermal conductivity of zirconia ceramics; blue diamonds: with 2.6 wt.% MgO, blue plusses: with 3.4 wt.% MgO, blue crosses: with 5 wt.% MgO, blue triangles: with 2.4 wt.% Y$_2$O$_3$, blue circles: with 4 wt.% Y$_2$O$_3$, blue squares: with 5.3 wt.% Y$_2$O$_3$ (Hasselman et al. 1987), red triangles: PSZ (35 % tetragonal, balance cubic with negligible monoclinic) (Williams et al. 1988), black triangles: with 3 mol.% Y$_2$O$_3$ (3Y-TSZ) (Stevens 1991), black squares: with 3 mol.% Y$_2$O$_3$ (3Y-TSZ) (Vassen et al. 2000), black circles: with 10 mol.% Y$_2$O$_3$ (Dura et al. 2010), green empty circles: Kingery's values corrected for porosity via the mixture rule (Kingery et al. 1954)), green full circles: Kingery's values corrected for porosity via the exponential relation.

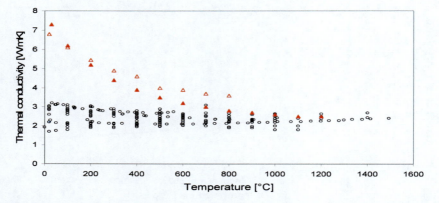

Figure 13. Comparison of the thermal conductivity of monoclinic zirconia [full triangles (Mévrel et al. 2004), empty triangles (Raghavan et al. 1998, 2001)) and of different types of doped zirconia (black empty circles; tetragonal and cubic phase with 5-17 wt.% Y$_2$O$_3$ or 2.6-5 wt.% MgO (An et al. 1999, Guo et al. 2002, Hasselman et al. 1987, Jang et al. 2004, Kingery et al. 1954, Mévrel et al. 2004, Raghavan et al. 1998, 2001, 2004, Schlichting et al. 2001, Stevens 1991, Su et al. 2001, Tanaka et al. 2001, Vassen et al. 2000, Williams et al. 1988, Wu et al. 2002)].

Table 4 lists thermal conductivity handbook data for titanates (BaTiO$_3$ and SrTiO$_3$) (Powell and Childs 1972, Lide 2010). It is evident that these materials exhibit relatively weak conductivity peaks at cryogenic temperatures (only approx. 20-25 W/mK at 30-40 K), while at room temperature their conductivity is 6-11 W/mK, comparable to silicates and the range of approx. 5-15 W/mK reported in (Childs et al. 1973). More recent work (Jezowski et al. 2007) has confirmed similar room temperature values even for materials with nanometric grain size (70-100 nm), but has shown that conductivity peaks of approx. 100 W/mK can be achieved (at temperatures of approx. 100 K). Moreover, by decreasing the grain size to 30 nm a glass-like behavior has been observed (without conductivity peak in the low-temperature region and with a conductivity continuously increasing with temperature), resulting in a conductivity of only 1.2 W/mK at room temperature (Jezowski et al. 2007).

Table 4. Thermal conductivity of titanates [in (W/mK)]
(Powell and Childs 1972, Lide 2010)

T (K)	BaTiO$_3$	SrTiO$_3$
5	4.2	2.4
30	24.0	21.0
40	25.0	19.2
100	12.0	18.5
250	4.8	12.5
300	6.2	11.2

Recent work has shown, that other piezoceramic materials (so-called PCR-type materials) may also have very low thermal conductivity, ranging from approx. 1.3 W/mK at room temperature to 2.0-2.8 W/mK at 500 °C (Gadzhiev et al. 2009).

The thermal conductivity of mixed yttrium-iron-aluminum-garnets (YIG-YAG) of composition Y$_3$Al$_x$Fe$_{5-x}$O$_{12}$ with x = 0, 0.7, 1.4 and 5.0 has been investigated for the temperature range from room temperature up to 1000 °C in (Padture and Klemens 1997). The shifted power fits of the temperature dependences of the thermal conductivity are (with T in °C):

$$k(T) = 23.7\,(T+67)^{-0.321} \tag{163}$$

for x ranging from 0 to 1.4 and

$$k(T) = 289.6\,(T+235)^{-0.629} \tag{164}$$

for x = 5.0. The thermal conductivities of these mixed-oxide garnets are similar to that of silicates, with average thermal conductivities of 5.6 ± 0.4 W/mK at room temperature, 4.5 ± 0.3 W/mK at 100 °C, 3.0 ± 0.1 W/mK at 500 °C and 2.7 ± 0.1 W/mK at 1000 °C for the low-x compositions. Low-temperature measurements of YAG (Yagi et al. 2007) have shown that at about 40 K peak conductivity values between approx. 60 and 100 W/mK are obtained for coarse-grained ceramics (with grain sizes of 3 and 7.5 μm, respectively), while the peak conductivity of YAG single crystals is approx. 800 W/mK, i.e. one order of magnitude higher (at 20-30 K). The values reported here for YAG ceramics and mixed YIG-YAG garnets are in good agreement with the data in (Childs et al. 1973), where the thermal conductivity of pure YIG garnets is in the range 4-7 W/mK at room temperature and peak conductivities are 70-140 W/mK at 20-30 K.

Another group of mixed oxides with very low thermal conductivity are the zirconates. In particular, a comparative study of barium zirconate ($BaZrO_3$), strontium zirconate ($SrZrO_3$) and lanthanum zirconate ($La_2Zr_2O_7$) has been performed in (Vassen et al. 2000). The thermal conductivity of lanthanum zirconate ($La_2Zr_2O_7$) is between 1.6 and 1.9 W/mK, i.e. the lowest of all zirconates (at the same time one of the lowest of all ceramics, together with zirconia, see above, and monazite-type rare earth phosphates (Du et al. 2009), which exhibit conductivities of 2.7-3.7 W/mK at room temperature but only 1.3-2.0 at 1000 °C, with a slight radiation contribution above 800 °C) and more or less constant from 200 to 1500 °C (average value 1.8 W/mK). The slightly increasing trend at temperature above 900 °C has to be attributed to the influence of radiation. The thermal conductivity data of barium and strontium zirconate can be nicely fitted by the shifted power fits

$$k(T) = 8.416 \cdot (T - 138)^{-0.135}, \tag{165}$$

$$k(T) = 5.476 \cdot (T - 60)^{-0.160}, \tag{166}$$

for the temperature ranges 200-1100 °C and 200-550 °C, respectively. In a recent paper (Chen et al. 2009) it is claimed that conductivities in the range 0.6-1.3 W/mK have been achieved in lanthanum zirconate ($La_2Zr_2O_7$). These values are among the very lowest thermal conductivities of all ceramic materials, comparable only to some silicate glasses and the yttrium silicates Y_2SiO_5 and γ-$Y_2Si_2O_7$ (Sun et al. 2008, 2009). Also in this work there seems to be no radiation influence below 900 °C. As a rough rule-of-thumb, the radiation contribution to heat transfer in polycrystalline ceramics can be assumed to be small below 1000 °C (Pampuch 2008). In (Du et al. 2009) it has been demonstrated that thermal conductivities around 1.1-1.2 W/mK for monazite-type rare earth phosphates ($REPO_4$) correspond to the amorphous limit (where the phonon mean free path is minimum). Theoretically calculated values for the minimum thermal conductivities (1.1 W/mK for $LaPO_4$ and 1.2 W/mK for $La_2Zr_2O_7$ (Sun et al. 2008, 2009)) are in good agreement with the experimentally measured values. Generally, this amorphous limit can be estimated via the aforementioned formula (Clarke 2003, Clarke and Phillpot 2005)

$$k_{min} \approx 0.87 k_B \rho^{1/6} E^{1/2} \left(\frac{n \cdot N_A}{M} \right)^{2/3}, \tag{167}$$

see Section 5. When using this formula, the amorphous limit is 2.9 W/mK for alumina (corundum, α-Al_2O_3), 2.6 W/mK for magnesia (MgO), 2.3 W/mK for magnesia-alumina spinel ($MgAl_2O_4$), 2.1 W/mK for titania (rutile, TiO_2) and 1.1-2.0 W/mK for the silicates considered there, viz. 1.1 W/mK for Y_2SiO_5, 1.4 W/mK for γ-$Y_2Si_2O_7$, 1.7 W/mK for mullite ($Al_6Si_2O_{13}$) and 2.0 for forsterite (Mg_2SiO_4) (Sun et al. 2009). The amorphous limit for zirconia has been estimated to be 1.1 W/mK (Cahill et al. 1992). These values can be recommended as an estimate of the grain boundary thermal conductivity for a precise phase mixture modeling of grain size effects (see Section 5). Of course, in systems were such precise information is lacking, it is reasonable to calculate an extremal estimate ("best-case" or "worst-case", whatever the purpose may be), based on the value of 1.1 W/mK, which is a reasonable estimate for the amorphous limit (minimum thermal conductivity) in the absence of a better alternative (Pabst 2009, Pabst and Gregorová 2010). The value of 1.1 W/mK is a good estimate also for the thermal conductivity of silicate glasses (range 0.6-1.4 W/mK, see below) and is sometimes recognized as a more or less universal value (Kingery 1960, Kingery et al. 1976).

The thermal conductivity of silica (SiO_2) in the form of low-quartz is between 7 W/mK (Schneider and Komarneni 2005) and 13 W/mK (Shackelford and Doremus 2008) at room temperature. This range is to be compared with the average value 7.6 ± 0.5 W/mK, calculated from the single-crystal values in Table A-1, the value of 8 W/mK listed in (Schlegel 1999) and, on the other hand, the value of 10.6 W/mK calculated from the single-crystal values in (Hellwege 1988). With respect to the low- and high-temperature values in (Gebhart 1993, Incropera and Dewitt 1991, Kleber 1990, Lide 2010, Powell and Childs 1972) we believe that the lower value (7.5 ± 0.5, say) is more reliable at room temperature. At 0 °C the thermal conductivity of low-quartz is approx. 8.9 W/mK, at 100 °C approx. 6.7 W/mK and at 500 °C approx. 3.5 W/mK (Gebhart 1993, Incropera and Dewitt 1991, Kleber 1990, Lide 2010, Powell and Childs 1972). According to (Lide 2010, Powell and Childs 1972) the low-temperature conductivity of low-quartz is approx. 450 W/mK at 20 K, but considerably higher peak conductivities of more than 1000 W/mK at 10-20 K have also been reported (Berman et al. 1950, Childs et al. 1973, Hellwege 1988, Ibach and Lüth 1988). At 100 K the thermal conductivity of low-quartz is in the range 3-50 W/mK (Childs et al. 1973), according to (Gebhart 1993, Incropera and Dewitt 1991) approx. 27 W/mK. The thermal conductivity of silica glass, silicates and silicate glasses is treated in Section 8. Sandstone as a natural rock is reported to have a conductivity of 2.9 W/mK at room temperature, while quartzite – its densified and recrystallized (diagenetically or metamorphically transformed) counterpart – has approx. 5.4 W/mK (Gebhart 1993, Incropera and Dewitt 1991).

The thermal conductivity of titania (TiO_2) in the form of rutile is approx. 8.3 W/mK at room temperature, as calculated from the single-crystal values in Table A-1, cf. also (Martienssen and Warlimont 2005). Again, also here a slightly higher value has been reported in the literature (approx. 10.4 W/mK) (Newnham 2005), but we prefer the former (lower value), because the latter is not consistent with the thermal conductivity at 0 °C, which is approx. 10.3 W/mK (Lide 2010, Powell and Childs 1972). The high-temperature thermal conductivity of rutile is approx. 6.6-7.0 W/mK at 100 °C, 3.8-4.0 W/mK at 500 °C and 3.2-3.3 W/mK at 1000 °C (Gebhart 1993, Incropera and Dewitt 1991, Kingery et al. 1954). Peak conductivities of TiO_2 in the low-temperature region (below 20 K) are between 80 and 2200 W/mK (Childs et al. 1973) (according to (Lide 2010, Powell and Childs 1972) approx. 170 W/mK at 4 K and 790 W/mK at 20 K).

Hafnia (HfO_2) has a conductivity of 1.1 W/mK at room temperature (Martienssen and Warlimont 2005), thoria (ThO_2) 14 W/mK at room temperature (Martienssen and Warlimont 2005), 10 W/mK at 100 °C, approx. 5 W/mK at 500 °C and approx. 3 W/mK at 1000 °C (Gebhart 1993, Incropera and Dewitt 1991, Salmang and Scholze 1982), urania (UO_2) 8-10 W/mK at room temperature (Childs et al. 1973, Martienssen and Warlimont 2005) and 3

W/mK at 500 °C (Salmang and Scholze 1982), yttria (Y$_2$O$_3$) 14 W/mK at 100 °C and 3 W/mK at 1000 °C (Salmang and Scholze 1982). The mixed oxide TiAl$_2$O$_5$ (sometimes called "tialite") is reported to have conductivity of 3.4 W/mK and fused-cast alumina-zirconia composite refractories 2.9-7.2 W/mK at room temperature (Bengisu 2001). The thermal conductivity of Y-Ba-Cu-superconductors (YBa$_2$Cu$_3$O$_{7-\delta}$), is approx. 2.7 W/mK at room temperature, decreasing to approx. 1-2 W/mK at 77 K according to (Martienssen and Warlimont 2005). According to other sources they exhibit a low-temperature maximum at around 50 K, which is of approx. 12 W/mK for single crystals but very flat (2-7 W/mK) or completely absent for polycrystalline samples, depending on the oxygen content and the degree of disorder (Nunez Regueiro and Castello 1991).

Table 5. Recommendable thermal conductivity values for oxide ceramics based on literature data (An et al. 1999, Bengisu 2001, Bisson et al. 2000, Burghartz and Schulz 1994, Guo et al. 2002, Hasselman et al. 1987, Jang et al. 2004, Kingery et al. 1954, Lee and Rainforth 1994, Martienssen and Warlimont 2005, Menčík 1992, Mévrel et al. 2004, NIST 2003, Padture and Klemens 1997, Pampuch 2008, Pierson 1996, Raghavan et al. 1998, 2001, 2004, Salmang and Scholze 1982, Schlichting et al. 2001, Shackelford and Doremus 2008, Slifka et al. 1998, Staroň and Tomšů 2000, Stevens 1991, Su et al. 2001, Tanaka et al. 2001, Vassen et al. 2000, Watanabe 1993, Williams et al. 1988, Wu et al. 2002), including calculated values from the fits obtained in this work (fit 0) as well as from fits based on previous literature surveys according to (Munro 1997a) (fit 1) and (Pabst and Gregorová 2007) (fit 2); "doped zirconia" means zirconia with dominating tetragonal (t) or cubic (c) phases, "undoped zirconia" means monoclinic (m)

Material	Room temperature	100 °C	500 °C	1000 °C
Al$_2$O$_3$	32.6 ± 4.5	26.5 ± 3.6	11.1 ± 0.7	7.3 ± 1.1
Al$_2$O$_3$ (fit 1)	33.4	26.3	11.4	7.2
Al$_2$O$_3$ (fit 2)	33.5	25.4	12.0	7.5
BaO	3			
BeO	285 ± 54	227 ± 14	75 ± 5	21 ± 2
BeO (fit 0)	350	234	75	35
CaO	20 ± 10	16 ± 1	9 ± 1	8 ± 1
CaCO$_3$ (calcite)	4.5	3.9		
MgO	51 ± 13	39 ± 6	15 ± 2	6 ± 2
MgO (fit 0)	49.4	36.1	15.1	8.7
MgAl$_2$O$_4$	20.4 ± 4.3	15.9 ± 0.8	9.4 ± 1.1	6.0 ± 1.0
MgAl$_2$O$_4$ (fit 0)	20.8	16.1	8.9	6.3
SiO$_2$ (β-quartz)	7.5 ± 0.5	6.7	3.5	-
SrO	10			
TiO$_2$ (rutile)	8.3	6.6	3.8	3.3
YBa$_2$Cu$_3$O$_{7-\delta}$	2.7			
YIG-YAG	5.6 ± 0.4	4.5 ± 0.3	3.0 ± 0.1	2.7 ± 0.1
ZnO	54			
ZrO$_2$ (doped)	2.4 ± 0.5	2.6 ± 0.5	2.4 ± 0.3	2.2 ± 0.2
ZrO$_2$ (t + c, fit 2)	2.4 ± 0.5	2.4 ± 0.5	2.3 ± 0.5	2.1 ± 0.4
ZrO$_2$ (undoped)	7.1 ± 0.3	6.2 ± 0.1	3.8 ± 0.3	2.6
ZrO$_2$ (m, fit 2)	7.2	6.1	3.8	2.8

According to calculations based on the single-crystal data in Table A-1 calcite (CaCO$_3$) has a thermal conductivity of approx. 4.5 W/mK at room temperature, which decreases to approx. 3.9 W/mK at 100 °C. The low-temperature conductivity of calcite is reported to be 20 W/mK at 83 K, approx. 6.8 W/mK at 194 K and approx. 4.9 W/mK at 273 K, i.e. 0 °C (Hellwege 1988, Lide 2010, Powell and Childs 1972). Limestone as a natural rock is reported to have a conductivity of approx. 2.2 W/mK at room temperature, while marble – its densified and recrystallized (diagenetically or metamorphically transformed) counterpart – has 2.8 W/mK (Gebhart 1993, Incropera and Dewitt 1991). The thermal conductivity of gypsum (CaSO$_4$ · 2H$_2$O, plaster) and anhydrite (CaSO$_4$) is considerably lower than that of limestone and marble, around 1 W/mK. Historically this has been a well-known criterion for readily distinguishing e.g. alabaster and marble statues. The thermal conductivity of ice (H$_2$O, hexagonal) at 0 °C is approx. 1.9-2.1 W/mK (Gebhart 1993, Gibson and Ashby 1997, Kleber 1990), with a low-temperature maximum of approx. 110-180 W/mK at 7-8 K (Childs et al. 1973), and low-temperature conductivities of 100-140 W/mK at 10 K and 5-8 W/mK at 10 K. For the thermal conductivity of liquid water and water vapor see Section 6 and Appendix C.

Table 5 lists thermal conductivity data for oxide ceramics based on en extensive literature search. Needless to say that the values have to be understood as typical values (the ± sign denotes an estimated uncertainty range rather than a standard deviation).

8. THERMAL CONDUCTIVITY OF SILICATES AND GLASSES

Silicates (including alumosilicates) are the most common constitutents of the Earth's crust, and as such not only historically the first but up to these days still the basic raw materials for the traditional ceramic industry (fine and coarse ceramics). Their applications range from floor tiles, wall tiles, roof tiles and stove tiles to bricks, refractories and electrical insulators. Crystalline silicate phases and silicate glasses (different from the original raw material composition) develop during the various technological processes and are thus common constituents of ceramic microstructures after firing. The relatively low thermal conductivity of silicates and glasses predetermines them for insulating applications. Therefore in many applications of silicates and silicate-based ceramics their thermal conductivity plays an important role and is often reduced in a controlled way by introducing porosity. Similar to ceramics *sensu strictu*, also inorganic glasses have irreplaceable functions in many areas of human life, including their key role in the current and future photovoltaic industry. Although most of these applications are based on the optical properties of glasses, the low thermal conductivity of glasses is a basic performance feature also in this class of materials.

This section deals with the thermal conductivity of polycrystalline silicates, mostly in the form of single-phase ceramics, and of silicate glasses, including pure silica glass (its crystalline counterpart, however, the pure oxide SiO$_2$ in the form of low-quartz, has been treated in Section 7 above). Figure 14 shows the classic results of Kingery and coworkers (Kingery et al. 1954), comparing the original results with the values corrected according to the exponential relation.

The difference between the two dependences is not insignificant, but it is evident that the thermal conductivity of the silicates forsterite (MgSiO$_4$), zircon (ZrSiO$_4$) and mullite (Al$_6$Si$_2$O$_{13}$) is very similar, i.e. their mutual differences are approximately of the same magnitude as the difference between two types of mullite (a batch-to-batch variation). Shifted-power fits of the temperature dependence of the thermal conductivity of silicate ceramics in the temperature range from 100 °C up to approx. 1400 °C (with T in °C) are listed in Table 6.

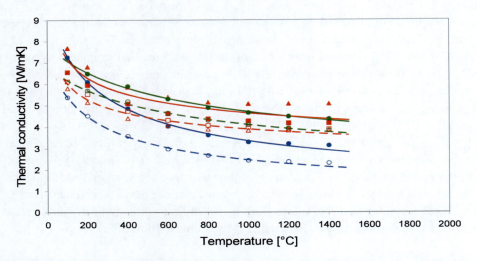

Figure 14. Thermal conductivity of silicate ceramics (Kingery et al. 1954); red: mullite (two types – squares and triangles), blue: forsterite, green: zircon; empty symbols (and dashed fit curves): corrected for porosity by the mixture rule, full symbols (and solid fit curves): corrected for porosity by the exponential relation.

Table 6. Shifted-power fits of the temperature dependence of the thermal conductivity of silicate ceramics; data from (Kingery et al. 1954), corrected for porosity by the mixture rule and the exponential relation, respectively

	Fit based on data corrected for porosity via the mixture rule	Fit based on data corrected for porosity via the exponential relation
Forsterite	$k(T) = 64.6 \cdot (T+107)^{-0.466}$	$k(T) = 88.2 \cdot (T+111)^{-0.466}$
Mullite 1	$k(T) = 17.3 \cdot (T+29)^{-0.213}$	$k(T) = 18.8 \cdot (T+33)^{-0.215}$
Mullite 2	$k(T) = 11.4 \cdot (T+32)^{-0.158}$	$k(T) = 15.3 \cdot (T+30)^{-0.161}$
Mullite average	$k(T) = 13.8 \cdot (T-6)^{-0.183}$	$k(T) = 16.4 \cdot (T-5)^{-0.183}$
Mullite average (2-par. fit)	$k(T) = 14.2 \cdot T^{-0.187}$	$k(T) = 16.9 \cdot T^{-0.187}$
Zircon	$k(T) = 42.2 \cdot (T+265)^{-0.326}$	$k(T) = 47.2 \cdot (T+257)^{-0.323}$

Literature values for the thermal conductivity of traditional silicate ceramics (hard porcelain, steatite, cordierite, fireclay products etc.) are generally low. Usually they lie in the range 1-4 W/mK at or around room temperature, with the highest values of 4 W/mK relating to high-strength alumina porcelain and dense forsterite (Martienssen and Warlimont 2005, Menčík 1992). Apart from the silicate glass phase, the properties of which are treated below, the most important phase of many silicate ceramics is mullite. The thermal conductivity of mullite is 5-7 W/mK at room temperature and decreases to 4-4.5 W/mK at 500 °C and to 3.5-4 W/mK above 1000 °C (Bengisu 2001, Dreyer 1974, Hayashi et al. 1998, Hildmann and Schneider 2005, Kingery et al. 1976, Lee and Rainforth 1994, Russell et al. 1996, Schneider and Komarneni 2005, Shackelford and Doremus 2008, Sivakumar et al. 2001). Single-crystal measurements (Hildmann and Schneider 2005) yield room temperature values of 5.4, 5.2 and 8 W/mK for the a-, b- and c-axes, respectively. Thus at room temperature the thermal conductivity of mullite is significantly lower than that of alumina (33 W/mK), but significantly higher than that of the silicate glass phase (approx. 1.1 W/mK). It is similar to the thermal conductivity of quartz, which is approx. 7.5 W/mK (see Section 7 and (Schlegel 1999, Schneider and Komarneni 2005)). At elevated temperature the conductivity difference between these phases (typical for many silicate ceramics and refractories) becomes generally smaller: at 500 °C we have 12 W/mK for alumina (Pabst and Gregorová 2007), 4 W/mK for mullite, 3.5 W/mK for quartz and more than 1.3 W/mK (possibly with a radiative contribution, resulting in apparent conductivities of up to 2.2-2.7 W/mK) (Schill 1993) for the silicate glass phase.

Reliable thermal conductivity values for pure magnesiosilicates and -alumosilicates (except for cordierite and steatite at room temperature with 1-3 W/mK and 2-4 W/mK, respectively (Bengisu 2001, Menčík 1992, Salmang and Scholze 1982), and forsterite at 1000 °C with approx. 3 W/mK (Staroň and Tomšů 2000)) and several other silicate phases (except for zircon with 4 W/mK at 1000 °C (Staroň and Tomšů 2000)) are rather rare in the engineering literature. The values range approximately from 0.7 W/mK for mica (muscovite and phlogopite) (Lide 2010), 0.7-1.5 W/mK for wollastonite ($CaSiO_3$) and CSH phase (i.e. the X-ray amorphous or "crypto-"crystalline $CaO \cdot SiO_2 \cdot H_2O$ "gel phase" in cements) (Schlegel 1999) to 2.9 W/mK for tourmaline (parallel to the c-axis), 6.4 W/mK for beryl, 17.7 W/mK for topaz (parallel to the c-axis) and 36 W/mK for Al-Fe-silicate garnet (Lide 2010, Powell and Childs 1972) (for the latter, however, the high value is questionable: values lower by one order of magnitude, namely 4-8 W/mK are given for natural garnets in (Childs et al. 1973)).

Most natural rock types (basalt, granite, limestone, sandstone etc.) are reported to have thermal conductivities in the range 1-3 W/mK (Gebhart 1993, Incropera and Dewitt 1991, Lide 2010, Weast 1988), but many of them (mainly sedimentary and metamorphic rocks) exhibiting microstructural anisotropy (e.g. slates are anisotropic due to the orientation texture of the constituent minerals, and their thermal conductivity is 2.5 W/mK within the plane and 1.4 W/mK normal to the plane).

At room temperature almost all silicate glasses exhibit thermal conductivities in the range 0.6-1.4 W/mK, with pure silica glass being among the most conductive and lead-containing glasses among the least conductive glasses. Increasing contents of alkali ions generally reduce the thermal conductivity of silicate glasses (Volf 1988). With increasing temperature the true thermal conductivity of these glasses increases only very slightly, but the influence of radiative heat transfer (depending on the optical properties of the glasses, especially the transparency in the near infrared region) can become significant at temperatures higher than about 300-400 °C in the case of colorless glasses (Fanderlík 1991, Volf 1990). According to (Volf 1990) the apparent conductivity caused by radiation can be as high as 3.0-4.4 W/mK at 600 °C, 11-16 W/mK at 800 °C, 28-48 W/mK at 1000 °C and 50-110 W/mK at 1200 °C. That means, for colorless glasses at approx. 800 °C the radiative conductivity is already about one order of magnitude higher than the true (phonon) conductivity.

Figures 15 and 16 indicate that the thermal conductivity of common silicate glasses is rather similar, around 1.1 W/mK at room temperature (around 1.05 W/mK for Na-Ca-silicate glasses and slightly higher, 1.15 W/mK, for borosilicate glasses (Volf 1990, Lide 2010)). Notable exceptions are pure, clear silica glass with approx. 1.4 W/mK, lead-containing flint glasses with approx. 0.7 W/mK (Volf 1990) (from 0.9 W/mK for a light flint with 25 % PbO down to 0.5 W/mK for a very heavy flint with 80 % PbO (Lide 2010)) and barium-containing crown glasses also with values down to 0.7 W/mK at room temperature (Lide 2010)).

At moderately elevated temperatures, up to about 200-400 °C (Fanderlík 1991, 1996, Schill 1993, Volf 1990), heat transfer is clearly dominated by true (phonon) conduction (Figure 15 shows that at 300 °C the radiative contribution is approx. 10 % of the apparent conductivity), but at higher temperatures the radiative (photonic) contribution to heat transfer becomes important (especially in colorless or, more precisely, non-heat absorbing glasses (Fanderlík 1996)) and at temperatures > 600 °C usually dominant (at 500-550 °C the radiative contribution to the apparent conduction is of the same order of magnitude as true conduction (Schill 1993), cf. also Figure 15). Above 800 °C the radiative contribution clearly dominates the heat transfer (Volf 1990), at around 1000 °C the true thermal conductivity is negligible (Volf 1988) and at melting temperatures the radiant heat flux can exceed true conduction by two orders of magnitude (Schill 1993), cf. also Figure 16. In glass melts, of course, convective heat transfer can play a major role, in addition to conductive and radiative heat transfer (Fanderlík 1991).

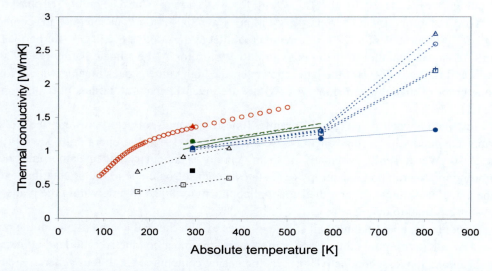

Figure 15. Thermal conductivity of silica glass (red circles (Fanderlík 1991) and red triangle – RT average value (Volf 1990)) and silicate glasses; green: borosilicate glasses (empirical fit curves from (Carwile and Hoge 1969, Stephens 1932, Volf 1990) and green circle – RT average value (Volf 1990)), black: lead silicate glasses (empty squares: very heavy flint glasses with 80 % PbO, empty triangles: light flint glasses with 25 % PbO (Lide 2010), full square – RT average value (Volf 1990) – black lines are guides to the eye), blue: soda-lime-silicate container glasses (empty triangles: white conserve glass, empty circles: white bottle glass, plusses: green bottle glass, empty squares: brown bottle glass – apparent conductivities with radiative components (Schill 1993), full circles: pure conduction without radiation for all soda-lime-silicate (Na-Ca-Si) glasses (Schill 1993) – blue lines are guides to the eye).

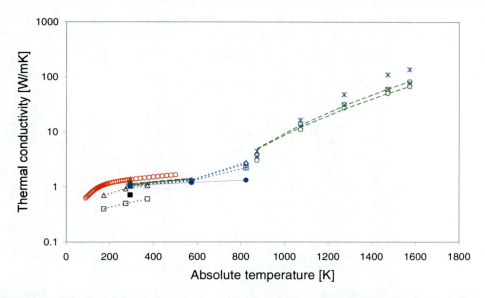

Figure 16. Thermal conductivity of silica glass and silicate glasses (below 900 K identical to Figure 15, see the figure caption above), including the influence of radiative heat transfer at higher temperatures (Volf 1990); green: two samples of borosilicate glasses (circles are data points, dashed curves are power-law fits), blue: soda-lime-silicate (Na-Ca-Si) glasses (crosses – container glass, stars – Float glasses).

The thermal conductivity of glasses in a certain temperature range around room temperature (from – 100 to +100 °C, say) can be estimated by empirical relations using additive factors. Two common relations are:

- the Sharp relation for room temperature (Volf 1988):

$$k_S = \frac{\sum_i w_i q_i}{\sum_i w_i s_i}, \qquad (168)$$

- the Ratcliffe relation for the temperatures – 100, 0 and +100 °C (Ratcliffe 1963):

$$k_R = \sum_i w_i r_i, \qquad (169)$$

where w_i are weight fractions of the oxides in the glass (melt) and q_i, s_i and r_i are empirical additive factors for each oxide. In the first case these factors are known for room temperature, in the second for the temperatures – 100, 0 and +100 °C. Table 7 lists the additive factors to be used in these relations in order to estimate the thermal conductivity of a silicate glass of known composition.

Table 7. Additive factors to estimate the thermal conductivity of common silicate glasses using the Ratcliffe or Sharp relations (Volf 1988), for resulting conductivities in units (cal/scmK), to by multiplied by 418.7 to obtain the SI units (W/mK); n.a. = not available

	q_i	s_i	r_i (−100 °C)	r_i (0 °C)	r_i (+100 °C)
SiO_2	130.4	0.4348	2.44	3.07	3.44
Al_2O_3	195.3	0.3125	3.23	3.72	2.14
B_2O_3	157.4	0.4255	1.09	1.59	2.49
Fe_2O_3	n.a.	n.a.	1.61	1.90	1.73
BaO	166.9	0.1408	0.39	0.46	0.75
CaO	225.6	0.2564	2.82	3.17	2.39
MgO	116.7	0.2564	6.37	5.92	4.53
PbO	117.0	0.1000	0.60	0.76	0.96
ZnO	146.6	0.1695	1.95	2.02	1.64
Na_2O	396.0	0.3448	− 1.24	− 1.29	− 0.67
K_2O	462.1	0.3448	0.54	0.58	0.39

The thermal conductivity at not too high temperatures (< 400 °C), i.e. the true thermal conductivity, is a material property that is determined largely by the bond strength (Volf 1988, 1990). The latter is strongest in silica glass (only Si-O-bonds), very strong also in borosilicate glasses (Si-O-, B-O- and Al-O-bonds) but somewhat weaker in alkali-lime-silicate glasses where non-bridging alkali metals are present. The presence of atoms with significantly different masses reduces thermal conductivity as well (Richerson 2006). This is the reason for the relatively low thermal conductivity of lead-containing silicate glasses. Between 0 and 300 °C the temperature dependence of borosilicate glasses is well described by the linear relation (with T in (°C))

$$k(T) = 1.1 + 0.001 \cdot T, \tag{170}$$

(Schill 1993, Volf 1990), and an analogous relation can be recommended for common soda-lime silicate glasses:

$$k(T) = 1.04 + 0.0005 \cdot T, \tag{171}$$

cf. Figures 15 and 16. At high temperatures (> 400 °C) the total heat flux and thus the apparent thermal conductivity of glasses,

$$k_{apparent} = k + k_{radiation}, \tag{172}$$

is determined by the optical properties (spectral transmittance and absorbance) and in the (visible and mainly in the) near-infrared region (Schill 1993, Volf 1988). Compared to the thermal conductivity of clear silica glass (1.4 W/mK; clear silica glass has the highest thermal conductivity of all technical glasses) that of opaque silica glass is somewhat lower (1.1 W/mK (Fanderlik 1991)), and the radiative contribution becomes important for the latter at higher temperatures than for the former. Similarly, the radiative contribution is higher for colorless glasses, or – more precisely – for glasses that are less absorbing (more transmissive) in the visible and near-infrared region. For example, in most glasses iron oxides have a distinct effect as coloring oxides; Fe^{2+} ions (FeO) absorb radiation in the near- infrared region (1.1 µm), while Fe^{3+} (Fe_2O_3) absorbs preferentially in the ultraviolet region (360 nm), so that its presence has no significant effect on the radiant heat flux (Volf 1990). For many glasses above 400 °C, when radiative heat flux becomes important, the (apparent) thermal conductivity (dominated by radiative transfer) *de facto* loses the character of a material property in the strict sense. In this case, of course, it cannot be calculated additively from the composition (Volf 1990).

The radiative contribution to the apparent thermal conductivity in a dense solid, e.g. a glass, increases with temperature according to a power law and can principally be estimated using the relation (Salmang and Scholze 1982, Schulle 1990)

$$k_{radiation} = \frac{16}{3} \sigma n^2 L_{photons} T^3, \qquad (173)$$

where n is the (real part of the) refractive index, σ the Stefan-Boltzmann constant (5.67·10^{-8} Wm^{-2}K^{-4}), $L_{photons}$ the mean free path for photons and T the absolute (Kelvin) temperature. For soda-lime-silicate container glasses values for $L_{photons}$ have been reported to be 56-93 µm at 20 °C, 620-990 µm at 300 °C and 2.2-3.7 mm at 550 °C (Schill 1993). For alumina single crystals at 750 °C it has been estimated to be approx. 10 cm (Lee and Kingery 1960). Note, however, that the experimentally determined temperature dependence is often stronger than predicted by this relation: temperature exponents from 3.5 to 5 have been reported (Richerson 2006, Schulle 1990).

The thermal conductivity of insulating bricks (based on diatomite, vermiculite, perlite etc.) with classification temperatures 900-950 °C and density 0.45-0.80 g/cm^3 is 0.10-0.20 W/mK at 200 °C, 0.12-0.21 W/mK at 400 °C and 0.19-0.24 W/mK at 800 °C (Salmang and Scholze 1982). The thermal conductivity of light-weight refractories (quartzite, fireclay, sillimanite and hollow-sphere alumina with porosities 45-80 %) with different classification temperatures in the range 1300-1870 °C and bulk densities in the range 0.8-1.2 g/cm^3 (1.4 g/cm^3 for hollow-sphere alumina) is between 0.3-0.6 W/mK (1.3 W/mK for hollow-sphere alumina) at 400 °C and 0.4-0.9 W/mK (1.1 W/mK for hollow-sphere alumina) at 1200 °C (Salmang and Scholze 1982). The thermal conductivity of fiber products (modules) at temperatures from room temperature up to 1200 °C is below 0.5 W/mK (Dietrichs and Krönert 1982, Salmang and Scholze 1982). There are indications that the conductivity of fiber products (with porosities typically > 90 %) is lower when the diameters of the fibers (and the pores between them) are smaller (Schupp 1981). It is reported that at temperatures above 800 °C radiative heat transfer dominates over convective heat transport (Salmang and Scholze 1982). When used as insulation materials in high-temperature aggregates like ovens, kilns and furnaces, the oven atmosphere can play a significant role. In particular, the insulation capability of fiber products decreases when the oven atmosphere is hydrogen (almost seven times higher conductivity than air) (Salmang and Scholze 1982).

9. THERMAL CONDUCTIVITY OF NON-OXIDES

The thermal conductivity of non-oxides and non-oxide ceramics is important from several points of view, ranging from theoretical interests (concerning e.g. the relation of the extremely high conductivity of diamond to its structure and its mechanical properties or the extremely high degree of conductivity anisotropy in graphite) to very practical concerns in common large-scale refractories, high-temperature structural materials, cutting tools and grinding media (e.g. silicon carbide, SiC, silicon nitride, Si_3N_4, sialons, $Si_{6-x}Al_xO_xN_{8-x}$, boron nitride, BN and tungsten carbide, WC), special high-temperature engineering materials (e.g. titanium boride, TiB_2, zirconium boride, ZrB_2, hafnium boride, HfB_2, and molybdenum silicide, $MoSi_2$) and modern electronic substrate materials (e.g. aluminum nitride, AlN). The comparison of literature data for polycrystalline non-oxide ceramics is generally more problematic than for oxides and other material groups, because non-oxides often require sintering aids in order to obtain more or less dense bodies (or even contain elemental phases, such as silicon in the case of reaction-bonded silicon carbide) and therefore they are often not available in pure form. After firing, these additives (mostly oxides) result in a grain boundary phase of a composition different from that of the crystalline bulk phase. The structure of this oxide-based (or hybrid oxide-nonoxide) grain boundary phase is of course different from that of the non-oxide crystallites: it is mostly amorphous or glassy (unless special annealing schedules are applied for recrystallization). For these reasons the uncertainty concerning true intrinsic thermal conductivity values for non-oxide phases is usually higher than for oxide and silicate ceramics. However, it may be said that the thermal conductivity of the most important non-oxides (AlN, SiC, Si_3N_4, WC and borides, not to speak of diamond, graphite and c-BN) at room temperature is higher than that of oxides (and of course silicates), the only notable exception being beryllia (BeO), see Section 7. Moreover, due to their nonionic bond character some non-oxides, especially so-called "interstitial" non-oxides (in contrast to "covalent" ones (Pierson 1996)), have features which are rather untypical for ceramics (e.g. good electric conductivity and thermal conductivity constant or even increasing with temperature). The reason for this behavior has to be sought in the partly metallic bond character and strong interactions between the metal atoms of these compounds.

A very special case of non-oxide materials, of course, are the carbon modifications – ranging from the traditionally well-investigated diamond and graphite types (extruded, pyrolytic etc.) to so-called diamond-like carbon and the more recent interest in carbon nanotubes (Krueger 2010, Yamamoto et al. 2008). It is well known that diamond is the solid bulk material with the highest thermal conductivity at and around room temperature (at this temperature its thermal conductivity is higher than that of any metal). Also its low-temperature behavior is extremely interesting from an academic point of view (and well investigated), because it shows very well the influence of isotopic effects on thermal conductivity and it can be extremely high (higher than 42000 W/mK at temperatures below 100 K (Wei et al. 1993)). On the other hand, a more practical aspect of the same coin, carbon nanotubes might pave the way to produce highly conducting polymer composites (Biercuk et al. 2002, Khare and Bose 2005, Kumari et al. 2008). For this reason we will give a brief account of diamond and graphite-based materials. However, a detailed treatment of the thermal conductivity of the different forms of carbon and polycrystalline carbon-based materials, mainly the extremely interesting nanotubes, is beyond the scope of this chapter. For more information on this specific topic the reader may consult e.g. (Pierson 1993, Krueger 2010, Yamamoto et al. 2008).

The class of non-oxides encompasses a large number of technically important compounds, including insulators, semiconductors and even materials with an electric conductivity comparable to metals. For structural and refractory applications these are mainly carbides, nitrides, borides and a few silicides, for functional applications mainly semiconductors, like gallium coumpounds (e.g. GaAs) and the chalkogenides of zinc and cadmium (ZnS, ZnSe, ZnTe, CdS, CdSe, CdTe). Table 8 lists thermal conductivity values for the latter group, but a more detailed analysis of this specific group is not intended here. Graphs of the temperature dependence of these values (in the low-temperature region) can be found in (Childs et al. 1973, Martienssen and Warlimont 2005).

Table 8. Thermal conductivity of some semiconducting compounds at room temperature (Martienssen and Warlimont 2005)

	k (W/mK)		k (W/mK)		k (W/mK)
GaN	130	ZnS	27	CdS	20
GaP	77	ZnSe	19	CdSe	9
GaAs	45.5	ZnTe	18	CdTe	7.1

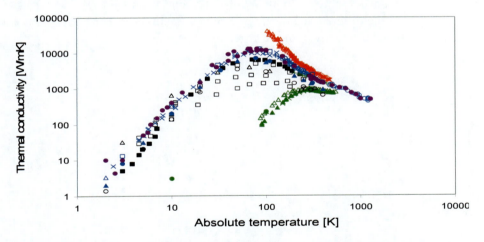

Figure 17. Thermal conductivity of diamond; isotopically enriched diamond single crystals – red (crosses: < 0.05 % ^{13}C (Onn et al. 1992), full squares: 0.07 % ^{13}C (Olson et al. 1993), empty triangles: 0.1 % ^{13}C (Wei et al. 1993)), polycrystalline diamond films – green (empty triangles: natural 1.1 % ^{13}C (Qiu et al. 1993), full triangles 0.04 % ^{13}C (Qiu et al. 1993), full circles: CVD (Ownby and Stewart 1991), crosses: CVD - low quality, plusses: CVD – high quality (Wort et al. 1994)), type I diamond – black (empty triangles (Slack 1964), empty circles (Lide 2010), type Ia – empty squares (Berman et al. 1975), type Ib – full squares (Berman et al. 1975)), type II diamond – blue (crosses (Ownby and Stewart 1991), type IIa – crosses (Berman et al 1975), plusses (Wei et al. 1993), empty circles (Olson et al. 1993), empty squares (Slack 1964), empty triangles (Lide 2010), type IIb – full triangles (Lide 2010), full circles (Slack 1964)) and unspecified diamond types – full violet circles (Adachi 2004, Naumann 2009, Onn et al. 1992, Ownby and Stewart 1991, Slack 1973).

Figures 17 and 18 show the temperature dependence of the thermal conductivity of diamond in the low- and high-temperature region, respectively. From Figure 18 it is evident that in the high-temperature region three types of results can be distinguished and assigned to different diamond types. The lowest conductivity is exhibited by so-called "type-I"-diamond and by polycrystalline diamond films. Its value of approx. 900-1000 W/mK is often reported as the room temperature value for the conductivity of diamond in handbooks and textbooks (Bengisu 2001, Martienssen and Warlimont 2005, Richerson 2006), although sometimes lower values, around 600 ± 50 W/mK, are cited as well (Martienssen and Warlimont 2005, Pierson 1996). Significantly higher, up to 2400 W/mK at room temperature, is the conductivity of "type-II"-diamond, corresponding to handbook values in the range 2000-2400 W/mK (Gebhart 1993, Martienssen and Warlimont 2005, Pierson 1996). Concerning all these values it must be emphasized, that errors in the absolute values can be quite large and may depend critically on the measurement technique used, especially in the upper conductivity range. A round robin test concerning the thermal conductivity of different CVD diamond samples at room temperature performed under the auspices of the US National Institute of Standards and Technology (NIST) in 1995 – measured were 10 commercial specimens at 6-10 experienced laboratories (Feldman 1995) – yielded results ranging from approx. 430 ± 80 W/mK to 1660 ± 560 W/mK (note the large relative errors). Nevertheless it is clear that diamond is the bulk solid with the highest thermal conductivity at room temperature. Moreover, Figure 18 shows that with isotopically purified diamond, in which the content of ^{13}C has been artificially reduced – compared to its natural ratio – even higher conductivity values may be attained (approx. 3600 W/mK at room temperature). As expected, the differences between the different diamond types increase with decreasing temperature, while at high temperature the conductivity of the different types converges.

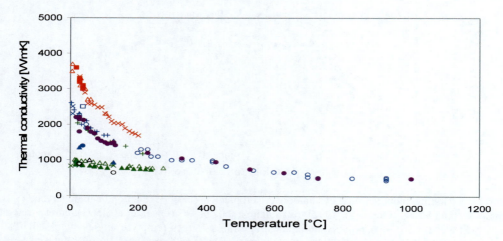

Figure 18. Thermal conductivity of diamond in the high-temperature region; isotopically enriched diamond single crystals – red (crosses: < 0.05 % ^{13}C (Onn et al. 1992), full squares: 0.07 % ^{13}C (Olson et al. 1993), empty triangles: 0.1 % ^{13}C (Wei et al. 1993)), polycrystalline diamond films – green (empty triangles: natural 1.1 % ^{13}C (Qiu et al. 1993), full triangles 0.04 % ^{13}C (Qiu et al. 1993), full circles: CVD (Ownby and Stewart 1991), crosses: CVD - low quality, plusses: CVD – high quality (Wort et al. 1994)), type I diamond – black (empty triangles (Slack 1964), empty circles (Lide 2010)), type II diamond – blue (crosses (Ownby and Stewart 1991), type IIa – crosses (Berman et al 1975), plusses (Wei et al. 1993), empty circles (Olson et al. 1993), empty squares (Slack 1964), empty triangles (Lide 2010), type IIb – full triangles (Lide 2010), full circles (Slack 1964)) and unspecified diamond types – full violet circles (Adachi 2004, Naumann 2009, Onn et al. 1992, Ownby and Stewart 1991, Slack 1973).

Peak values of the low-temperature thermal conductivity commonly reported for diamond are around 13000 W/mK at temperatures of 70-90 K (Childs et al. 1973, Martienssen and Warlimont 2005), but much higher values have been measured for isotopically refined diamond (Wei et al. 1993), see also Figure 17. For "type-I" diamond (and polycrystalline films) the conductivity is approx. 800-900 W/mK at 100 °C, for "type-II" diamond approx. 1500 W/mK at 100 °C and 800-900 W/mK at 500 °C. Interestingly, the in-plane conductivity of pyrolytic graphite is approx. 2000 W/mK (Bengisu 2001, Gebhart 1993, Lide 2010, Richerson 2006). This value is of course determined by the in-plane conductivity of the graphene sheet itself, and therefore similar to the axial thermal conductivity of carbon nanotubes (Yamamoto et al. 2008). For pyrolytic graphite the ratio between in-plane and perpendicular conductivity can be higher than 200, see Table A-1. Of course, pyrolytic graphite is structurally close to a single crystal, whereas usual graphite produced e.g. by extrusion has a much lower degree of anisotropy: in-plane values of 350-400 W/mK and perpendicular values of order 8-80 W/mK are commonly reported, resulting in average values of 200-250 W/mK or lower for densely pressed graphite products, cf. Table A-1 and (Martienssen and Warlimont 2005, Richerson 2006, Salmang and Scholze 1982). Note that amorphous (glassy) and badly graphitized carbon is reported to have a much lower conductivity (approx. 1-5 W/mK (Gebhart 1993, Salmang and Scholze 1982)) and coal (anthracite) is even more insulating, with approx. 0.26 W/mK (Gebhart 1993); of course, carbon foams, fiber felts, mats, blankets and modules can exhibit conductivities of order 0.01 W/mK (Salmang and Scholze 1982), depending on porosity and microstructure. A more detailed treatment of the other forms of carbon is beyond the scope of this treatise. The interested reader may refer to (Krueger 2010, Pierson 1993).

The most important non-oxide ceramics are based on silicon carbide (SiC) or silicon nitride (Si_3N_4), followed by boron nitride (BN), boron carbide (B_4C), aluminum nitride (AlN), tungsten carbide (WC), titanium boride (TiB_2), zirconium boride (ZrB_2) and molybdenum silicide ($MoSi_2$), the nuclear materials uranium carbide (UC) and uranium nitride (UN), as well as the interstitial compounds such as titanium carbide (TiC), tantalum carbide (TaC), titanium nitride (TiN), which are mainly used as powders or coatings. We will begin with carbides and nitrides and then briefly summarize the rest.

From the viewpoint of thermal conductivity it is useful to distinguish between covalent and interstitial carbides and nitrides (Pierson 1996). The covalent ones (e.g. AlN, B_4C, BN, SiC, Si_3N_4, WC) exhibit the typical thermal conductivity decrease at high temperatures and their electric conductivity is low (they are at best semiconductors), while the interstitial ones (e.g. TiC, TaC, TiN) show an increase of thermal conductivity with temperature (with a thermal conductivity similar to that of the host metals) and are good electrical conductors. This is attributed to the metallic character of these compounds. Therefore also the electronic contribution to their thermal conductivity is significant (Pierson 1996).

Figures 19 and 20 show the thermal conductivity of silicon carbide (SiC). According to these data the shifted-power fit for the temperature range from room temperature up to 1500 °C (with T in °C) is

$$k(T) = 1794 \, (T + 232)^{-0.538}. \tag{174}$$

The majority of literature data for densely sintered SiC at room temperature are in the wide range from 30 to 170 W/mK, with an estimated average of approx. 100 ± 70 W/mK, but a much higher value (270 W/mK) has been reported as well (Takeda 1998, Watari et al. 2003)), which is in good agreement with the values around 250 W/mK that have been extrapolated for SiC particles from measurements of SiC-reinforced composites (Barea et al. 2003, Hasselman et al. 1992, Kawai 2001). Of course, single crystal values are much higher (290-490 W/mK) (Müller et al. 1998, Slack 1964, 1973), see Table 9 and a theoretical limit of as high as 700 W/mK has been predicted for the thermal conductivity of defect-free SiC single crystals at room temperature (Slack 1964). Typical values at 1000 °C are in the range 31-53 W/mK (estimated average 40 ± 10 W/mK), for single crystals around 61 W/mK (Müller et al. 1998). It seems that a significant difference between (hexagonal) α- and (cubic) β-SiC has not been observed to date; many literature sources (especially data bases) do not specify the modification and polytype. The low-temperature peak conductivity of SiC ceramics ranges between 390 W/mK at 130 K (Nemoto et al. 1985) and 5200 W/mK at 50 K (Slack 1964, 1973), cf. also (Childs et al. 1973, Naumann 2009).

Another empirical fit for the temperature dependence of thermal conductivity (with T in °C) of sintered polycrystalline SiC (α-SiC-6H) in the temperature range between room temperature and 1600 °C has been given in (Munro 1997b):

$$k(T) = \frac{52000 \cdot \exp(1.24 \cdot 10^{-5} T)}{T + 437}. \tag{175}$$

For SiC single crystals (α-SiC-6H) the following fit relation has been proposed from room temperature up to approx. 2000 °C (Müller et al. 1998):

$$k(T) = 451700 \cdot (T + 273)^{-1.29}. \tag{176}$$

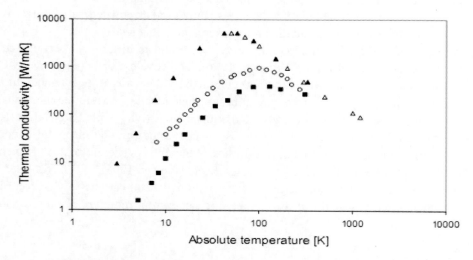

Figure 19. Thermal conductivity of SiC in the low-temperature region; data from (Nemoto et al. 1985) (full squares), (Slack 1964, 1973) (full and empty triangles) and (Adachi 2004, Naumann 2009) (empty circles).

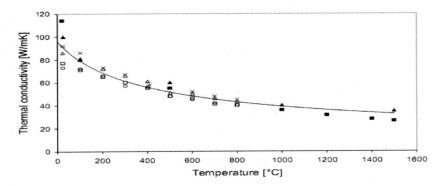

Figure 20. Thermal conductivity of SiC in the high-temperature region; data from (Morell 1985) (full triangles), (Munro 1997b) (full squares), and (Liu and Lin 1996) (empty tringles: pure α-SiC, empty squares: pure β-SiC, empty circles: 50 wt.% α- and 50 wt % β-SiC, crosses: 20 wt.% α and 80 wt.% β-SiC).

Figure 21 shows the temperature dependence of the thermal conductivity of aluminum nitride (AlN) ceramics sintered with yttria as a sintering aid. According to these data the shifted-power fit for the temperature range from room temperature up to 500 °C (with T in °C) is

$$k(T) = 5152 \cdot (T + 168)^{-0.655}. \tag{177}$$

Literature data for the thermal conductivity of AlN at room temperature range from approx. 100 to 260 W/mK (Bengisu 2001, Boey and Tok 2003, Chen et al. 1994, Jackson et al. 1997, Jarrige et al. 1997, Kanai et al. 1992, Kim et al. 1996, Kuramoto et al. 1989, NIST 2003, Pampuch 2008, Pezzotti et al. 2000a, 2000b, Pierson 1996, Raghavan 1991, Richerson 2006, Shaffer et al. 1989, Slack 1973, Watari et al. 1996), followed by a gap and then the value 320 W/mK, the intrinsic thermal conductivity of oxygen-free AlN single crystals (Pierson 1996, Martienssen and Warlimont 2005). Aluminum nitride is a promising material for electronic substrates (partly replacing the currently dominating alumina Al_2O_3), because due to its high thermal conductivity and an electric resistivity comparable to the best insulators it is optimally suited for redistributing dissipative heat in modern high-density integrated circuits and other electronic devices (Bengisu 2001, Richerson 2006). It is highly stable, but prone to hydrolysis by atmospheric humidity if not fully sintered (Boey and Tok 2003, Pezzotti et al. 2000a, 2000b, Pierson 1996). The low-temperature thermal conductivity of AlN single crystals can achieve extremely high values (70000 W/mK at 30 K) (Slack et al. 1987), although the low-temperature peak values of polycrystalline AlN ceramics are between only 150 W/mK at 180 K (Ivanov et al. 1997) and 2300 W/mK at 45 K (Slack et al. 1987), cf. also (Slack 1973).

Figure 22 shows the temperature dependence of the thermal conductivity of silicon nitride (Si_3N_4) ceramics sintered with yttria as a sintering aid. According to these data the shifted-power fit for the temperature range from room temperature up to 1400 °C (with T in °C) is

$$k(T) = 2807 \cdot (T + 264)^{-0.662}. \tag{178}$$

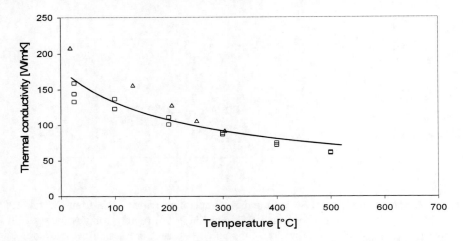

Figure 21. Thermal conductivity of aluminum nitride (AlN); triangles: with 4 wt.% Y_2O_3 (Jackson et al. 1990), squares: with 3 wt.% Y_2O_3 (De Baranda et al. 1993).

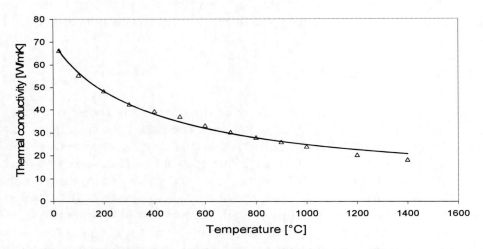

Figure 22. Thermal conductivity of silicon nitride (Si_3N_4); hot isostatically pressed with 6 wt.% Y_2O_3 as a sintering aid (empty triangles (Keyes 1992, NIST 2003)), full curve: shifted power fit.

The majority of literature data for the thermal conductivity of silicon nitride (Si_3N_4) ceramics at room temperature range from approx. 15 to 50 W/mK ((Bengisu 2001, Lee and Rainforth 1994, Martienssen and Warlimont 2005, Menčík 1992, Pampuch 2008, Pierson 1996, Salmang and Scholze 1982, Skopp et al. 1990, Staroň and Tomšů 2000), see also the temperature dependence in (Martienssen and Warlimont 2005), which starts at approx. 32 W/mK at room temperature), i.e. significantly lower than the value of 67 W/mK indicated by Figure 22 and this fit relation. Even lower (approx. 10-30 W/mK) is the thermal conductivity of sialons ($Si_{6-x}Al_xO_xN_{8-x}$), which are formed by adding alumina (Al_2O_3) to silicon nitride (Si_3N_4) and are based on the fact that the lattice of the latter can accommodate other atoms (e.g. Al and O) (Bengisu 2001, Lee and Rainforth 1994, Pampuch 2008). Silicon nitride (and sialon) occurs in two different hexagonal modifications (α below 1600 °C and β above 1600 °C). The latter seems to have a slightly higher thermal conductivity (e.g. α-Si_3N_4 17 W/mK and β-Si₃N₄ 28 W/mK (Martienssen and Warlimont 2005), α-sialon 11 W/mK and β-sialon 22 W/mK (Bengisu 2001, Lee and Rainforth 1994)), but the difference is not well investigated. On the other hand, some authors have determined much higher values for Si₃N₄, ranging from 102-122 W/mK for sintered materials (Watari et al. 1999, 2006) to 180 W/mK for the conductvity of β-Si₃N₄ along the c-axis (Li et al. 1999), see also (Kushan et al. 2007, Sivakumar et al. 2009).

For Si₃N₄ a strong grain size dependence of the thermal conductivity has been reported, which can be modeled using a conductivity value of 0.6 W/mK for the grain boundary phase (Kushan et al. 2007). Also for SiC ceramics a significant grain size dependence of the thermal conductivity has been reported in the range 2-8 μm (60-90 W/mK) (Jang and Sakka 2008).

Figure 23 gives an overall comparison of the temperature dependence of the thermal conductivity of interstitial borides, carbides and nitrides. It is evident that borides have usually higher thermal conductivities than interstitial carbides and nitrides. Only titanium and hafnium boride (TiB₂ and HfB₂) exhibit the typical decrease of thermal conductivity with temperature, while interstitial carbides and nitrides commonly exhibit a slight increase of conductivity with temperature, which is attributed to the metal-like character of these compounds (Pierson 1996). The low-temperature behavior of NbC, TiC and ZrC, is characterized by a gradual increase from approx. 3-12 W/mK at 10 K to approx. 10-30 W/mK at 100 K, a clear maximum at low temperatures being absent, see (Childs et al. 1973).

Table 9 gives an overview of the thermal conductivity of non-oxides at room temperature and at higher temperature (where available), based on an extensive literature research. Needless to say that these values have to be considered as typical, since the scatter of experimental (and handbook) data is usually larger than in the case of oxide ceramics. Nevertheless, the values have been critically checked and therefore we believe them to provide realistic estimates for practical purposes.

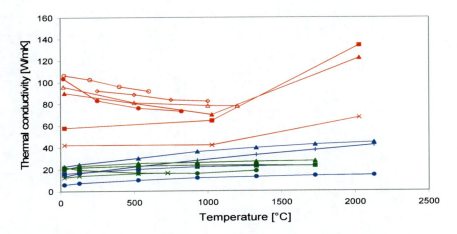

Figure 23. Thermal conductivity of interstitial borides (red), carbides (blue) and nitrides (green); TiB₂ (red full triangles (Cutler 1991), red empty triangles (Munro 2000)), ZrB₂ (red squares (Cutler 1991)), HfB₂ (red full circles (Gasch et al. 2008), red empty circles (Opeka et al. 1999), red empty diamonds (Loehman et al. 2006), corrected for 31 % porosity via the exponential relation), VB₂ (red crosses (Cutler 1991)), TiC (blue triangles (Pierson 1996)), ZrC (blue squares (Pierson 1996)), HfC (blue circles (Pierson 1996)), NbC (blue plusses (Pierson 1996)), TiN (green triangles (Pierson 1996)), ZrN (green squares (Pierson 1996)), HfN (green circles (Pierson 1996)), VN (green crosses (Pierson 1996)).

Table 9. Thermal conductivity of non-oxides (recommended orientational values), based on (Bengisu 2001, Childs et al. 1973, Cutler 1991, Gasch et al. 2008, Gibson and Ashby 1997, Jang and Sakka 2008, Keyes 1992, Kinoshita et al. 2001, Kleber 1990, Komeya and Matsui 2005, Kopitzki 1989, Kushan et al. 2007, Lee and Rainforth 1994, Li et al. 1999, Liu and Lin 1996, Liu et al. 1995b, Martienssen and Warlimont 2005, Munro 1997b, 2000, Neshpor 1968, NIST 2003, Opeka et al. 1999, Pampuch 2008, Pierson 1996, Richerson 2006, Menčík 1992, Salmang and Scholze 1982, Sigl 2003, Sivakumar et al. 2009, Slack 1964, 1973, Staroň and Tomšů 2000, Watari et al. 1989, 1999, 2003, 2006, Zhou et al. 2004, Zimmermann et al. 2008); values calculated from fits are calculated from the fits in this work (fit 0) and from (Munro 1997b) (fit 1)

	Room temperature	100 °C	500 °C	1000 °C
C (type I diamond)	950 ± 400	850		
C (type II diamond)	2400	1500	850	
C (graphite)	for random orientation < 200-250; for anisotropic materials from 400 (in-plane) / 100 (normal) to 2000 (in-plane) / 5 (normal)			
AlN	170 ± 50, most values in the range 120-220 (260), single crystal 320	140	110 (single crystal 180)	90 (single crystal 150)
AlN (fit 0)	165	132	73	
B_4C	27 ± 6			
BN (hexagonal)	25 ± 10	34	28	19.5
BN (cubic)	800 ± 500 (hot-pressed 300–600, single crystal 1300)			
SiC (hexagonal α-SiC – polytypes 6H, 4H, 2H and 15 R, cubic β-SiC – polytype 3C)	100 ± 70, most values in the range 30-170 (270) (single crystal 290-490)	70	55	40 ± 10 (single crystal 61)
SiC (fit 0)	91	79	52	39
SiC (fit 1 for α-SiC with equiaxed SiC(6H) grains)	113	97	55	36
Si_3N_4	70 ± 50 (α 17, β 28), values in the range 20-120 (single crystal 180)	29.5	24	18.5
Si_3N_4 (fit 0)	66	57	35	25
Sialon ($Si_{6-x}Al_xO_xN_{8-x}$)	20 ± 10 (α 11, β 22)	12 ± 4	8.5 ± 2.5	6 ± 1
TiN	30 ± 10	23	25	26
ZrN	20 ± 1	22	22.5	23.5
HfN	22	20	16.5	16
VN	11	13.5	15.5	
NbN	4			

	Room temperature	100 °C	500 °C	1000 °C
TaN	9			
TiC	24 ± 6	28.5	34	37.5
ZrC	18 ± 3	17	20	22
HfC	13 ± 7	7	10	12
VC	25			
NbC	14	16	22	29
TaC	27 ± 5			
WC	120			
UC	23			
UN	17 ± 4			
TiB_2	75 ± 50 (values in the range 25-125)	92	81	76 ± 2
ZrB_2	65 ± 45 (values in the range 20-110, single crystal 98 parallel and 140 perpendicular to the c-axis)	61	64	64
HfB_2	80 ± 30 (values in the range 50-110)		83 ± 8	
VB_2	42			42
LaB_6	48			
$MoSi_2$	45 ± 15			
$TaSi_2$	22			

10. SUMMARY AND COMPARISON OF DIFFERENT MATERIAL CLASSES

The topic of this chapter has been the thermal conductivity of ceramics and their typical phases, i.e. oxides, silicates (including glasses) and non-oxides, including some forms of carbon (diamond and graphite). Ceramics can be defined as inorganic non-metallic solid materials. In this section we compare the thermal conductivity of ceramics with other material classes, in particular with metals and organic polymers (for gases and liquids see Section 6, Appendix C and (Touloukian et al. 1970)).

The thermal conductivity of metals at room temperature is higher than that of traditional ceramics. The reason is that in the electrically conductive metals, in addition to lattice vibrations (phonons), also the free electrons contribute to the thermal conductivity, and at high temperatures the latter usually dominates the former. Moreover, traditional silicate-based ceramics have typically high glass contents. This does not mean, however, that the thermal conductivity of electrically insulating materials (dielectrics) is generally lower than that of metals. In fact, many metals have thermal conductivities comparable to oxide ceramics, and many non-oxide ceramics have higher thermal conductivities at room temperature. At low temperatures, even oxide ceramics with moderate thermal conductivity may become highly thermally conducting, e.g. alumina. On the other hand, the thermal conductivity of organic polymers is generally lower than that even of the least conducting ceramics and glasses. While conductivity values lower than approx. 1 W/mK cannot be achieved by dense ceramics and metals, some polymers can attain values lower by one order of magnitude (approx. 0.1 W/mK). Values lower than this can only achieved by introducing porosity. Depending on the porosity, conductivity values approaching the pore-filling gas can be achieved (for very small, i.e. nanoscale, pore size even lower values), and when the pore space is void (vacuous) the thermal conductivity can be arbitrarily close to zero (vacuum). In principle, the dependence of thermal conductivity on microstructure (porosity and grain size) is the same for all classes of materials.

This chapter was intended to provide the necessary tools for estimating the thermal conductivity of ceramic materials, considered as multiphase mixtures. Together with the material data given in Sections 7-9 it should enable the reader to give realistic estimates of the effective thermal conductivity for all common polycrystalline ceramic materials, including composites, for which the phase composition is known. In particular, with the help of this chapter the reader should be able to calculate the effective thermal conductivity of dense isotropic polycrystalline ceramics when single-crystal values (components of the thermal conductivity tensor) are available (Section 2). Moreover, the effective thermal conductivity for composites of any composition can be predicted by calculating upper and lower bounds (Section 3) and applying model relations (Section 4), which also allow a realistic assessment of the influence of porosity. On the other hand, the phase mixture model presented here (Section 5) represents a rational way of estimating the influence of grain size on the effective thermal conductivity. The mechanisms of heat transfer have been treated partly from a traditional atomistic point of view, in order to show the standard way how to take into account or estimate the contributions of convection and radiation to the (apparent) thermal conductivity (Section 6).

Beyond the scope of this chapter are questions concerning the atomistic origin of the different (intrinsic, i.e. limited only by anharmonicity effects) thermal conductivity values for different (pure) materials (defect-free single-crystals) and their temperature dependence, as well as the influence of defects, isotope composition and sample size on the thermal conductivity, especially at low (cryogenic) temperatures. Of course, also outside of the scope of this chapter are questions concerning the thermal conductivity of single-phase solid mixtures in the form of mixed crystals (crystalline solid solutions), alloys and intermetallic compounds. Since the thermal conductivity is closely related to propagating lattice vibrations (phonons), it depends on the mass of the vibrating atoms (chemical composition and isotopes) and is indirectly related to other properties where the chemical bond strength plays a role, e.g. elastic moduli and other mechanical properties. The reader may consult e.g. (Berman 1976, Grimvall 1999) for more information on these important topics.

A general conclusion that can be drawn from low-temperature literature data of thermal conductivity (and partly from the data presented in Sections 7-9) is that at low temperature the thermal conductivity of many materials can vary by more than one order of magnitude and is determined more by its defect structure (scattering of phonons by point defects, line defects, planar defects and volume defects) and sample size than by the chemical composition of the material itself. Therefore low-temperature conductivity data from the literature should always be taken as very rough (order-of-magnitude) estimates only, and sometimes – e.g. when sample size effects come into play – their value as material properties is questionable anyway. This conclusion is immediately confirmed by a glance into the comprehensive data bases on this topic, e.g. (Childs et al. 1973, NIST 2003).

Most crystalline solids exhibit a thermal conductivity maximum at temperatures below 100 K (as a rule-of-thumb this maximum can be expected at approx. 0.1 of the Debye temperature (Grimvall 1999)), and the height of this maxima covers more than three orders of magnitude, ranging from approx. 15 W/mK (for silicates like natural garnets) to approx. 20000-25000 W/mK (for diamond, alumina and some metals like Ag, Al, Cu and Ga) (Childs et al. 1973, Gebhart 1993, Martienssen and Warlimont 2005). In singular cases even higher values are reported, e.g. 84500 W/mK for pure metallic gallium at 1.8 K (Kittel 1988), 70000 W/mK for aluminum nitride at 30 K (Slack et al. 1987) or more than 42000 W/mK for isotopically refined (0.1 % ^{13}C) diamond (type II single crystal) at temperatures below 100 K (Wei et al. 1993). Some materials, in particular polymers, metallic alloys (e.g. steel) and amorphous inorganic dielectrics (glasses), do not exhibit any maximum at low temperature.

The thermal conductivity of ceramic materials at room temperature ranges from approx. 1 W/mK for silicate glasses to 2000-3000 W/mK for some carbon modifications (type-II diamond, as well as pyrolytic graphite and carbon nanotubes in the in-plane direction of the graphene sheet). Type-II diamond has the highest thermal conductivity of all isotropic solids (bulk materials) at room temperature (approx. 2400 W/mK, while type-I diamond has approx. 950 W/mK), followed by cubic boron nitride (c-BN) single crystals (1300 W/mK, in contrast to hexagonal boron nitride, h-BN, with only 25 W/mK), silicon carbide (hexagonal α-SiC) single crystals (490 W/mK perpendicular to the c-axis) and aluminum nitride (AlN, 320 W/mK). Note, however, that the effective thermal conductivity of polycrystalline c-BN, α-SiC and AlN ceramics are much lower than these values (approx. 800 W/mK, 100 W/mK and 170 W/mK, respectively) and that the thermal conductivity of most other non-oxide ceramics is in the range 20-100 W/mK at room temperature, see Table A-1. Some of these, especially the so-called interstitial carbides and nitrides, do not exhibit the hyperbolic conductivity decrease with temperature (some even exhibit an increase). This behavior is more typical for metals than for ceramics and has to do with the electronic contribution to thermal conductivity in metals and metal-like compounds with quasi-free electrons.

Beryllia (BeO) has the highest conductivity of all oxides at room temperature (around 200-400 W/mK), while doped zirconia (ZrO$_2$) has the lowest conductivity of all simple oxides (around 2 W/mK), matched only by some mixed oxides (zirconates), phosphates and silicates (La$_2$Zr$_2$O$_7$, LaPO$_4$, Y$_2$SiO$_5$, γ-Y$_2$Si$_2$O$_7$), which can have thermal conductivities comparable to silicate glasses (around 1 W/mK). Above room temperature the majority of these oxides exhibit decreasing thermal conductivity with increasing temperature (at least up to the temperature, at which radiation affects heat transfer, which ranges from approx. 1400-1500 °C for simple oxides to 800-900 °C for zirconates and other mixed oxides). Only silicate glasses and materials with very defective structures, such as doped zirconia, exhibit temperature-independence or even a slight increase of the thermal conductivity with temperature (note, however, that in transparent glasses radiation can become important at temperature as low as approx. 300 °C, whereas in doped zirconia the radiation influence can be neglected up to approx. 1400 °C). As mentioned above, at low temperatures the thermal conductivity of oxides is very sensitive to defects and impurities. Except for defect-free alumina (α-Al$_2$O$_3$, corundum) single crystals, which have the highest peak thermal conductivity of all oxides, the peak conductivities of other oxides are much lower and do not exceed 5000 K (e.g. 3400 W/mK for MgO at 27 K, 2200 W/mK for TiO$_2$ at 12 K, 1600 W/mK for BeO at 80 K, 1200 W/mK for SiO$_2$ at 11 K, 110 W/mK for H$_2$O at 8 K, 22 W/mK for BaTiO$_3$ at 30 K and 16 W/mK for almandine garnet at 30 K) (Berman et al. 1950, Childs et al. 1973, Hellwege 1988, Ibach and Lüth 1988, Martienssen and Warlimont 2005). Note, however, that even these typical peak conductivities of oxides and oxide ceramics are comparable to the highest peak conductivities of other inorganic non-metallic compounds, e.g. the electrically insulating halogenides (NaCl with more than 2000 W/mK at 20 K (Kittel 1988) and CaF_2 with 3000 W/mK at 13 K (Childs et al. 1973)) and the semiconducting chalkogenides (e.g. GaAs with almost 4000 W/mK at 10-20 K (Martienssen and Warlimont 2005)).

The room temperature values of the thermal conductivity of many metals are higher than those of ceramics (including the highly conductive ones, such as AlN, BeO and SiC). The highest conductivity is exhibited by silver Ag (429 W/mK), followed by copper Cu (400 W/mK), gold Au (317 W/mK), aluminum Al (237 W/mK), beryllium Be (210 W/mK), and others. Also the thermal conductivity of elemental semiconducting silicon (Si) can attain values approaching 200 W/mK (Childs et al. 1973) although the average value is around 140 W/mK and values as low as 40 W/mK have also been reported for some types of Si (Childs et al. 1973). On the other hand, titanium (Ti), a metal used in many structural today, has a rather low thermal conductivity (22 W/mK at room temperature), and most rare earth metals (except for ytterbium (β-Yb) which has a conductivity of approx. 37 W/mK) have even lower conductivities, ranging from approx. 10 W/mK to 17 W/mK. Bismuth (Bi), manganese (Mn) and mercury (Hg) have the lowest thermal conductivity of all pure metals (7-9 W/mK at room temperature), together with some actinides (e.g. Np 6 W/mK, Pu 8 W/mK) (Lide 2010) (see Table 10).

The thermal conductivity of pure crystalline metals exhibits a steep increase at low temperatures, usually with a peak at temperatures below 100 K, see Figure 24. At low (cryogenic) temperatures the thermal conductivity of silver attains peak values of approx. 17200-19300 W/mK (at 5-7 K), whereas for copper and aluminum values of 24300-24900 W/mK and 23500-23900 W/mK, respectively, are reported (at 9-10 K) (Gebhart 1993, Lide 2010). The peak values for semiconducting Ge and Si are much lower, only up to approx. 4000-5000 W/mK at 10-30 K (Childs et al. 1973, Martienssen and Warlimont 2005), respectively, and highly dependent on the isotopic purity (Hellwege 1988, Lide 2010). Probably the highest value reported in the literature is 84500 W/mK for metallic gallium (Ga) at 1.8 K (Kittel 1988).

The reason for the high thermal conductivity of metals is that, in addition to lattice vibrations (phonons), also the free electrons contribute to the thermal conductivity in metals, and at high temperatures the latter usually dominates the former. At room temperature the electronic contribution to heat transfer (determined by the Wiedemann-Franz law) can exceed the phonon contribution by two orders of magnitude (Hellwege 1988, Grimwall 1999), and also at very low temperature the thermal conductivity of metals is intimately related to their electric resistivity (Hellwege 1988). Moreover, at low temperatures isotope effects can be significant, for metals as well as for elemental semiconductors such as Ge (Hellwege 1988). At higher temperature (for gold e.g. above 70 K (Hellwege 1988)) the thermal conductivity of metals is more or less constant, see Figure 24.

Similar to nonmetallic solid solutions (mixed crystals), for which the thermal conductivities are below those predicted by the (linear) mixture rule (Eucken 1932, Hellwege 1988), the thermal conductivities of metallic alloys and mixtures of elemental semiconductors (e.g. Si-Ge) (Abeles 1963) are always lower than predicted by the linear mixture rule (molar-fraction-weighted arithmetic mean) (Grimvall 1999, Klemens 1960).

Table 10. Thermal conductivity (W/mK) of pure metals and elemental semiconductors at room temperature (Aylward and Findlay 1986, Childs at al. 1973, Gebhart 1993, Incropera and Dewitt 1991, Kopitzki 1989, Lide 2010, Martienssen and Warlimont 2005, Pierson 1996, Weast 1988)

	k (W/mK)		k (W/mK)		k (W/mK)
Ag	429	In	84	Ru	112
Al	237	Ir	59-147	Sb	25
As	50	K	102	Sc	16
Au	317	La	14	Se	0.2-2
B	27	Li	70-85	Si	40-200
Ba	18	Lu	16	Sm	13
Be	210	Mg	164	Sn	67
Bi	8-9	Mn	8	Sr	35
Ca	130-200	Mo	140	Ta	60
Cd	100	Na	141	Tb	11
Ce	11	Nb	57	Tc	185
Co	100	Nd	16	Te	2-5
Cr	94	Ni	88	Th	41-65
Cs	36	Np	6	Ti	22
Cu	401	Os	88	Tl	46
Dy	11	Pb	36	Tm	17
Er	14	Pd	75	U	25
Eu	14	Pm	17	V	30
Fe	80	Po	20	W	170-180
Ga	40	Pr	12	Y	16
Gd	10	Pt	72	Yb	37
Ge	60	Pu	7-9	Zn	120
Hg	8.3-8.4	Rb	60	Zr	22
Hf	23	Re	60		
Ho	16	Rh	89-150		

The thermal conductivities of polymers lie in the surprisingly narrow range of 0.1-0.5 W/mK at room temperature (Childs et al. 1973), and if more specific information is lacking a value of 0.3 ± 0.1 W/mK is commonly recommended, see e.g. (Gibson and Ashby 1997). Polystyrene (PS), polyvinylchloride (PVC), polypropylene (PP), polymethylmethacrylate (PMMA = "plexiglass") and rubber are at the lower end of this range (approx. 0.1-0.2 W/mK), while high-density polyethylene (HDPE) and some epoxies are at the upper end (0.5 W/mK), and teflon, polyurethane (PU) and many other polymers are intermediate (approx. 0.3 W/mK) (Gebhart 1993, Gibson and Ashby 1997). Also the temperature dependence of the thermal conductivity is weak for polymers (less than 10 % between 0 and 100 °C) (Gibson and Ashby 1997) and in most cases there is no conductivity maximum at low temperatures (Childs et al. 1973). The thermal conductivity of partially crystalline polymers is, however, strongly dependent on the degree of crystallinity and on the degree of stretching. The strong covalent bond of the C-C-backbone is responsible for a higher conductivity, while the weak van-der-Waals bonds in the directions perpendicular to the backbone (polymer chain) reduce the overall conductivity. Thus the thermal conductivity of amorphous polymers is between those of organic liquids (approx. 0.1 W/mK) and inorganic glasses (approx. 1 W/mK)

(Hellwege 1988). The thermal conductivity of dry wood, leather, paper and dry foodstuffs is approx. 0.1-0.2 W/mK, while fruits, meat and human soft tissue are approx. 0.5 W/mK, mainly due to their water content (the thermal conductivity of water at room temperature is 0.6 W/mK) (Gebhart 1993, Incropera and Dewitt 1991).

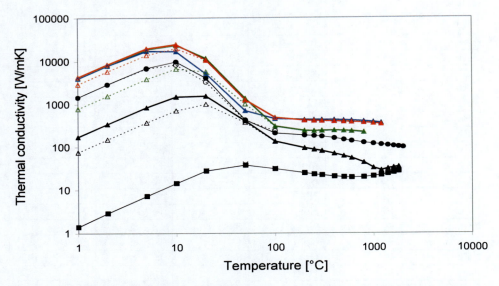

Figure 24. Thermal conductivity of selected metals according to (Gebhart 1993) (empty symbols and dotted lines) and (Lide 2010) (full symbols and full lines); Ag (blue), Al (green), Cu (red), Fe (black triangles), Ti (black squares) and W (black circles).

APPENDIX A. THERMAL CONDUCTIVITY VALUES OF SINGLE CRYSTALS AND THE CORRESPONDING POLYCRYSTALLINE MATERIALS

It is evident that the effective medium model prediction lies always within the Hashin-Shtrikman bounds, whereas the weakly anisotropic model does not. Thus, from a principal point of view, the former should be recommended to estimate the thermal conductivity of statistically isotropic materials. The difference of the two predictions, however, is very small (probably smaller than experimental errors of measurement) even when the degree of anisotropy is high. In the case of graphite true single crystals of sufficient size are not available, but pyrolytic graphite approaches single-crystal behavior to some degree. The in-plane thermal conductivity of the graphene sheet is probably the highest thermal conductivity value for any material at room temperature (at the same time it represents the strongest chemical bond) and determines the axial thermal conductivity of carbon nanotubes, which is usually given in the range 2000-3000 W/mK (Berber et al. 2000, Krueger 2010, Yamamoto et al. 2008), well comparable to the value of 2400 W/mK for type-II diamond. It should be mentioned, however, that some authors report even higher values for individual carbon nanotubes, namely 6600 W/mK for single-wall carbon nanotubes (SWCNT) and > 3000 W/mK for multi-wall carbon nanotubes (MWCNT), cf. (Baughman et al. 2000, Biercuk et al. 2002, Kumari et al. 2008).

Table A-1. Components of the thermal conductivity tensor for non-cubic uniaxial single crystals (and highly oriented graphite materials) at or around room temperature (24 ± 6 °C) and at 0 °C (for ice) according to various sources (Chojnacki 1979, Grimvall 1999, Haussühl 1983, Kleber 1990, Lide 2010, Nye 1985, Pabst and Gregorová 2007, Salmang and Scholze 1982), and effective thermal conductivities of the corresponding statistically isotropic polycrystalline materials calculated according to the micromechanical bounds and model approximations (Grimvall 1999, Torquato 2002); the anisotropy parameter ε is defined in Section 2

Material	$k_\parallel$ (W/mK) (∥ c-axis, out-of-plane)	$k_\perp$ (W/mK) (⊥ c-axis, in-plane)	$k_\perp / k_\parallel$	$\lvert\varepsilon\rvert$	Molyneux bounds Upper	Molyneux bounds Lower	Hashin-Shtrikman bounds upper	Hashin-Shtrikman bounds lower	Effective-medium model	Weakly aniso-tropic model
Alumina (α-Al$_2$O$_3$)	38.9	31.2	0.80	0.25	33.8	33.4	33.7	33.6	33.6	33.5
Bismuth	6.65	9.24	1.39	0.28	8.38	8.18	8.319	8.305	8.315	8.27
	5	9	1.80	0.44	7.67	7.11	7.52	7.45	7.50	7.40
Cadmium	83	104	1.25	0.20	97.0	95.9	96.7	96.6	96.7	96.4
Calcite	5.0	4.2	0.84	0.19	4.47	4.44	4.457	4.455	4.456	4.444
Graphite (extruded)	89	355	3.99	0.75	266	178	249	222	243	237
	8	400	50	0.98	269	23	233	49	208	212
Graphite (pyrolytic)	10	2000	200	0.995	1337	30	1148	67	1010	1043
	5.7	1950	342	0.997	1302	17	1117	39	981	1015
Ice at 0 °C (hexagonal H$_2$O)	2.3	1.9	0.83	0.21	2.03	2.02	2.03	2.03	2.03	2.02
Quartz (trigonal SiO$_2$ / low-quartz)	10	6	0.60	0.67	7.33	6.92	7.21	7.16	7.18	6.94
	10.4	6.2	0.60	0.68	7.6	7.16	7.47	7.42	7.44	7.18
	11.3	6.5	0.58	0.74	8.1	7.57	7.94	7.87	7.90	7.57
Rutile (t-TiO$_2$)	10	7	0.70	0.43	8.0	7.78	7.93	7.91	7.92	7.81
	10.4	7.4	0.71	0.41	8.4	8.19	8.33	8.32	8.32	8.22
	12.9	9.3	0.72	0.39	10.5	10.25	10.42	10.40	10.41	10.29
Tin (β-Sn)	52	74	1.43	0.30	66.7	64.9	66.1	66.0	66.1	65.7
Zinc (Zn)	124.2	120.4	0.97	0.03	121.67	121.64	121.66	121.66	121.66	121.65
Zirconia (t-ZrO$_2$)	2.92	2.89	0.99	0.01	2.90	2.90	2.90	2.90	2.90	2.90

APPENDIX B. SINGLE-INCLUSION SOLUTIONS

All correct model relations for the effective thermal conductivity of two-phase materials are based on single-inclusion solutions or reduce to the latter in the limit of low volume fractions (of a second phase). This subject has a long tradition, beginning with spherical inclusions and going back to Maxwell (Maxwell 1873) and Rayleigh (Rayleigh 1892) (as well as Einstein in the field of suspension rheology (Einstein 1906)). Ellipsoidal inclusion models and randomly oriented fiber and platelet models are of a younger date, going back to the works of Polder and van Santen (Polder and van Santen 1946) in conductivity context,

Eshelby (Eshelby 1957) in elasticity context and Jeffery in viscosity context (Jeffery 1922), see (Torquato 2002). The single-inclusion problem for thermal conductivity (with Fourier's law as the constitutive equation) is completely analogous to the single-inclusion problem for electrical conductivity (with Ohm's law as the constitutive equation). The latter have been solved within the theory of electrodynamics, based on the pioneering work of Maxwell (Maxwell 1873), see e.g. (Jackson 1983, Stratton 1961). A complete derivation of the spherical- and spheroidal- and ellipsoidal-inclusion problem can be found in (Torquato 2002).

The problem, generally formulated, is to find the temperature field for an ellipsoidal inclusion of conductivity k_2 in an infinite matrix of conductivity k_1 in which the temperature gradient imposed from the outside is a constant vector. The solution of this problem consists in solving the Laplace equation ($\text{div grad } T = 0$) under appropriate boundary conditions, see (Jackson 1983, Torquato 2002). The result is that the first-order coefficients (for spherical inclusions the so-called polarizability) in Maxwell-type approximations and cluster expansions are given in terms of so-called depolarization tensors (or the corresponding scalar factors in the case of statistically isotropic microstructures).

The local field inside an ellipsoidal inclusion (i.e. the internal temperature gradient $\text{grad } T$) is uniform and is linearly related to the externally applied field (i.e. the temperature gradient imposed from outside via the surrounding matrix). For a spherical inclusion the proportionality coefficient of this linear relation is a second-order tensor $\mathbf{R}$ ("field concentration tensor") given by

$$\mathbf{R} = (1 - \beta_{21})\mathbf{1} = \frac{3 k_1}{k_2 + 2 k_1}\mathbf{1},$$

when the inclusion is isotropic (Markov 2000). Also the polarization, defined as $(k_2 - k_1)\text{grad } T$, is uniform within the inclusion, i.e. linearly dependent on the imposed field with the proportionality coefficient being a second-order tensor $\mathbf{M}$ ("polarization concentration tensor") given for spherical inclusions by

$$\mathbf{M} = (k_2 - k_1)\mathbf{R} = 3 k_1 \beta_{21} \mathbf{1}.$$

The generalized d – dimensional analogues of the above equations are (Torquato 2002):

$$\mathbf{R} = (1 - \beta_{21})\mathbf{1} = \frac{d k_1}{k_2 + (d-1) k_1}\mathbf{1},$$

$$\mathbf{M} = (k_2 - k_1)\mathbf{R} = d k_1 \beta_{21} \mathbf{1}.$$

with the polarizability β_{21} which is for spherical inclusions

$$\beta_{21} = \frac{k_2 - k_1}{k_2 + 2k_1}.$$

We emphasize that the tensorial prefactors multiplying the imposed field vector are in general shape-dependent. Uniformity of the field and polarization inside the inclusions (sometimes called the "Eshelby property") is only guaranteed for inclusion shapes without singularities (i.e. edges, vertices or cusps), i.e. for ellipsoids (Jackson 1983, Stratton 1961). It is absent for non-ellipsoidal inclusions (Lubrada and Markenscoff 1998). Following (Markov 2000, Milton 2002, Torquato 2002) for triaxial ellipsoids one obtains the general relation

$$\mathbf{R} = \left(1 + \mathbf{A}\frac{k_2 - k_1}{k_1}\right)^{-1},$$

where $\mathbf{A}$ is the (symmetric, second-order) depolarization tensor of the d-dimensional (hyper-) ellipsoid, which in the principal axes frame has i, j (= 1, 2, 3 ... d) positive eigenvalues ("depolarization factors") given by the elliptic integrals (Torquato 2002)

$$A_i = \left(\prod \frac{a_j}{2}\right) \int_0^\infty \frac{dt}{(t + a_i^2)\sqrt{\prod(t + a_j^2)}},$$

where a_i is the semiaxis of the (hyper-) ellipsoid along the x_i direction. The corresponding elliptic integrals in 3D (for ellipsoids with semiaxes a, b, c) are (Markov 2000, Milton 2002)

$$A_1 = \frac{1}{2}abc \int_0^\infty \frac{dt}{(t + a^2)\sqrt{(t + a^2)(t + b^2)(t + c^2)}},$$

and analogously for A_2 and A_3. Since the depolarization tensor has unit trace (tr $\mathbf{A} = 1$, i.e. in 3D $A_1 + A_2 + A_3 = 1$) (Markov 2000, Milton 2002) and its eigenvalues are positive, we have $0 \leq A_i \leq 1$ (Milton 2002, Torquato 2002). For the special case of a (hyper-) sphere we have $A_i = 1/3$ for all i. Depolarization factors for general triaxial ellipsoids have been summarized and tabulated in several papers (Osborn 1945, Stoner 1945) and books (Kellogg 1953), for the special case of spheroids see already (Maxwell 1873).

For spheroidal inclusions (i.e. inclusions in the shape of "rotational", "rotary" or "biaxial" ellipsoids, i.e. ellipsoids with one axis of rotational symmetry, usually chosen to be c-axis) aligned along the x_3-axis, the depolarization tensor is

$$A = \begin{pmatrix} A_\perp & 0 & 0 \\ 0 & A_\perp & 0 \\ 0 & 0 & 1-2A_\perp \end{pmatrix} = \begin{pmatrix} \frac{1}{2}(1-A_{axial}) & 0 & 0 \\ 0 & \frac{1}{2}(1-A_{axial}) & 0 \\ 0 & 0 & A_{axial} \end{pmatrix},$$

see (Markov 2000, Milton 2002, Torquato 2002). The depolarization factors for prolate spheroids with aspect ratio $c/a > 1$ are

$$A_{axial}(prolate) = \frac{1-\varepsilon^2}{\varepsilon^2}\left[\frac{1}{2\varepsilon}\ln\left(\frac{1+\varepsilon}{1-\varepsilon}\right)-1\right],$$

whereas for oblate spheroids with aspect ratio $c/a < 1$ we have

$$A_{axial}(oblate) = \frac{1}{\varepsilon^2}\left[1 - \frac{\sqrt{1-\varepsilon^2}}{\varepsilon}\arcsin\varepsilon\right],$$

(an error in (Torquato 2002) has been tacitly corrected here), where the eccentricity ε of a prolate spheroid is defined as

$$\varepsilon_{prolate} \equiv \sqrt{1-\left(\frac{a}{c}\right)^2}$$

and that of an oblate spheroid as

$$\varepsilon_{oblate} \equiv \sqrt{1-\left(\frac{c}{a}\right)^2}.$$

The following special and extreme cases are of great practical interest:

- Spherical inclusions ($c/a = 1$, $A_\perp = 1/3$) $\rightarrow$ $A_1 = A_2 = A_3 = \frac{1}{3}$,
- Needle-shaped inclusions ($c/a = \infty$, $A_\perp = 1/2$) $\rightarrow$ $A_1 = A_2 = \frac{1}{2}$, $A_3 = 0$,
- Disk-shaped inclusions ($c/a = 0$, $A_\perp = 0$) $\rightarrow$ $A_1 = A_2 = 0$, $A_3 = 1$.

Although the aforementioned "Eshelby property" is missing for that case, the depolarization factors can be evaluated explicitly also for elliptical cylinders (Milton 2002), for which $c \rightarrow \infty$ and a and b are the semiaxes of the cross-section ellipse, with $a < b$. In this case

$$A_1 = \frac{b}{a+b}, \quad A_2 = \frac{a}{a+b}, \quad A_3 = 0$$

The polarization and temperature gradient fields within the ellipsoid, averaged over all orientations, are easily obtained from the isotropic averages of the tensors **M** and **R**, i.e.

$$\mathbf{M} = M\,\mathbf{1},$$

$$\mathbf{R} = \Omega\,\mathbf{1},$$

where the scalars M and Ω are given by

$$M = (k_2 - k_1)\Omega,$$

$$\Omega = \frac{\operatorname{tr}\mathbf{R}}{3} = \frac{1}{3}\sum \frac{1}{1 + A_i \frac{k_2 - k_1}{k_1}}.$$

Thus, for randomly oriented spheroids we have (Markov 2000)

$$\Omega = \frac{1}{3}\left(\frac{2k_1}{k_1 + (k_2 - k_1)A_\perp} + \frac{k_1}{k_1 + (k_2 - k_1)A_{axial}} \right),$$

and for the special and extreme cases of spherical and randomly oriented needle-shaped and disk-shaped inclusions we have, respectively,

$$\Omega = \frac{3k_1}{2k_1 + k_2}, \quad \text{(sphere)}$$

$$\Omega = \frac{5k_1 + k_2}{3(k_1 + k_2)}, \quad \text{(needle)}$$

$$\Omega = \frac{k_1 + 2k_2}{k_2}. \quad \text{(disk)}$$

APPENDIX C. THERMAL CONDUCTIVITY AND MEAN FREE PATH OF GASES AND VAPORS

Table C-1. Thermal conductivity (W/mK) and mean free path (nm) of gases and vapors (at p = 1 atm if not indicated otherwise, RT = room temperature, FCHC = fluoro-chloro-hydrocarbons, HFCHC = hydrogenated FCHC) (Aylward and Findlay 1986, Gebhart 1993, Kutzendörfer and Máša 1991, Moore 1981, Schlegel 1999)

	k (W/mK) at 0 °C	k (W/mK) at RT	k (W/mK) 100 °C	k (W/mK) 200 °C	k (W/mK) 500 °C	k (W/mK) 1000 °C	L_{gas} (nm)
H_2	0.172	0.182	0.216	0.258	0.384	0.593	112
He	0.142	0.150	0.171	-	0.294	-	180
Ne		0.049					
Ar	0.016	0.019	0.022	0.026	0.038	0.051	64
Kr		0.009					
Xe		0.006					
N_2	0.024	0.025	0.032	0.038	0.055	0.082	60
O_2	0.025	0.026	0.033	0.040	0.061	0.089	65
Air	0.024	0.026	0.032	0.038	0.055	0.076	
H_2O vapor	0.016	0.017	0.025	0.034	0.070	0.147	
CO	0.023	0.025	-	0.037	0.053	0.077	58
CO_2	0.014	0.017	0.020	0.031	0.055	0.090	40
CH_4 (methane)	0.031	0.031	0.046	0.064	0.122	-	-
C_2H_6 (ethane)	0.019	0.019	0.032	0.047	0.108	-	-
C_3H_8 (propane)	0.015	0.015	0.026	0.040	0.095	-	-
C_4H_{10} (butane)	0.013	0.013	0.023	0.036	0.090	-	-
C_2H_4 (ethene = ethylene)	0.017						35
NH_3 (ammonia)	0.022	0.024					44
Cl_2	0.008						29
FCHC (R-11, R-12, R-22)		0.008					
HFCHC (R-123)		0.010					
Ethanol vapor	0.023	-	0.023	0.035	0.086	-	-

APPENDIX D. EMISSIVITIES OF SELECTED MATERIALS

Table D-1. Total normal emissivities of selected materials (Kutzendörfer and Máša 1991, Modest 1993, Weast 1988) (AS = alumina-silica); values for metals are for pure metals with unoxidized, polished surfaces if not indicate differently (unusual values are in brackets)

Material	T (°C)	ε	Material	T (°C)	ε
Al_2O_3	−25	0.80	Al	20-500	0.03-0.08
Al_2O_3	275-500	0.63-0.42	... alloys	20-500	0.03-0.22
Al_2O_3	500-825	0.42-0.26	... oxidized	95-600	0.11-0.31
			... anodized	−25	0.79-0.84
Al_2O_3 (10 µm)	1010-1565	0.30-0.18	Antimony Sb	35-260	0.28-0.31
Al_2O_3 (50 µm)		0.39-0.28	Be	150-600	0.18-0.30
Al_2O_3 (100 µm)		0.50-0.40	... anodized	150-600	0.90-0.82
AS (with Fe_2O_3)		0.43-0.78	Bi	75	0.05-0.34
Asbestos board (paper)	23-370	0.96-0.93	Brass	22-375	0.03-0.22
Brick (red)	20	0.93	... oxidized	200-600	0.61-0.59
	1000-1100	0.29-0.75	Bronze	50	0.1
Carbon (filament)	20-1400	0.98-0.53	Co	500-1000	0.13-0.23
Concrete	38-1000	0.63-0.94	Cr	50-1000	0.08-0.38
CuO	800-1100	0.66-0.54	Cu	50-100	0.02-0.15
Enamel white (vitreous)	20	0.90-0.95	... oxidized	25-600	0.16-0.88
Fireclay (chamotte) and fireclay bricks	1000-1220	0.60-0.75	Gold Au	20-600	0.02-0.14
Iron oxide (FeO and Fe_2O_3)	500-1200	0.85-0.89	Iron Fe (cast or wrought)	20-1000	0.05-0.25 (0.97)
Glass (pyrex, lead, soda)	20-540	0.95-0.85	... oxidized	40-1200	0.64-0.95
Graphite	0-3600	0.70-0.98	Mercury Hg	25-100	0.10-0.12
Gypsum, plaster	0-200	0.90-0.93	Mg	35-260	0.07-0.13
Ice	0	0.97-0.99	Mo	35-1000	0.05-0.13
				2750	0.29
MgO	375-1705	0.55-0.20	Ni	20-25	0.02-0.03
				100	0.06
				200-400	0.07-0.09
				500	0.12
				1000	0.19
Magnesite brick	1000	0.38	Ni - oxidized	200-600	0.37-0.48
Marble	20	0.93	Ni-alloys	50-1300	0.06-0.98
NiO	650-1255	0.59-0.86	Ni-alloys, oxidized	20-600	0.26-0.98
Paper	20-35	0.91-0.95	Pb	125-225	0.05-0.08
Porcelain (glazed)	20	0.92-0.93	Pb - oxidized	25-200	0.28-0.63
SiO_2 (quartz)	20-840	0.93-0.41	Pt	25	0.04
				100	0.05
				200	0.06
				500	0.10
				1000	0.15
				1400	0.18
				1650	0.19
Refractory fibers (AS)	500-900	0.19-0.34	Silver Ag	20-625	0.02-0.05
SiO_2 (10 µm)	1010-1565	0.42-0.33	Stainless steel	0-1050	0.35-0.8 (0.07-0.98)
SiO_2 (70-600 µm)		0.62-0.46	... liquid		0.28
Sandstone	35-260	0.83-0.90	Ta	1340-3000	0.19-0.31
				1500	0.21
				2000	0.26
Serpentine	23	0.90	Tin Sn	25-100	0.04-0.08
Silica glass	35	0.84	Ti	−25-425	0.10-0.73

Table D-1. (Continued)

Material	T (°C)	ε	Material	T (°C)	ε
Silica brick	1000-1100	0.80-0.85	Tungsten W	25-3300	0.02-0.39
				25	0.02
				100	0.03
				500	0.07
				1000	0.15
				1500	0.23
				2000	0.28
				3300	0.39
SiC	150-1400	0.96-0.81	Zn	25-400	0.04-0.05 (0.28)
Sillimanite brick	1390	0.29			
Slate	35	0.67-0.80			
Soot (candle soot, lamp black)	0-1000	0.91-0.95			
Water	0-100	0.95-0.96			
Wood	20-70	0.94-0.75			
ThO_2	275-825	0.58-0.21			
$ZrSiO_4$	240-830	0.92-0.52			

ACKNOWLEDGMENT

This work was part of the project IAA401250703 "Porous ceramics, ceramic composites and nanocomposites" (Grant Agency of the Academy of Sciences of the Czech Republic) and the frame research program MSM6046137302 "Preparation and research of functional materials and material technologies using micro- and nanoscopic methods" (Ministry of Education, Youth and Sports of the Czech Republic). The support is gratefully acknowledged. J. H. acknowledges financial support from specific university research grants (MSMT no. 21/2010 and 2011).

REFERENCES

Abeles, B. *Phys. Rev.* 1963, *131*, 1906-1911.
Adachi, S. *Handbook of Physical Properties of Semiconductors;* Springer: New York, NY, 2004.
Adams, J. B.; Wolfer, W. G.; Foiles, S. M. *Phys. Rev. B* 1989, *40*, 9479-9484.
Aifantis, E. C. *Mater. Sci. Forum* 1993, *123-125*, 553-557.
Aifantis, E. C. *J. Mech. Behavior Mater.* 1994, *5*, 355-359.
An, K.; Ravichandran, K.; Dutton, R. E.; Semiatin, S. L. *J. Am. Ceram. Soc.* 1999, *82*, 399-406.
Archie G. E. *Trans. AIME* 1942, *146*, 54-61.
Ashcroft, N. W.; Mermin, N. D. *Solid State Physics;* Saunders College: Philadelphia, PA, 1976, pp 421-468, 495-505.
Avellaneda, M.; Cherkaev, A. V.; Lurie, K. A.; Milton, G. W. *J. Appl. Phys.* 1988, *63*, 4989-5003.
Aust, K. T.; Hibbard, G.; Palumbo, G. In *Encyclopedia of Nanoscience and Nanotechnology*; Nalwa, H. S.; Ed.; American Scientific Publishers: Valencia, CA, 2004; Vol. 4, pp. 489-498.

Aylward, G. H.; Findlay, T. J. V. *Datensammlung Chemie in SI-Einheiten* (in German, second edition); VCH-Physik-Verlag: Weinheim, Germany, 1986, pp 4-87.

Bakker, K.; Kwast, H.; Cordfunke, E. H. P. *J. Nucl. Mater.* 1995, *223*, 135-142.

Barea, R.; Osendi, M. I.; Miranzo, P.; Ferreira, J. M. F. *J. Am. Ceram. Soc.* 2005, *88*, 777-779.

Baughman, R. H.; Zakhidov, A. A.; de Heer, W. A. *Science* 2000, *297*, 787-792.

Beasley, J. D.; Torquato, S. *J. Appl. Phys.* 1986, *60*, 3576-3581.

Bengisu, M. *Engineering Ceramics;* Springer: Berlin, Germany, 2001; pp 287-301, 472-541.

Benguigui, L. *Phys. Rev B* 1986, *34*, 8176-8178.

Benveniste, Y. *Z. Angew. Math. Phys.* 1986, *37*, 696-713.

Beran, M. J. *Nuovo Cimento* 1965, *38*, 771-782.

Beran, M. J. *Statistical Continuum Theories;* Wiley-Interscience: New York, NY, 1968; pp 1-424.

Berber, S.; Kwon, Y.-K.; Tománek, D. *Phys. Rev. Lett.* 2000, *84*, 4613-4616.

Berman, R. *Adv. Phys.* 1953, *2*, 103-140.

Berman, R. *Thermal Conductivity in Solids;* Oxford University Press: Oxford, UK, 1976.

Berman, R.; Klemens, P. G.; Simon, F. E.; Fry, T. M. *Nature* 1950, *166*, 864-866.

Berman, R.; Hudson, P. R. W.; Martinez, M. *J. Phys. C: Solid State Phys.* 1975, *8*, L430-L434.

Berryman, J. G. *J. Phys. D* 1985, *18*, 585-597.

Berryman, J. G.; Milton, G. W. *J. Appl. Phys. D: Appl. Phys.* 1988, *21*, 87-94.

Biercuk, M. J.; Llaguno, M. C.; Radoslavijevic, M.; Hyun J. K.; Johnson, A. T. *Appl. Phys. Lett.* 2002, *80*, 2767-2769.

Bisson, J.-F.; Fournier, D.; Poulain, M.; Lavigne, O.; Mévrel, R. *J. Am. Ceram. Soc.* 2000, *83*, 1993-1998.

Boey, F. Y. C.; Tok, A. I. Y. *J. Mater. Process. Technol.* 2003, *140*, 413-419.

Brown, W. F. *J. Chem. Phys.* 1955, *23*, 1514-1517.

Bruggeman, D. A. G. *Ann. Phys. Leipzig* 1935, *24*, 636-679.

Bruno, O. P. *Proc. Roy. Soc. Lond. A* 1991, *433*, 353-381.

Burghartz, S.; Schulz, B. *J. Nucl. Mater.* 1994, *212*, 1065-1068.

Cahill, D. G.; Watson, S. K.; Pohl, R. O. *Phys. Rev.* 1992, *B46*, 6131-6140.

Carwile, L. C. K.; Hoge, H. J. *Glastech. Ber.* 1969, *42*, 100.

Carsley, J. E.; Ning, J.; Milligan, W. W.; Hackney, S. A.; Aifantis, E. C. *Nanostruct. Mater.* 1995, *5*, 441-448.

Cernuschi, F.; Ahmaniemi, S.; Vuoristo, P.; Mäntylä, T. *J. Eur. Ceram. Soc.* 2004, *24*, 2657-2667.

Chaim, R. *J. Mater. Res.* 1997, *12*, 1828-1836.

Chaim, R.; Hefetz, M. *J. Mater. Sci.* 2004, *39*, 3057-3061.

Charvat, F. R.; Kingery, W. D. *J. Am. Ceram. Soc.* 1957, *40*, 306-315.

Chen, C. F.; Perisse, M. E.; Ramires, A. F. *J. Mater. Sci.* 1994, *29*, 1595-1600.

Chen, H.; Gao, Y.; Liu, Y.; Luo, H. *J. Alloys Compounds* 2009, *480*, 843-848.

Cherkaev, A. V.; Gibiansky, L. V. *Proc. Roy. Soc. Edinburgh A* 1992, *122*, 93-125.

Cherkaev, A. V.; Gibiansky, L. V. *J. Mech. Phys. Solids* 1993, *41*, 937-980.

Childs, G. E.; Ericks, L. J.; Powell, R. L. *Thermal Conductivity of Solids at Room Temperature and Below;* US Department of Commerce / National Bureau of Standards: Boulder, CO, 1973; pp 1-536.

Chojnacki, J. *Základy chemické a fyzikální krystalografie* (Fundamentals of chemical and physical crystallography, in Czech); Acadmia: Prague, Czechoslovakia, 1979; pp 377-387.

Choy, T. C.; Alexopoulos, A.; Thorpe, M. F. *Proc. Roy. Soc. Lond. A* 1998, *454*, 1973-1992.

Christensen, R. M. *Mechanics of Composite Materials;* Wiley: New York, NY, 1979; pp 1-348.

Clarke, D. R. *Surf. Coat. Technol.* 2003, *163-164*, 67-74.

Clarke, D. R.; Phillpot, S. R. *Mater. Today*, 2005, *8*, 22-29.

Coble, R. L.; Kingery, W. D. *J. Am. Ceram. Soc.* 1956, *39*, 377-385.

Corson, P. B. *J. Appl. Phys.* 1974, *45*, 3159-3182.

Cutler, R. A. Engineering properties of borides, in Schneider, S. J. (Ed.) *Engineered Materials Handbook, Volume 4;* ASM International: Russell Township, OH, 1991; pp 787-803.

De Baranda, P. S.; Knudsen, A. K.; Ruh, E. *J. Am. Ceram. Soc.* 1993, *76*, 1751-1771.

DeVera, A. L.; Strieder, W. *J. Phys. Chem.* 1977, *81*, 1783-1790.

Dietrichs, P.; Krönert, W. *Interceram* 1982, *31*, 293-306.

Douglas, J. F.; Garboczi, E. J. *Adv. Chem. Phys.* 1995, *91*, 85-153.

Dreyer, W. *Materialverhalten anisotroper Festkörper;* Springer: Wien, Austria, 1974.

Du, A.; Wan, C.; Qu, Z.; Pan, W. *J. Am. Ceram. Soc.* 2009, *92*, 2687-2692.

Dura, O. J.; Bauer, E.; Vazquez, L.; de la Torre, M. A. *J. Appl. Phys. D* 2010, *43*, 105407.

Ehre, D.; Chaim, R. *J. Mater. Sci.* 2008, *43*, 6139-6143.

Einstein, A. *Ann. Phys.* 1906, *19*, 289-306.

Elbel, H.; Vollath, D. *J. Nucl. Mater.* 1988, *153*, 50-58.

Epperson, J. E.; Siegel, R. W.; White, J. W.; Klippert, T. E.; Narayanasamy, A.; Eastman, J. A.; Trouw, F. *Mater. Res. Soc. Symp. Proc.* 1989, *132*, 15-18.

Eshelby, J. D. *Proc. Roy. Soc. Lond. A* 1957, *241*, 376-396.

Estrin, Y.; Kim H. S.; Bush, M. B. In *Encyclopedia of Nanoscience and Nanotechnology*; Nalwa, H. S.; Ed.; American Scientific Publishers: Valencia, CA, 2004; Vol. 8, pp. 489-498.

Eucken, A. *Ceram. Abstr.* 1932, *11*, 576.

Eucken, A. *Ceram. Abstr.* 1933, *12*, 231.

Fanderlik, I. *Silica Glass and its Application;* Elsevier: Amsterdam, The Netherlands, 1991; pp 206-230.

Fanderlik, I. *Vlastnosti skel* (Properties of Glasses, in Czech); Informatorium: Prague, Czech Republic, 1996; pp 29-31.

Feldman, A. Round robin thermal conductivity measurements on CVD diamond, in Feldman, A.; Tzeng, Y.; Yarbrough, W.; Yoshikawa, M.; Murakawa, M. (Eds.) *Applications of Diamond Films and Related Materials;* National Institute of Standards and Technology: Gaithersburg, MD, 1995; pp 627-630.

Feng, S.; Halperin, B. I.; Sen, P. N. *Phys. Rev. B* 1987, *35*, 197-214.

Francl, J.; Kingery, W. D. *J. Am. Ceram. Soc.*, 1954, *34*, 99-107.

Fricke, J. *J. Non-Cryst. Solids* 1988, *100*, 169-173.

Fricke, J.; Tillotson, T. *Thin Solid Films* 1997, *297*, 212-223.

Gadzhiev, G. G.; Omarov, Z. M.; Abdullaev, K. K.; Reznichenko, L. A.; Kravchenko, O. Y. *Bull. Russ. Acad. Sci.* 2009, *73*, 1128-1129.

Gasch, M.; Johnson, S.; Marschall, J. *J. Am. Ceram. Soc.* 2008, *91*, 1423-1432.

Gebhart, B. *Heat Conduction and Mass Diffusion;* McGraw-Hill: New York, NY, 1993; pp 579-607.

Gibson, L. J.; Ashby, M. F. *Cellular Solids – Structure and Properties* (second edition); Cambridge University Press: Cambridge, UK, 1997, pp 52-92, 283-308.

Gleiter, H. *Prog. Mater. Sci.* 1989, *33,* 223-315.

Gordon, F. H.; Turner, S. P.; Taylor, R.; Clyne, T. W. *Composites* 1994, *25,* 583-592.

Green, D. J. *An Introduction to the Mechanical Properties of Ceramics;* Cambridge University Press: Cambridge, UK, 1998; pp 1-336.

Greenwood, N. N.; Earnshaw, A. *Chemie prvků* (Chemistry of the Elements, Czech translation); Informatorium: Prague, Czech Republic, 1993, pp 1381.

Grimvall, G. *Thermophysical Properties of Materials* (second edition); Elsevier: Amsterdam, The Netherlands, 1999, pp 70-166, 255-376.

Guo, H.; Xu, H.; Bi X.; Gong, S. *Mater. Sci. Eng. A* 2002, *325,* 389-393.

Halpin, J. C.; Kardos, J. L. *Polymer Eng. Sci.* 1976, *16,* 344-352.

Hashin, Z. *ASME J. Appl. Mech.* 1962, *29,* 143-150.

Hashin Z., Shtrikman S.: *J. Appl. Phys.* 1962, *33,* 3125-3131.

Hashin, Z.; Shtrikman, S. *Phys. Rev.* 1963, *130,* 129-133.

Hasselman, D. P. H. *Mater. Sci. Eng.* 1985, *71,* 251-264.

Hasselman, D. P. H.; Johnson, L. F. *J. Compos. Mater.* 1987, *21,* 508-515.

Hasselman, D. P. H.; Johnson, L. F.; Bentsen, L. D.; Syed, R.; Lee, H. L.; Swain, M. V. *Am. Ceram. Soc. Bull.* 1987, *66,* 799-806.

Hasselman, D. P. H.; Donaldson, K. Y.; Geiger, A. L. *J. Am. Ceram. Soc.* 1992, *75,* 3137-3140.

Haussühl, S. *Kristallphysik;* Physik-Verlag / Verlag Chemie: Weinheim, Germany, 1983; pp 128-130, 418.

Hayashi, K.; Kyaw, T. M.; Okamoto, Y. *High Temp. High Press.* 1998, *30,* 283-290.

Hellwege, K. H. *Einführung in die Festkörperphysik* (Introduction to Solid State Physics, in German, third edition); Springer: Berlin, Germany, 1988; pp 136-137, 565-580.

Helsing, J. *J. Math. Phys.* 1994, *35,* 1688-1692.

Helsing, J.; Helte, A. *J. Appl. Phys.* 1991, *69,* 3583-3588.

Hetherington, J. H.; Thorpe, M. F. *Proc. Roy. Soc. Lond. A* 1992, *438,* 591-604.

Hildmann, B.; Schneider, H. *J. Am. Ceram. Soc.* 2005, *88,* 2879-2882.

Hill, R. *Proc. Phys. Soc. A* 1952, *65,* 349-354.

Hu, W.; Guan, H.; Sun, X.; Li, S.; Fukumoto, M.; Okane, I. *J. Am. Ceram. Soc.* 1998, *81,* 2209-2212.

Ibach, H.; Lüth, H. *Festkörperphysik* (Solid State Physics, second edition, in German); Springer: Berlin, Germany, 1989, pp 69-86.

Incropera, F. P.; Dewitt, D. P. *Fundamentals of Heat and Mass Transfer* (third edition); John Wiley and Sons: New York, NY, 1991.

Ivanov, S. N.; Popov, P. A.; Egorov, G. V.; Sidorov, A. A.; Kornev, B. I.; Zhukova, L. M.; Ryabov, V. P. *Phys. Solid State* 1997, *39,* 81-83.

Jackson, J. D. *Klassische Elektrodynamik* (Classical Electrodynamics, German translation, second edition); de Gruyter: Berlin, Germany, 1983; pp 69-158.

Jackson, T. B.; Donaldson, K. Y.; Hasselman, D. P. H. *J. Am. Ceram. Soc.* 1990, *73,* 2511-2514.

Jackson, T. B.; Virkar, A. V.; More, K. L.; Dinwiddie R. D.; Cutler, R. A. *J. Am. Ceram. Soc.* 1997, *80*, 1421-1435.

Jang, B. K.; Yosiha, M.; Yamaguchi, N.; Matsubara, H. *J. Mater. Sci.* 2004, *39*, 1823-1825.

Jang, B. K.; Sakka, Y. *J. Alloys Compounds* 2008, *463*, 493-497.

Jarrige, J.; Lecompte, J. P.; Mullot, J.; Muller, G. *J. Eur. Ceram. Soc.* 1997, *17*, 1891-1895.

Jeffery, G. B. *Proc. Roy. Soc. Lond. A* 1922, *102*, 161-179.

Jeffrey, D. J. *Proc. R. Soc. Lond. A* 1973, *335*, 355-367.

Jezowski, A.; Mucha, J.; Pazik, R.; Strek, W. *Appl. Phys. Lett.* 2007, 90, 1141041-1141043.

Jiang, B; Weng, G. J. *Int. J. Plast.* 2004, *20*, 2007-2026.

Kanai, T.; Ando, A.; Tanemoto, K. *Jap. J. Appl. Phys.* 1992, *31*, 1426-1427.

Kaviany, M. *Principles of Heat Transfer in Porous Media* (2nd edition); Springer: New York, NY, 1995, pp 1-572.

Kawai, C. *J. Am. Ceram. Soc.* 2001, *84*, 896-898.

Kellogg, O. D. *Foundations of Potential Theory;* Springer: Berlin, Germany, 1953.

Kerner, E. H. *Proc. Phys. Soc. (London) B* 1956, *69*, 802-813.

Keyes, B. L. P. *Ceramic Technology Project Database*: September 1992 Summary Report (ORNL Report Number: ORNL/M-2775); Oak Ridge National Laboratory: Oak Ridge, TN, 1992, pp 1-171.

Khadar, M. A.; Biju, V.; Inoue; A. *Mater. Res. Bull.* 2003, *38*, 1341-1349.

Khare, R.; Bose, S. *J. Miner. Mater. Charact. Eng.* 2005, *4*, 31-46.

Kim, H. S. *Scripta Mater.* 1998, *39*, 1057-1061.

Kim, I. C.; Torquato, S. *J. Appl. Phys.* 1992, *71*, 2727-2735.

Kim, W. J.; Kim, D. K.; Kim, C. H. *J. Am. Ceram. Soc.* 1996, *79*, 1066-1072.

Kim, H. S.; Bush, M. B. *Nanostruct. Mater.* 1999, *11*, 361-367.

Kim, H. S.; Bush, M. B.; Estrin, Y. *Mater. Sci. Eng. A* 2000, *276*, 175-185.

Kim, H. S.; Estrin, Y.; Bush, M. B. *Acta Mater.* 2000, *48*, 493-504.

Kim, H. S.; Estrin, Y.; Bush, M. B. *Mater. Sci. Eng. A* 2001, *316*, 195-199.

Kingery, W. D. *Introduction to Ceramics;* John Wiley and Sons: New York, NY, 1960, pp 461-510.

Kingery, W. D. The thermal conductivity of ceramic dielectrics, in Burke, J. E. (Ed.) *Progress in Ceramic Science, Volume 2;* Pergamon Press: Oxford, UK, 1962, pp 182-235.

Kingery, W. D.; McQuarrie, M. C. *J. Am. Ceram. Soc.* 1954, *37*, 67-72.

Kingery, W. D.; Francl, J.; Coble, R. L.; Vasilos, T. *J. Am. Ceram. Soc.* 1954, *37*, 107-110.

Kingery, W. D.; Bowen, H. K.; Uhlmann, D. R. *Introduction to Ceramics* (second edition); John Wiley and Sons: New York, NY, 1976, pp 583-645.

Kinoshita, H.; Otani, S.; Kamiyama, S.; Amano, H.; Akasaki, I.; Suda, J.; Matsunami, H. *Jpn. J. Appl. Phys.* 2001, *40*, L1280-L1282.

Kirchheim, R.; Mütschele, T.; Kieneinger, W. *Mater. Sci. Eng.* 1988, *99*, 457-462.

Kittel, C. *Einführung in die Festkörperphysik* (Introduction to Solid State Physics, 7th edition, in German); Oldenbourg: München, Germany, 1988; pp 130-158, 577-580, 689-693.

Kleber, W. *Einführung in die Kristallographie* (Introduction to Crystallography, 17th edition, in German); Bautsch, H.-J.; Bohm, J.; Kleber, I.; Eds.; Verlag Technik: Berlin, Germany, 1990; pp 237-240, 253-256.

Klemens, P. G. *Phys. Rev.* 1960, *119*, 507-509.

Klemens, P. G. *High Temp. - High Press.* 1991, *23,* 241.

Komeya, K.; Matsui, M. High temperature engineering ceramics, in *Structure and Properties of Ceramics;* Swain, M. W.; Ed. (= Volume 11 of Materials Science and Technology – A Comprehensive Treatment; Cahn, R. W.; Haasen, P.; Kramer, E. J.; Eds.) Wiley-VCH: Weinheim 2005 (paperback reprint of the 1994 edition); pp 517-565.

Kopitzki, K. *Einführung in die Festkörperphysik* (Introduction to Solid State Physics, second edition, in German); Teubner: Stuttgart, Germany, 1989; pp 64-90, 393.

Krueger, A. *Carbon Materials and Nanotechnology;* Wiley-VCH: Weinheim, Germany, 2010; pp 186-217.

Kumari, L.; Zhang, T.; Du, G. H.; Li, W. Z.; Wang, Q. W.; Datye, A.; Wu, K. H. *Compos. Sci. Technol.* 2008, *68,* 2178-2183.

Kuneš, J. *Modelování tepelných procesů* (Modelling of Thermal Processes, in Czech); SNTL: Prague, Czechoslovakia, 1989, pp 13-220.

Kuramoto, N.; Taniguchi, H.; Aso, I. *Am. Ceram. Soc. Bull.* 1989, *68,* 883-887.

Kushan, S. R.; Uzun, I.; Dogan, B.; Mandal, H. *J. Am. Ceram. Soc.* 2007, *90,* 3902-3907.

Kutzendörfer, J. Máša, Z. *Žárovzdorné tepelně isolační materiály* (Thermally Insulating Refractory Materials, in Czech); Informatorium: Prague, Czechoslovakia 1991, pp 1-36, 205-255.

Lado, F., Torquato S.: *Phys. Rev. B* 1986, *33,* 3370-3378.

Landauer, R. *J. Appl. Phys.* 1952, **23**, 779-784.

Landauer, R. Electrical conductivity in inhomogeneous media, in *Electrical, Transport and Optical Properties of Inhomogeneous Media;* Garland, J. C.; Tanner, D. B.; Eds.; American Institute of Physics: New York, NY, 1978; pp 2-43.

Lee, D. W.; Kingery, W. D. *J. Am. Ceram. Soc.* 1960, *43,* 594-607.

Lee, W. E.; Rainforth, W. M. *Ceramic Microstructures – Property Control by Processing;* Chapman and Hall: London, UK, 1994, pp 263-556.

Li, B. C.; Pottier, L.; Rogre, J. P.; Fournier, D.; Watari, K.; Hirao, K. *J. Eur. Ceram. Soc.*, 1999, *19,* 1631-1639.

Lide, D. R. (Ed.) *CRC Handbook of Chemistry and Physics* (90th edition, internet version); CRC Press / Taylor and Francis: Boca Raton, FL, 2010, pp 4-119 - 4-124, 12-72 - 12-211.

Liu, D. M.; Lin, B. W. *Ceram. Intern.* 1996, *22,* 407-414.

Liu, D. M.; Tuan, W. H.; Chiu, C. C. *Mater. Sci. Eng. B* 1995, *31,* 287-291.

Liu, D. M.; Chen, C. J.; Ray Lee, R. R. *J. Appl. Phys.*, 1995, *77,* 494-496.

Loeb, A. L. *J. Am. Ceram. Soc.* 1954, *37,* 96-99.

Loehman, R.; Corral, E.; Dumm, H.-P.; Kotula, P.; Tandon, R. Ultra high temperature ceramics for hypersonic vehicle applications, project SAND2006-2925: Albuquerque, NM, 2006.

Lu, X.; Caps, R.; Fricke, J.; Alviso, C. T.; Pekala, R. W. *J. Non-Cryst. Solids* 1995, *188,* 226-234.

Lu, T. J.; Levi, C. G.; Wadley, H. N. G.; Evans, A. G. *J. Am. Ceram. Soc.* 2001, *84,* 2937-2946.

Lubrarda, V. A.; Markenscoff, X. *Int. J. Solids Structures* 1998, *35,* 3405-3411.

Makhlouf, S. A.; Kassem, M. A.; Abdel-Rahim, M. A. *J. Mater. Sci.* 2009, *44,* 3438-3444.

Markov, K. Z. *J. Mech. Phys. Solids* 1998, *46,* 357-388.

Markov, K. Z. Elementary micromechanics of heterogeneous media, in Markov, K. Z.; Preziosi, L. (Eds.) *Heterogeneous Materials – Micromechanics Modeling Methods and Simulations;* Birkhäuser: Boston, MA, 2000; pp 1-162.

Martienssen, W.; Warlimont, H. (Eds.) *Springer Handbook of Condensed Matter and Materials Data;* Springer: Berlin, Germany, 2005, pp 1-1120.

Maxwell, J. C. *A Treatise on Electricity and Magnetism;* Clarendon Press: London, UK, 1873 (reprint Dover: New York, NY, 1954).

McLachlan, D. S. *J. Phys. C - Solid State Phys.* 1985, *18,* 1891.

McLachlan, D. S. *Solid State Commun.* 1986, *60,* 821-825.

McLachlan, D. S.; Blaszkiewicz, M.; Newnham, R. E. *J. Am. Ceram. Soc.* 1990, *73,* 2187-2203.

McLachlan, D. S.; Cai, K.; Sauti, G. *Int. J. Refract. Metals Hard Mater.* 2001, *19,* 437-445.

Menčík, J. *Strength and Fracture of Glass and Ceramics;* Elsevier: Amsterdam, The Netherlands, 1992, pp 89-98.

Mévrel, R.; Laizet, J.-C.; Azzopardi, A.; Leclercq, B.; Poulain, M.; Lavigne O.; Demange, D. *J. Eur. Ceram. Soc.* 2004, *24,* 3081-3089.

Miller, M. N. *J. Math. Phys.* 1969, *10,* 1988-2004.

Miller, C. A.; Torquato, S. *J. Appl. Phys.* 1990, *68,* 5486-5493.

Milton, G. W. *J. Appl. Phys.* 1981, *52,* 5286-5304.

Milton, G. W. *J. Mech. Phys. Solids* 1982, *30,* 177-191.

Milton, G. W. Correlation of the electromagnetic and elastic properties of composites and microgeometries corresponding with effective medium approximations, in Johnson, D. L.; Sen, P. N.; Eds.; *Physics and Chemistry of Porous Media;* American Institute of Physics: New York, NY, 1984.

Milton, G. W. *Comm. Math. Phys.* 1985, *99,* 463-500.

Milton, G. W. *The Theory of Composites;* Cambridge University Press: Cambridge, UK, 2002; pp 1-719.

Miyayama, M.; Koumoto, K.; Yanagida, H. Engineering properties of single oxides, in Schneider, S. J. (Ed.) *Engineered Materials Handbook, Volume 4;* ASM International: Russell Township, OH, 1991; pp 748-757.

Modest, M. F. *Radiative Heat Transfer;* McGraw-Hill: New York, NY, 1993, pp 75-382, 759-774.

Mogro-Campero, A.; Johnson, C. A.; Bednarczyk, P. J.; Dinwiddie, R. B.; Wang, H. *Surf. Coat. Technol.* 1997, *94-95,* 102-105.

Molyneux, J. E. *J. Math. Phys.* 1970, *11,* 1172-1184.

Mooney, M. *J. Colloid Sci.* 1951, *6,* 162-170.

Moore, W. J. *Fyzikální chemie* (Physical chemistry, in Czech); SNTL: Prague, Czechoslovakia, 1981, p 154-180.

Mori, T.; Tanaka, K. *Acta Metall.* 1973, *21,* 571-574.

Müller, S. G.; Eckstein, R.; Fricke, J.; Hofmann, D.; Hofmann, R.; Horn, R.; Meling, H.; Nilsson, O. *Mater. Sci. Forum* 1998, *264-268,* 623-636.

Munro, R. G. *J. Am. Ceram. Soc.* 1997, *80,* 1919-1928.

Munro, R. G. *J. Phys. Chem. Ref. Data* 1997, *26,* 1195-1203.

Munro, R. G. *J. Res.. Nat. Inst. Stand. Technol.* 2000, *105,* 709-720.

Mütschele, T.; Kirchheim, R. *Script. Metall.* 1987, *21,* 1101-1104.

Nair, S. S.; Khadar, M. A. *Sci. Technol. Adv. Mater.* 2008, *9,* 1-4.

Naumann, R. J. *Introduction to the Physics and Chemistry of Materials;* CRC Press: Boca Raton, FL, 2009; pp 311-338.

Nemat-Nasser, S.; Hori M. *Micromechanics – Overall Properties of Heterogeneous Materials* (second edition); North-Holland / Elsevier: Amsterdam, NL, 1999; pp 1-786.

Nemoto, T.; Sasaki, S.; Hakuraku Y. *Cryogenics* 1985, *25*, 531-532.

Neshpor, V. S. *J. Eng. Phys. Thermophys.* 1968, *15*, 321-325.

Nesi, V.; Milton, G. W. *J. Mech. Phys. Solids* 1991, *39*, 525-542.

Newnham, R. E. *Properties of Materials – Anisotropy, Symmetry, Structure;* Oxford University Press: Oxford, UK, 2005; pp 203-210.

Nielsen, L. E. *J. Appl. Polym. Sci.* 1973, *17*, 3819-3825.

Nielsen, L. E. *Ind. Eng. Chem. Fundam.* 1984, *13*, 17-28.

Nieto, M. I.; Martínez, R.; Mazerolles, L.; Baudín, C. *J. Eur. Ceram. Soc.* 2004, *24*, 2293-2301.

Nikolopoulos, P.; Ondracek, G. *J. Am. Ceram. Soc.* 1983, *66*, 238-241.

NIST (National Institute of Standards and Technology) *NIST Standard Reference Database No. 30 / Web SCD Database* (www.ceramics), version April 2003; US Department of Commerce / NIST: Gaithersburg, MD, 2003.

Norris, A. N. *J. Appl. Mech.* 1989, *56*, 83-88.

Nunez Regueiro, M. D.; Castello, D. *Int. J. Mod. Phys. B* 1991, *5*, 2003-2035.

Nye, J. F. *Physical Properties of Crystals* (first paperback edition); Oxford University Press: Oxford, UK, 1985; pp 195-214.

Öchsner, A.; Murch, G. E.; de Lemos, M. J. S. (Eds.) *Cellular and Porous Materials – Thermal Properties Simulation and Prediction;* Wiley-VCH: Weinheim, Germany, 2008, pp 1-422.

Okaz, A. M.; El-Messih, S. A.; El-Osairy, M. *Powder Technol.* 1986, *47*, 35-38.

Olson, J. R.; Pohl, R. O.; Vandersande, J. W.; Zoltan, A.; Anthony, T. R.; Banholzer, W. F. *Phys. Rev. B* 1993, *47*, 14850-14856.

Ondracek, G. *Metall.* 1982, *36*, 523-531.

Ondracek, G. *Mater. Chem. Phys.* 1986, *15*, 281.

Ondracek, G. *Rev. Powder Metall. Phys. Ceram.* 1987, *3*, 205-322.

Ondracek, G.; Schulz, B. *J. Nucl. Mater.* 1973, *46*, 253-258.

Onn, D. G.; Witek, A.; Qiu, Y. Z.; Anthony, T. R.; Banholzer, W. F. *Phys. Rev. Lett.* 1992, *68*, 2806-2809.

Opeka, M. M.; Talmy, I. G.; Wuchina, E. J.; Zaykoski, J. A.; Causey, S. J. *J. Eur. Ceram. Soc.* 1999, *19*, 2405-2414.

Osborn, J. A. *Phys. Rev.* 1945, *67*, 351-357.

Ownby, P. D.; Stewart, R. W. Engineering properties of diamond and graphite, in Schneider, S. J. (Ed.) *Engineered Materials Handbook, Volume 4;* ASM International: Russell Township, OH, 1991; pp 821-834.

Pabst, W. *Ceram.-Silik.* 2005, *49*, 254-263.

Pabst, W. *J. Mater. Sci.* 2005, *40*, 2667-2669.

Pabst, W. Steps across the border – from micromechanics to the properties of nanoceramics, in Tseng T.-Y.; Nalwa, H. S. (Eds.) *Handbook of Nanoceramics and Their Based Devices, Volume 3: Characterization and Properties;* American Scientific Publishers: Stevenson Ranch, CA, 2009, pp 207-228.

Pabst, W.; Gregorová, E. *Ceram.-Silik.* 2004, *48*, 14-23.

Pabst, W.; Gregorová, E. *J. Mater. Sci.* 2004, *39*, 3213-3215.
Pabst, W.; Gregorová, E. *J. Mater. Sci.* 2004, *39*, 3501-3503.
Pabst, W.; Gregorová, E. Effective elastic moduli of alumina, zirconia and alumina-zirconia composite ceramics, in Caruta, B.M. (Ed.) *Ceramics and Composite Materials: New Research;* Nova Science Publishers: New York, NY, 2006; pp. 31-100.
Pabst, W.; Gregorová, E. *Ceram. Intern.* 2006, *32*, 89-91.
Pabst, W.; Gregorová, E. *Ceramika-Ceramics* 2006, *97*, 71-84.
Pabst, W.; Gregorová, E. Effective thermal and thermoelastic properties of alumina, zirconia and alumina-zirconia composite ceramics, in Caruta, B. M. (Ed.) *New Developments in Materials Science Research;* Nova Science Publishers: New York, NY, 2007; pp 77-137.
Pabst, W.; Gregorová, E. Exponential porosity dependence of thermal conductivity, in Koenig, J. R.; Ban, H. (Eds.) *Thermal Conductivity 29 / Thermal Expansion 17;* DEStech Publications: Lancaster, PA, 2008; pp 487-498.
Pabst, W.; Gregorová, E. *Phase Mixture Models for the Properties of Nanoceramics;* Nova Science Publishers: New York, NY, 2010; pp 1-74.
Pabst, W.; Hostaša, J. *Adv. Sci. Technol.* 2010, *63*, 68-73.
Pabst, W.; Tichá, G.; Gregorová, E. *Ceram.-Silik.* 2004, *48*, 41-48.
Pabst, W.; Gregorová E.; Tichá, G. *J. Eur. Ceram. Soc.* 2007, *27*, 479-482.
Pabst, W.; Gregorová, E.; Hostaša, J. *AIP Conf. Proc.* 2009, *1145*, 109-112.
Padture, N. P.; Klemens, P. G. *J. Am. Ceram. Soc.* 1997, *80*, 1018-1020.
Palumbo, G.; Thorpe, S. J.; Aust, K. T. *Scripta Met. Mater.* 1990, *24*, 1347-1350.
Pampuch, R. *ABC of Contemporary Ceramic Materials;* Techna Group: Faenza, Italy, 2008, pp 71-85.
Pezzotti, G.; Kamada, I.; Miki, S. *J. Eur. Ceram. Soc.* 2000, *20*, 1197-1203.
Pezzotti, G.; Nakahira, A.; Tajika, M. *J. Eur. Ceram. Soc.* 2000, *20*, 1319-1325.
Phan-Thien, N.; Pham, D. C. *Int. J. Eng. Sci.* 2000, *38*, 73-88.
Phillpot, S. R.; Wolf, D.; Gleiter, H. *Scripta Metall. Mater.* 1995, *33*, 1245-1251.
Pierson, H. O. *Handbook of Carbon, Graphite, Diamond, and Fullerenes*; Noyes Publications: Park Ridge, NJ, 1993.
Pierson, H. O. *Handbook of Refractory Carbides and Nitrides: Properties, Characteristics, Processing, and Applications*; William Andrew Publishing: Westwood, NJ, 1996; pp 55-99, 137-155, 181-208, 223-247.
Ping, D. H.; Li, D. X.; Ye, Q. *J. Mater. Sci. Lett.* 1995, *14*, 1536-1540.
Polder, D.; Van Santen, J. H. *Physica* 1946, *12*, 257-271.
Pouchon, M. A.; Degueldre, C.; Tissot, P. *Thermochim. Acta* 1998, *323*, 109-121.
Powell, R. L.; Childs, G. E. in Gray, D. E. (Ed.) *American Institute of Physics Handbook* (third edition); McGraw-Hill: New York, NY, 1972.
Prager, S. *J. Chem. Phys.* 1969, *50*, 4305-4312.
Qiu, Y. Z.; Witek, A.; Onn, D. G.; Anthony, T. R.; Banholzer, W. F. *Thermochim. Acta* 1993, *218*, 257-258.
Raghavan, N. S. *Mater. Sci. Eng. A* 1991, *148*, 307-317.
Raghavan, S.; Wang, H.; Dinwiddie, R. B.; Porter, W. D.; Mayo, M. J. *Scripta Mater.* 1998, *39*, 1119-1125.
Raghavan, S.; Wang, H.; Porter, W. D.; Dinwiddie, R. B.; Mayo, M. J. *Acta Mater.* 2001, *49*, 169-179.

Raghavan, S.; Wang, H.; Dinwiddie, R. B.; Porter, W. D.; Vassen, R.; Stöver, D.; Mayo, M. J. *J. Am. Ceram. Soc.* 2004, *87*, 431-437.

Ratcliffe, E. H. *Glass Technol.* 1963, *4*, 113-128.

Rayleigh, Lord (Strutt, J. W.) *Philos. Mag.* 1892, *34*, 481-502.

Reeves, A. J.; Taylor, R.; Clyne, T. W. *Mater. Sci. Eng. A* 1991 *141*, 129-138.

Reuss, A. *Z. Angew. Math. Mech.* 1929, *9*, 49-58.

Richerson, D. W. *Modern Ceramic Engineering* (third edition); Taylor and Francis: Boca Raton, FL, 2006; pp 189-198.

Russell, L. M.; Donaldson, K. Y.; Hasselman, D. P. H.; Ruh, R.; Adams, J. W. *J. Am. Ceram. Soc.* 1996, *79*, 2767-2770.

Sahimi, M. *Applications of Percolation Theory;* Taylor and Francis: London, UK, 1994; pp 1-258.

Sahimi, M. *Heterogeneous Materials I – Linear Transport and Optical Properties;* Springer: New York, NY, 2003; pp 1-290.

Sakai, S.; Tanimoto, H.; Mizubayashi, H. *Acta Mater.* 1999, *47*, 211-217.

Salmang, H.; Scholze, H. *Keramik* (Ceramics, in German, fifth edition in two volumes); Springer: Berlin, Germany, 1982, pp 227-234 (Vol.1), 90-232 (Vol.2).

Santos, W. N. D. *J. Eur. Ceram. Soc.* 2003, *23*, 745-755.

Santos, W. N.; Filho, P. I.; Taylor, R. *J. Eur. Ceram. Soc.* 1998, *18*, 807-811.

Schelling, P. K.; Phillpot, S. R. *J. Am. Ceram. Soc.* 2001, *84*, 2997-3007.

Schill, F. *Chlazení skla* (Cooling of Glass, in Czech); Informatorium: Prague, Czech Republic, 1993; pp 143-152.

Schlegel, E. *Wärmedämmstoffe für den Wärmeschutz von Gebäuden und Anlagen* (Thermally insulating materials for thermal protection of buildings and industrial equipment, in German; Sitzungsber. Sächs. Akad. Wiss. Leipzig, Technikwissenschaftliche Klasse, Volume 1, Issue 2); Hirzel: Stuttgart, Germany, 1999, pp 1-25.

Schlichting, K. W.; Padture, N. P.; Klemens, P. G. *J. Mater. Sci.* 2001, *36*, 3003-3010.

Schneider, H.; Komarneni, S. *Mullite;* Wiley-VCH: Weinheim, Germany, 2005; pp 149-156, 322.

Schulgasser, K. *J. Appl. Phys.* 1976, *47*, 1880-1886.

Schulgasser, K. *J. Phys. C* 1977, *10*, 407-417.

Schulle, W. *Feuerfeste Werkstoffe* (Refractories, in German); Deutscher Verlag für Grundstoffindustrie: Leipzig, Germany, 1990, pp 79-101.

Schulz, B. *High Temp. - High Pressures* 1981, *13*, 649.

Schulz, B.; Wedemeyer, H. *J. Nucl. Mater.* 1986, *139*, 35-41.

Schupp, M. *Gas Wärme Internat.* 1981, *30*, 350-356.

Sen, P. N.; Scala, C.; Cohen, M. H. *Geophys.* 1981, *46*, 781-795.

Shackelford, J. F.; Doremus, R. H. *Ceramic and Glass Materials;* Springer: New York, NY, 2008, pp 14-110.

Shaffer, P. T. B.; Schorr, J. R.; Hexemer, R. L. *Ceramic Industry* 1989, *132*, 25-27.

Sigl, L. S. *J. Eur. Ceram. Soc.* 2003, *23*, 1115-1122.

Sivakumar, R.; Doni Jayaseelan, D.; Nishikawa, T.; Honda, S.; Awaji, H. *Ceram. Intern.* 2001, *27*, 537-541.

Sivakumar, R.; Aoyagi, K.; Akiyama, T. *Ceram. Intern.* 2009, *35*, 1391-1395.

Skopp, A.; Woydt, M.; Habig, K. H. *Trib. Intern.* 1990, *23*, 189-199.

Slack, G. A. *J. Appl. Phys.* 1964, *35*, 3460-3466.

Slack, G. A. *J. Phys. Chem. Solids* 1973, *34*, 321-335.
Slack, G. A.; Austerman,. S. B. *J. Appl. Phys.* 1971, *42*, 4713-4717.
Slack, G. A.; Tanzill, R. A.; Pohl, R. O.; Vandersande, J. W. *J. Phys. Chem. Solids* 1987, *48*, 641-647.
Slifka, A. J.; Filla, B. J.; Phelps, J. M. *J. Res. National Inst. Stand. Technol.* 1998, *103*, 357-363.
Smith, L. N.; Lobb, C. J. *Phys. Rev. B* 1979, *20*, 3653-3658.
Smith, D. S.; Grandjean, S.; Absi, J.; Kadiebu, S.; Fayette, S. *High Temp. – High Press.* 2003, *35-36*, 93-99.
Smith, D. S.; Fayette, S.; Grandjean, S.; Martin, R.; Telle, R.; Tönnessen, T. *J. Am. Ceram. Soc.* 2003, *86*, 105-111.
Staff, N. N. *Ceram. Industr.* 1989, *132*, 34-35.
Staroň, J.; Tomšů, F. *Žiaruvzdorné materiály – výroba, vlastnosti a použitie* (Refractories – Production, Properties and Application, in Slovakian); Slovmag: Lubeník, Slovak Republic, 2000; pp 11-13, 154-333.
Stauffer, D.; Aharony, A. *Introduction to Percolation Theory* (second edition); Taylor and Francis: London, UK, 1985; pp 1-181.
Stephens, R. W. B. *Philos. Mag.* 1932, *14*, 897-914.
Stevens, R. Engineering properties of zirconia, in Schneider, S. J. (Ed.) *Engineered Materials Handbook, Volume 4;* ASM International: Russell Township, OH, 1991; pp 775-786.
Stoner, E. C. *Philos. Mag.* 1945, *36*, 803-820.
Stratton, J. A. *Teorie elektromagnetického pole* (Electromagetic Theory, Czech translation); SNTL: Prague, Czechoslavkia, 1961; pp 164-224.
Su, Y. J.; Wang, H.; Porter, W. D.; De Arellano Lopez, A. R.; Faber, K. T. *J. Mater. Sci.* 2001, *36*, 3511-3518.
Sugawara, A.; Yoshizawa, Y. *J. Appl. Phys.* 1962, *33*, 3135-3138.
Sun, Z.; Zhou, Y.; Wang, J.; Li, M. *J. Am. Ceram. Soc.* 2008, *91*, 2623-2629.
Sun, Z.; Li, M.; Zhou, Y. *J. Eur. Ceram. Soc.* 2009, *29*, 551-557.
Takeda, Y. *Am. Ceram. Soc. Bull.* 1988, *67*, 1961-1963.
Tanaka, H.; Sawai, S.; Morimoto, K.; Hisano, K. *J. Therm. Anal. Calorim.* 2001, *64*, 867-872.
Taylor, R. E. *J. Am. Ceram. Soc.*, 1962, *45*, 74-78.
Tessier-Doyen, N.; Grenier, X.; Huger, M.; Smith, D. S.; Fournier, D.; Roger, J. P. *J. Eur. Ceram. Soc.* 2007, *27*, 2635-2640.
Thomas, G. J.; Siegel, R. W.; Eastman, J. A. *Scripta Metall. Mater.* 1990, *24*, 201-206.
Thorvert, J. F.; Kim, I. C.; Torquato, S.; Acrivos, A. *J. Appl. Phys.* 1990, *67*, 6088-6098.
Tichá, G.; Pabst, W.; Smith, D. S. *J. Mater. Sci.* 2005, *40*, 5045-5047.
Torquato, S. *J. Chem. Phys.* 1984, *81*, 5079-5088.
Torquato, S. *J. Appl. Phys.* 1985, *58*, 3790-3797.
Torquato, S. *J. Chem. Phys.* 1985, *83*, 4776-4785.
Torquato, S. *J. Chem. Phys.* 1986, *84*, 6345-6359.
Torquato, S. Connection between the morphology and effective properties of heterogeneous materials, in *Macroscopic Behavior of Heterogeneous Materials from the Microstructure* (AMD Volume 147); Torquato, S.; Krajcinovic, D. (Eds.); American Society of Mechanical Engineers: New York, NY, 1992; pp. 53-65.
Torquato, S. *Random Heterogeneous Materials – Microstructure and Macroscopic Properties*; Springer: New York, NY, 2002; pp 1-701.

Torquato, S.; Rubinstein J. *J. Appl. Phys.* 1991, *69*, 7118-7125.
Torquato, S.; Stell, G. *J. Chem. Phys.* 1983, *79*, 1505-1510.
Torquato, S.; Stell, G. *Int. J. Eng. Sci.* 1985, *23*, 375-383.
Torquato, S.; Lado, F.; Smith, P. A. *J. Chem. Phys.* 1987, *86*, 6388-6393.
Touloukian, Y. S.; Liley, P. E.; Saxena, S. C. *Thermal Conductivity – Nonmetallic Liquids and Gases* (Thermophysical Properties of Matter Vol. 3); Plenum: New York, NY, 1970.
Truesdell, C. *Rational Thermodynamics* (second edition); Springer: New York, NY, 1984; pp 365-395.
Vassen, R.; Cao, X.; Tietz, F.; Basu, D.; Stöver, D. *J. Am. Ceram. Soc.* 2000, *83*, 2023-2028.
Voigt, W. *Ann. Physik Leipzig (Wiedemann)* 1889, *38*, 573-587.
Voigt, W. *Lehrbuch der Kristallphysik* (reprint of the 1910 edition); Teubner: Leipzig, 1928; pp 954-964.
Volf, M. B. *Mathematical Approach to Glass*; Elsevier: Amsterdam, The Netherlands, 1988; pp 254-265.
Volf, M. B. *Technical Approach to Glass*; Elsevier: Amsterdam, The Netherlands, 1990, pp 194-202.
Walker, F. J.; Anderson, A. C. *Phys. Rev. B* 1984, *29*, 5881-5890.
Wang, C.-C. On the symmetry of the heat-conduction tensor, in Truesdell, C. *Rational Thermodynamics* (second edition); Springer: New York, NY, 1984; pp 396-401.
Wang, N.; Palumbo, G.; Wang, Z.; Erb, U.; Aust, K. T. *Scripta Met. Mater.* 1993, *28*, 253-256.
Wang, N.; Wang, Z.; Aust, K. T.; Erb, U. *Acta Metall. Mater.* 1995, *43*, 519-528.
Wang, J.; Wolf, D.; Phillpot, S. R.; Gleiter, H. *Philos. Mag. A* 1996, *73*, 517-555.
Wang, N.; Wang, Z.; Aust, K. T.; Erb, U. *Mater. Sci. Eng. A* 1997, *237*, 150-158.
Watanabe, H. *Thermochim. Acta* 1993, *218*, 365-372.
Watari, K.; Seki, Y.; Ishizaki, K. *J. Ceram. Soc. Jpn.* 1989, *97*, 56-62.
Watari, K.; Valecillos, M. C.; Brito, M. E.; Toriyama, M.; Kanzaki, S. *J. Am. Ceram. Soc.* 1996, *79*, 3103-3108.
Watari, K.; Hirao, K.; Toriyama, M.; Ishizaki, K. *J. Am. Ceram. Soc.* 1999, *82*, 777-779.
Watari, K.; Nakano, H.; Sato, K.; Urabe, K.; Ishizaki, K.; Cao, S.; Mori, K. *J. Am. Ceram. Soc.* 2003, *86*, 1812-1814.
Watari, K.; Hirao, K.; Brito, M. E.; Toriyama, M.; Ishizaki, K. *J. Mater. Online* 2006, *2*, 1-17.
Weast, R. C. *CRC Handbook of Chemistry and Physics* (first student edition); CRC Press: Boca Raton, FL, 1988; pp E-2 – E-15, E-25 – E-27, E-323 – E-324.
Wei, L.; Kuo, P. K.; Thomas, R. L.; Anthony, T. R.; Banholzer, W. F. *Phys. Rev. Lett.* 1993, *70*, 3764-3767.
Weissberg, H. L. *J. Appl. Phys.* 1963, *34*, 2636-2639.
Wen, S.; Yan, D. *Ceram. Int.* 1995, *21*, 109-112.
Wiener, O. *Abh. Math.-Phys. Klasse Königl. Sächs. Gesellsch. Wissensch.* 1912, *32*, 509-604.
Williams, R. K.; Bates, J. B.; Graves, R. S.; McElroy, D. L.; Weaver, F. J. *Int. J. Thermophysics* 1988, *9*, 587-598.
Willis, J. R. *J. Mech. Phys. Solids* 1977, *25*, 185-202.
Wort, C. J. H.; Sweeney, C. G.; Copper, M. A.; Sussmann, R. S. *Diamond and Related Materials, Volume 3;* Elsevier Science: Amsterdam, The Netherlands, 1994; pp 1158-1167.

Wu, J.; Wei, X.; Padture, N. P.; Klemens, P. G.; Gell, M.; García, E.; Miranzo, P.; Osendi, M. I. *J. Am. Ceram. Soc.* 2002, *85*, 3031-3035.

Yagi, H.; Yanagitani, T.; Numazawa, T.; Ueda, K. *Ceram. Intern.* 2007, *33*, 711-714.

Yamamoto, T.; Watanabe, K.; Hernández, E. R. Mechanical properties, thermal stability and heat transport in carbon nanotubes, in Jorio, A.; Dresselhaus, M. S.; Dresselhaus, G. (Eds.) *Carbon Nanotubes – Advanced Topics in the Synthesis, Structure, Properties and Applications;* Springer: Berlin, Germany, 2008; pp 165-194.

Yang, H.-S.; Bai, G.-R., Thompson, L. J.; Eastman J. A. *Acta Mater.* 2002, *50*, 2309-2317.

Yeheskel, O; Chaim, R.; Shen, Z.; Nygren, M. *J. Mater. Res.* 2005, *20*, 719-725.

Zallen, R. *The Physics of Amorphous Solids;* Wiley: New York, NY, 1983; pp 49-59, 135-204.

Zhang, Z. M. *Nano / Microscale Heat Transfer;* McGraw-Hill: New York, NY, 2007; pp 1-479.

Zhou, Y.; Hirao, K.; Watari, K.; Yamauchi, Y.; Kanzaki, S. *J. Eur. Ceram. Soc.* 2004, *24*, 265-270.

Zimmermann, J. W.; Hilmas, G. E.; Fahrenholtz, W. G.; Dinwiddie, R. B.; Porter, W. D.; Wang, H. *J. Am. Ceram. Soc.* 2008, *91*, 1405-1411.

Živcová, Z.; Gregorová, E.; Pabst, W.; Smith, D. S.; Michot, A.; Poulier, C. *J. Eur. Ceram. Soc.* 2009, *29*, 347-353.

In: Advances in Materials Science Research, Volume 7 ISBN 978-1-61209-821-0
Editor: Maryann C. Wythers © 2012 Nova Science Publishers, Inc.

Chapter 2

RECENT ADVANCED IN TIN DIOXIDE MATERIALS: DEVELOPMENTS IN THIN FILMS, NANOWIRES AND NANORODS

Z. W. Chen, Z. Jiao, M. H. Wu, C. H. Shek, C. M. L. Wu, and J. K. L. Lai

Shanghai Applied Radiation Institute, Institute of Nanochemistry
and Nanobiology, School of Environmental and Chemical Engineering,
Shanghai University, Shanghai 200444, People's Republic of China,
and Department of Physics and Materials Science,
City University of Hong Kong, Tat Chee Avenue, Kowloon Tong, Hong Kong

1. INTRODUCTION

Tin oxide is a unique material of widespread technological applications, particularly in the field of gas sensors [1], dye-based solar cells [2], transparent conducting electrodes [3], and catalyst supports [4]. New assessment strategies for tin dioxide (SnO_2) functional materials are of fundamental importance in the development of micro/nano-devices [5]. However, the as-grown SnO_2 materials typically possess a high density of defects [6], which would degrade their properties. Therefore, the synthesis of defect-free SnO_2 materials is of great interest. In order to provide guidance for the search of better SnO_2 functional materials with suitable optical and electrical properties, it is necessary to investigate the temperature effects of SnO_2 thin film and nanostructured materials. It was found that the influence of annealing temperature on material properties is especially remarkable [7]. However, some challenges are still remained in the motivation to clarify the intricate aspects of SnO_2 thin film and nanostructured materials as well as the applications. This chapter summarizes the recent new research in our group concerning the microstructures and properties of SnO_2 materials, including thin films, nanowires and nanorods. The present work mainly focuses on the synthesis, characterization and applications of SnO_2 thin films and nanowires by using pulsed laser deposition techniques as well as the SnO_2 nanorods by using micro-emulsion method. It is an interdisciplinary work that integrates the areas of physics, chemistry and materials science. The results may enable novel SnO$_2$ functional materials with appropriate microstructures to be tailor made for a large number of applications and provide new opportunities for future study of SnO$_2$ architectures with the goal of optimizing functional material properties for specific applications.

2. TIN DIOXIDE THIN FILMS

2.1. Fractal Assessment Strategies of Tin Dioxide Thin Films

Semiconductor oxides are fundamental to the development of smart and functional materials, devices, and systems [8-11]. These oxides have two unique structural features: mixed cation valences and an adjustable oxygen deficiency, which are the bases for creating and tuning many novel material properties, from chemical to physical [12-15]. Due to the increasing importance of air pollution and the need to monitor concentration levels of gases such as CO, CO$_2$, NO$_x$, O$_3$, SO$_2$ etc., the development of many kinds of sensors and control systems has been jolted into action in recent years. Tin dioxide (SnO$_2$) has been used as a gas sensor material to detect combustible and toxic gases such as CO, NH$_3$, NO$_2$, H$_2$S and CH$_4$. Commercial sensors typically use sintered SnO$_2$ powders, but thin films SnO$_2$ are gaining increasing popularity [16, 17]. With the advent of advanced thin film technology more cost-effective, reproducible devices can be constructed with a reduction in device size and a concomitant increase in the speed of response by using SnO$_2$ thin films.

It is known that SnO$_2$ is used as a gas sensor because the number of electrons in the conduction band is affected by the adsorption of gaseous species on its surface [18]. Reducing gas molecules, such as CO, react with the oxygen species (O$_2^-$, O$^-$) on the semiconductor surface [19]. This lowers the height of the Schottky barrier and increases the conductance of the material [20, 21]. Gas sensors using SnO$_2$ are widely used due to its high sensitivity to humidity and inflammable gases. In this type of sensors, gas concentration is related to the material's electrical impedance due to the adsorption of gas molecules on the SnO$_2$ surface. In a pure air environment, SnO$_2$ adsorbs oxygen that captures its electrons, thereby raising its resistivity. When a reducing gas is present, it competes for the adsorbed oxygen and hence the SnO$_2$ resistivity decreases. However, the electrical properties of SnO$_2$ are strongly dependent on material fabrication parameters. Since gas sensing is based on adsorption mechanisms on the SnO$_2$ grain surface, for high sensitivity a small grain size is desirable in order to achieve a high specific area, i.e. adsorption area per unit volume [22, 23].

The surface conductance of semiconducting oxide is affected by the concentration of ambient gases. Resistive gas sensors are based on this principle and the nature of the sensing mechanism is related to the electrical response of gas sensors to reactive gases. The change in the sensor resistance provides an indication of the gas concentration [24]. These sensors can be quite versatile as they may be used to detect oxygen, flammable gases and common toxic gases. Their mechanism of operation is complex, involving interactions between gaseous molecules and defects on the surface and grain boundaries. The sign of a change in resistance depends on whether the solid has *n*-type or *p*-type conductivity [25, 26]. Structural properties such as grain size, grain geometry as well as specific surface area can significantly affect the gas sensing properties of semiconducting SnO$_2$. In order to control these structural characteristics, the microstructure evolution of SnO$_2$ thin films should be understood. Fractal method is a potentially powerful technique to characterize microstructures, and we are applying this technique to SnO$_2$ thin films for the first time in this study. Besides showing some examples of geometric structures of SnO$_2$ thin films, we shall discuss in detail the applicability and relevance of fractal theory to studying the microstructure and gas sensing behavior of SnO$_2$ based environmental functional materials.

An integrated device for different gas species is highly desirable for versatile advanced applications. Despite the high sensitivity of SnO$_2$ to many gases, it is often susceptible to electrical drift which requires long stabilization periods, as well as permanent poisoning after extended periods of operation. New fractal assessment strategies for this material are of fundamental importance in the development of micro-devices. In this section, we report on new insight on fractal assessment strategies on SnO$_2$ thin films prepared by the pulsed laser deposition (PLD) technique. We report new results on the experimental preparation of SnO$_2$ thin films at different substrate temperatures with interesting fractal features. The microstructure evolution of SnO$_2$ thin films has been investigated using X-ray diffraction and scanning electron microscopy and its structure has been evaluated by fractal methodology for the first time. The dependence of fractal dimensions on substrate temperature in the SnO$_2$ thin films has been characterized by fractal theory. Experimental evidence indicated that fractal clusters with various sizes, densities, and fractal dimensions formed in SnO$_2$ thin films prepared under different substrate temperatures. This formation of significant fractal features is rather unusual. It was found that these fractal structures were sensitively dependent upon the substrate temperature, which was a key parameter affecting the gas sensing behavior. Our findings may enable novel SnO$_2$ environmental functional materials with appropriate fractal structures to be tailor made for a large number of applications such as the monitoring of environmental harmful gases and provide new opportunities for future study of fractal structure SnO$_2$ architectures, with the goal of optimizing environmental functional material properties for specific applications.

In order to obtain the sintered SnO$_2$ target for pulsed laser deposition (PLD), we synthesized a pure nanocrystalline SnO$_2$ powder by the sol-gel method [27]. The fabrication method is described in the following. Meta-stannic acid sol (parent sol) was precipitated by treating a cold ethanol solution of SnCl$_4$ (27 %) with an aqueous ammonia solution (28 %) until a suitable pH value was reached. Dry powder with average grain size of about 4 nm was obtained by drying the parent sol, which had been washed repeatedly with de-ionized water. The SnO$_2$ discs, 15 mm in diameter and 4 mm in thickness, were prepared by compacting the powder under uniaxial pressure of 0.4 GPa, and sintered at 1150 °C for 2 h. The sintered disc consisted of high-purity cassiterite structure SnO$_2$ (99.8 %).

SnO$_2$ thin film was prepared by PLD techniques using the above sintered SnO$_2$ disc [10]. The target was cleaned with methanol in an ultrasonic cleaner before installation to minimize contamination. The laser was a KrF excimer laser (Lambda Physik, LEXtra 200, Germany) producing pulse energies of 350 mJ at a wavelength of 248 nm and a frequency of 10 Hz. The duration of every excimer laser pulse was 34 ns. The laser energy was transmitted onto the target in a high-vacuum chamber through an ultraviolet (UV)-grade fused silica window using an UV-grade fused silica lens. During the experiment, the target was kept rotating at a rate of 15 rpm to avoid drilling. The fluence was set at 5 J/cm^2 per pulse, corresponding to a total of approximately 1.5×10^5 laser pulses. The growth rate was estimated to be about 0.3 nm/s (or about 1 μm/h). The ablated substance was collected on a Si (100) substrate mounted on a substrate holder 4 cm away from the target. The high vacuum in the deposition chamber was achieved by using a cryopump (Edwards Coolstar 800). The base pressure prior to laser ablation was about 1×10^{-6} mbar, and the oxygen partial pressure during laser ablation was set about 3×10^{-2} Pa. All deposition processes were carried out by in-situ operation on the substrate at temperatures of 300 °C, 350 °C, 400 °C, and 450 °C.

X-ray diffraction (XRD) was performed with a Philips X'pert diffractometer using Cu K$_\alpha$ radiation (1.5406 Å) in reflection geometry. A proportional counter with an operating voltage of 40 kV and a current of 40 mA was used. XRD patterns were recorded at a scanning rate of $0.05°s^{-1}$ in the 2θ ranges from 20° to 60°. Scanning electron micrographs were obtained using a JEOL-JSM6335F scanning electron microscope (SEM). SEM images were digitized by using the Fractal Images Process Software (FIPS). These digitized images were divided into boxes of 360 × 360 size and then processed by the fractal theory [28]. Four intact fractal patterns were selected from these digitized images. The average value of the fractal dimensions (D), the fractal density and the average size of the fractal clusters for these digitized fractal patterns were obtained by using the box-counting method [29]. The carbon monoxide (CO) gas sensing property in the sensor was measured by a simple electrical measuring system. The test CO gas was introduced in the chamber by an injector with variable volume which facilitated control of gas concentration in the range 25-500 ppm. After the sensor was stabilized, the process was repeated by injecting a higher amount of the CO gas.

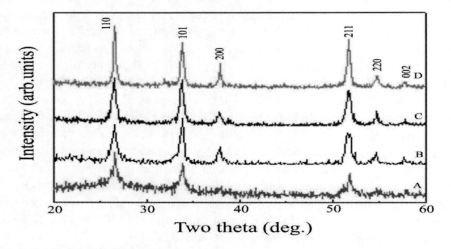

Figure 1. X-ray diffraction (XRD) patterns of SnO2 thin films prepared on Si (100) substrate at temperatures of (A) 300 oC; (B) 350 oC; (C) 400 oC; and (D) 450 oC.

It is known that SnO$_2$ has a tetragonal rutile crystalline structure (known in its mineral form as cassiterite) with point group D_{4h}^{14} and space group $P4_2/mnm$. The unit cell consists of two metal atoms and four oxygen atoms. Each metal atom is situated amidst six oxygen atoms which approximately form the corners of a regular octahedron. Oxygen atoms are surrounded by three tin atoms which approximate the corners of an equilateral triangle. The lattice parameters are a = 4.7382(4) Å, and c = 3.1871(1) Å. Figures 1A, B, C, and D show X-ray diffraction (XRD) patterns of the SnO_2 thin films prepared on Si (100) substrate at 300 °C, 350 °C, 400 °C, and 450 °C respectively. The major diffraction peaks of some lattice planes can be indexed to the tetragonal unit cell structure of SnO_2 with lattice constants $a = 4.738$ Å and $c = 3.187$ Å, which are consistent with the standard values for bulk SnO_2 (International Center for Diffraction Data (ICDD), PDF File No. 77-0447). The (hkl) peaks observed are (110), (101), (200), (211), (220), and (002). No characteristic peaks belonging to other tin oxide crystals or impurities were detected. The high intensity of these peaks suggests that these thin films mainly consist of the crystalline phase. As the substrate temperature increased, the crystallinity of the thin films was enhanced as manifested by the intensity and sharpness of the XRD peaks of the SnO_2 thin films. The substrate temperature dependence can be interpreted mainly by the mobility of the atoms in the thin films. At low substrate temperatures, the vapor species have a low surface mobility and are located at different positions on the surface. The low mobility of the species will prevent full crystallization of the thin films. However at high substrate temperatures the species with high enough mobility will arrange themselves at suitable positions in the crystalline cell [30-32]. The SnO_2 average grain sizes were calculated using the Scherrer formula: $D = K\lambda / \beta \cos\theta$, where D is the diameter of the nanoparticles, $K = 0.9$, λ (Cu K_α) = 1.5406 Å, and β is the full-width-at-half-maximum of the diffraction lines. The results show that the average grain sizes of the SnO_2 nanoparticles at different substrate temperatures are in the range of 25.3-27.8 nm. SnO_2 nanoparticle size increases from 25.3 nm at 300 °C to 26.2 nm at 350 °C. It then increases to 27.0 nm at 400 °C and finally to 27.8 nm at 450 °C. In fact, SnO_2 nanostructures can work as sensitive and selective chemical sensors. SnO_2 nanostructure sensor elements can be configured as resistors whose conductance can be modulated by charge transfer across the surface or as a barrier junction device whose properties can be controlled by applying a potential across the junction. Functionalizing the surface further offers a possibility to improve their sensing ability based on a better understanding of the influence of significant microstructural features, for example, the development of gas sensors for the detection of environmentally harmful gases.

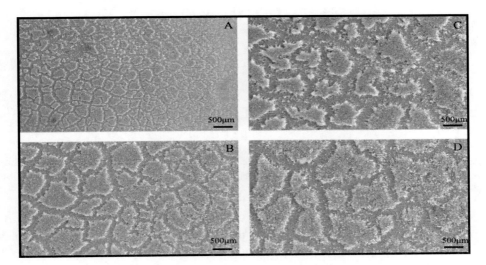

Figure 2. Scanning electron microscopy (SEM) images of SnO_2 thin films prepared on Si (100) substrate at temperatures of (A) 300 °C; (B) 350 °C; (C) 400 °C; and (D) 450 °C.

Figure 2 presents scanning electron microscopy (SEM) images of SnO_2 thin films prepared on Si (100) substrate at temperatures of (A) 300 °C, (B) 350 °C, (C) 400 °C, and (D) 450 °C respectively. The SEM observation indicated that all thin films produced under different substrate temperatures exhibited self-similar fractal patterns. It can be seen from Figure 2 that the fractal patterns are open and loose structure with increasing substrate temperature. The average sizes of the fractal patterns (or clusters) are about 0.307 μm (Figure 2A), 0.906 μm (Figure 2B), 1.202 μm (Figure 2C) and 1.608 μm (Figure 2D). The average sizes of the fractal clusters for four thin films were estimated by measurement on the fractal regions. The measuring procedure is as follows: for each SEM image, we chose ten fractal patterns at random to get an average value. The average sizes of the fractal patterns were obtained by averaging the values of SEM images with different orientations. It was found that the average sizes of the fractal clusters increase with increasing substrate temperature.

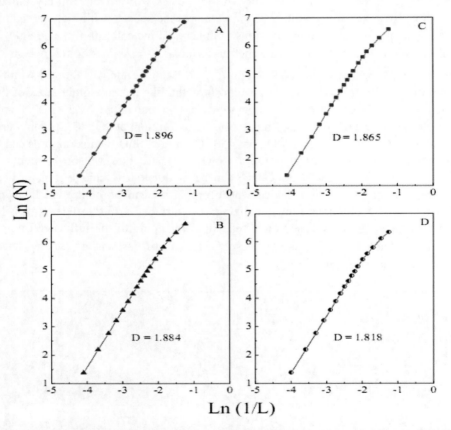

Figure 3. Plots of ln(N) versus ln(1/L) of the fractal cluster regions in Figure 2, where L is the box size and N is the number of boxes occupied by the SnO_2 crystalline structure for substrate temperatures at (A) 300 °C; (B) 350 °C; (C) 400 °C; and (D) 450 °C.

Figure 3 shows that the plots of ln(N) versus ln(1/L) of the fractal cluster regions in Figure 2, where L is the box size and N is the number of boxes occupied by the SnO_2 clusters. It can be seen that all plots show good linearity, which means that the morphologies of SnO_2 clusters have scale invariance within these ranges. So the SnO_2 clusters can be regarded as fractals. In order to obtain the fractal dimension (D), we fit a linear relationship for the function ln(N) versus ln(1/L). The results show that the fractal dimension (D) is 1.896 at 300 °C as shown in Figure 3A, 1.884 at 350 °C as shown in Figure 3B, 1.865 at 400 °C as shown in Figure 3C, and 1.818 at 450 °C as shown in Figure 3D. We found that the fractal dimension (D) decreases with increasing substrate temperature. The smaller fractal dimension means that the SnO$_2$ thin films are composed of the open and loose fractal structure with finer branches. Figures 4A to C show the distribution of the fractal average size, fractal dimension and fractal density for different substrate temperatures. It can be seen that there is an obvious increase in average fractal size (Figure 4A), and the fractal dimension generally decreases (Figure 4B) with increasing substrate temperature. In general, the fractal density is determined by the initial nucleation probability of the core crystal. From Figure 4C, the fractal density was calculated to be 18, 6, 3, and 2 mm^{-2} at 300 °C, 350 °C, 400 °C, and 450 °C respectively. It was found that the fractal density gradually decreases with increasing substrate temperature. In the present work, the initial increase in nucleation probability was due to strain relaxation caused by the low short-range temperature field at 300 °C, so that the fractal density and their occupation area were high. With the increase of substrate temperature, the higher long-range temperature field may promote new nuclei and subsequent growth, which leads to the fractal growth of the fine branches and a lower fractal density. This fractal structure may lead to improvement in the design of gas sensors for the monitoring of environmental pollutants.

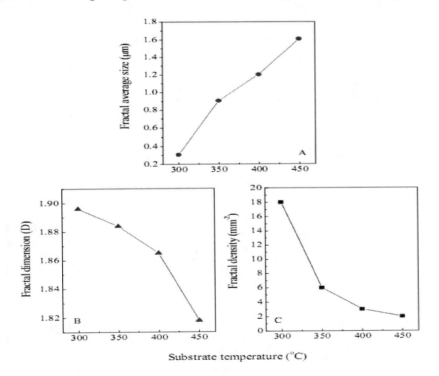

Figure 4. (A) The fractal average size; (B) the fractal dimension; (C) the fractal density versus the substrate temperature.

On the basis of our experimental observation, the formation process of SnO$_2$ nanocrystals and fractal clusters could be reasonably described by a novel model, and be separated into eight steps, which illustrated in detail in Figure 5.

(i). Operation of the KrF excimer laser at a repetition rate of 10 Hz at an incident angle of 45° to the polished sintered cassiterite SnO$_2$ target rotating at a rate of 15 rpm to avoid drilling.

(ii). Production of the high-temperature and high-pressure SnO$_2$ plasma at the solid-liquid interface quickly after the interaction between the pulsed laser and SnO$_2$ target.

(iii). Subsequent expansion of the high-temperature and high-pressure SnO$_2$ plasma leading to cooling of the SnO$_2$ [33-36]. In our case, the interval between two successive pulses is much longer than the life of the plasma. Therefore, the next laser pulse had no interaction with the former plasma.

(iv). Deposition of the SnO$_2$ plume on the Si (100) substrate after the disappearance of the plasma, inducing the initial nucleation of SnO$_2$ nanocrystals.

(v). Grain rotation culminating in a low-energy configuration. This process is directly related to the reduction of surface energy, aimed at minimizing the area of high-energy interfaces [37, 38].

(vi). Possible formation of a coherent boundary between grains due to grain rotation, with the consequence of removing the common grain boundary and culminating in a single larger SnO$_2$ nanocrystal. This is the coalescence process.

(vii). Growth of SnO$_2$ nanocrystals along preferred crystallographic directions which could be predicted by an analysis of the surface energy in several crystallographic orientations.

(viii). Formation of the fractal structure as SnO$_2$ crystallizes and nucleates at high energy interfaces such as grains boundaries.

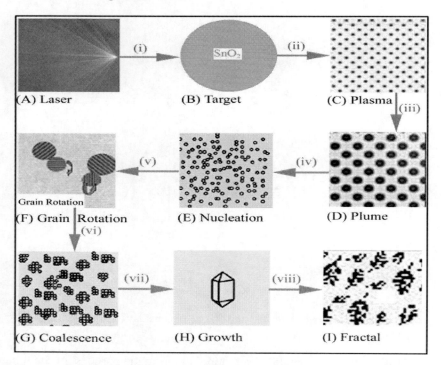

Figure 5. The formation process of SnO$_2$ nanocrystals and fractal clusters. (A) Laser; (B) Target; (C) Plasma; (D) Plume; (E) Nucleation; (F) Grain Rotation; (G) Coalescence; (H) Growth; and (I) Fractal.

According to the fractal theory [39, 40], the heat released by crystallization leads to a local temperature rise in the surrounding area and this temperature field can propagate quickly and stimulate new nuclei appearing randomly in nearby regions. The stimulated nuclei of the next generation can also cause a local temperature rise and repeat the above process many times until SnO$_2$ fractal patterns are formed. Based on the above proposed formation mechanism, we characterize the formation processes of SnO$_2$ nanocrystals and fractal structure in Figure 5A~I. We believe that laser ablation technique is an appropriate method to synthesize a series of environmental functional materials with controlled composition, morphology and nanocrystal size, which are of important in the study of the sensitivity of SnO$_2$ thin films.

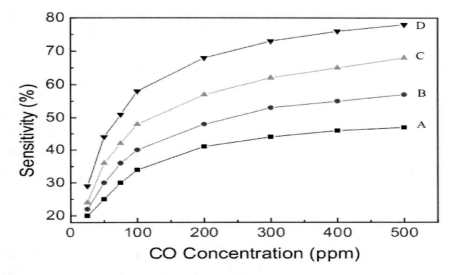

Figure 6. The CO gas sensing behavior of SnO$_2$ thin films prepared on Si (100) substrate at temperatures of (A) 300 °C; (B) 350 °C; (C) 400 °C; and (D) 450 °C.

To verify the gas sensing behavior of these SnO$_2$ thin films with interesting features of the fractal structure, we investigate the sensitivity dependence on carbon monoxide (CO) concentration, so as to achieve the aim of monitoring environmental pollutants. Figure 6 shows the CO gas sensing behavior of the SnO$_2$ thin films prepared on Si (100) substrate at (A) 300 °C, (B) 350 °C, (C) 400 °C, and (D) 450 °C respectively. The measurement was performed at room temperature with CO concentrations of 25, 50, 75, 100, 200, 300, 400, and 500 ppm. We observe that the sensitivity increases with increasing CO concentration and substrate temperature. Similarly Cooper and Cicera found their SnO$_2$ thin film sensor possessed higher sensitivity to CO by using different procedures [41, 42]. Further advancement of this gas sensor fabricated by the SnO$_2$ thin films with fractal structure to detect environmental harmful gases such as CO requires a clear understanding of its gas sensing mechanism. Our experimental results show that the CO gas sensing behavior clearly depends on the fractal dimension, fractal density, and average sizes of the fractal clusters (Figures. 4 and 6). We propose a Random Tunneling Junction Network (RTJN) mechanism to explain this gas sensing behavior. After the fractal formation, the fractal clusters consist of the SnO$_2$ grains with the morphology of fine dendrite-like nanocrystals incorporating many tunneling junctions of varying sizes. From the view of electron transport, the whole thin film is made up of a series of tunneling junctions. For the SnO$_2$ thin films deposited at different substrate temperatures, the sizes of the fractal branches with different fractal dimensions are different, leading to differences in the height of the Schottky barrier of the tunneling junctions, with the consequence that the breakdown voltages are also different. During the measurement of the gas sensitivity, the reducing gas molecules such as CO react with the oxygen species (O$_2^-$, O$^-$) ionized on the surface of the SnO$_2$ particles. This lowers the height of the Schottky barrier, and increases the conductance [19-21]. For example, for the SnO$_2$ thin film deposited at the lower substrate temperature (e.g. at 300 °C) with the larger fractal dimension, the junction i will have the higher resistance state due to the thicker fractal branches, so the external voltage V_i cannot lower the Schottky barrier S_i and the junction i cannot be broken. The gas sensitivity is then lowered (Figure 6A). Conversely, for the SnO$_2$ thin film deposited at the higher substrate temperature (e.g. at 450 °C) with the smaller fractal dimension, the junction i will have the lower resistance state due to the finer fractal branches, so the external voltage V_i can lower the Schottky barrier S_i and the junction i will be broken. Therefore, the gas sensitivity would be higher (Figure 6D). As mentioned above, there is a relationship between the fractal dimension and the size of the fractal branches in that the number of the fine branches increases with decreasing fractal dimension. Therefore, the smaller the fractal dimension, the larger the number of junctions with the smaller Schottky barrier S_i and lower resistance state. The present findings reveal new opportunities for future study of fractal structure SnO$_2$ architectures, with the goal of optimizing environmental functional material properties for specific applications.

2.2. Annealing Effects of Tin Dioxide Thin Films

Tin dioxide (SnO$_2$) is an important compound semiconducting material due to its interesting properties such as high electrical conductivity (8 × 10^{-4} Ωcm), high transparence over shorter wavelength range, high chemical stability, direct band gap [43-45], etc. SnO$_2$ is an n-type semiconductor (E$_g$ = 3.6 eV at 300 K) that is widely used under various forms in a broad range of important applications, ranging from solid-state gas sensors to liquid crystal displays, photovoltaic cells and transparent conducting electrodes [10, 46-48], etc. SnO$_2$ is a promising candidate for such applications due to its high sensitivity to various gases, high transparency in the visible wavelength range, high stability and low cost [49-52].

Since the properties of thin films strongly depend on its microstructure, composition, and crystal defects, the influence of annealing temperature on material properties is especially remarkable. In recent years, SnO$_2$ thin films have attracted much attention due to their high surface area [53-55]. There are many different techniques used for depositing SnO$_2$ thin films, for example, RF-sputtering [56], DC-magnetron sputtering [57], thermal evaporation [58, 59], ion beam deposition [60], chemical vapour deposition [61-65], spray pyrolysis [66], successive ionic layer deposition (SILD) [67], and other chemical methods [51, 68], etc. Sberveglieri has presented a review of the techniques applied for tin oxide thin films deposition [69]. As it is shown there, all methods discussed require high annealing temperature in order to fabricate good quality polycrystalline films. High temperature, however, damages the surface of the thin films and increases the interface thickness, which has negative effect on the optical properties especially on the waveguiding.

Pulsed laser deposition (PLD) is a growth technique in which photonic energy is coupled to the bulk starting material via electronic processes [70, 71]. An intense laser pulse passes through an optical window of a vacuum chamber and is focused onto a target. Above a certain power density, significant material removal occurs in the form of an ejected luminous plume. The threshold power density needed to produce such a plume depends on the target material, its morphology, and the laser pulse wavelength and duration, but might be of the order of 10-500 MW/cm^2 for ablation using ultraviolet (UV) excimer laser pulses of 10 ns duration. Material from the plume is then allowed to recondense on a substrate, where thin film growth occurs. The growth process may be supplemented by a passive or reactive gas or ion source, which may affect the ablation plume species in the gas phase or the surface reaction [72]. The diversity of thin films grown using PLD is enormous and perhaps recommends its flexibility more persuasively than anything else. The rationale for using PLD in preference to other deposition techniques lies primarily in its pulsed nature, the possibility of carrying out surface chemistry far from thermal equilibrium, and under favorable conditions, the ability to reproduce in thin films the same elemental ratios of even highly chemically complex bulk ablation targets [73]. The physical processes in PLD are highly complex and interrelated, and depend on the laser pulse parameters and the properties of the target material [74]. Laser ablation for thin film growth has many advantages: (i) the energy source (laser) is outside the vacuum chamber which, in contrast to vacuum-installed devices, provides a much greater degree of flexibility in materials use and geometrical arrangements; (ii) almost any condensed matter material can be ablated; (iii) the pulsed nature of PLD means that thin film growth rates may be controlled to any desired amount; (iv) the amount of evaporated source material is localized only to that area defined by the laser focus; (v) under optimal conditions, the ratios of the elemental components of the bulk and thin films are the same, even for chemically complex systems; (vi) the kinetic energies of the ablated species lie mainly in a range that promotes surface mobility while avoiding bulk displacements; (vii) the ability to produce species with electronic states far from chemical equilibrium opens up the potential to produce novel or metastable materials that would be unattainable under thermal conditions. PLD techniques were successfully applied for growing of quality tin oxide thin films. They were produced by ablation of either Sn target in oxidizing oxygen atmosphere [52] or SnO$_2$ target [10, 55, 75]. PLD offered many advantages of reduced contamination due to the use of laser light, control of the composition of deposited structure and in-situ doping. It is a versatile and powerful tool for production of nanoparticles with desired size and composition only by varying the experimental deposition conditions [70].

This section attempts to explore the changes in the microstructural and morphological properties, and root-mean-square (RMS) surface roughness of SnO$_2$ thin films as a result of annealing. The microstructural properties of SnO$_2$ thin films deposited on glass substrates by pulsed laser deposition techniques were investigated before and after annealing in the temperature range from 50 to 550 °C with a step of 50 °C, respectively. The influence of the annealing temperature on the microstructural and morphological properties of SnO$_2$ thin films was investigated using X-ray diffraction, scanning electron microscopy, transmission electron microscopy, and selected area electron diffraction. Experimental measurements showed that the amorphous microstructure almost transformed into a polycrystalline SnO$_2$ phase and preferred orientation related to the (110), (101) and (211) crystal planes with increasing annealing temperatures. We found that the thin film annealed at 200 °C demonstrated the best crystalline properties, viz. optimum growth conditions. However, the thin film annealed at

100 °C revealed the minimum average root-mean-square roughness of 20.6 nm with average grain size of 26.6 nm.

The target for pulsed laser deposition (PLD) was a sintered SnO_2 disc. The method employed to synthesize the pure SnO_2 powder is described by the direct oxidation reaction at 1050 °C, viz.: $Sn + O_2 \rightarrow SnO_2$, carried out in a horizontal quartz tube. The circular target was consisted of high-purity cassiterite SnO_2 (99.8 %). The size of the target was about $\phi 15$ mm × 4 mm, and it was cleaned with methanol in an ultrasonic cleaner before installation to minimize contamination. The laser used was a KrF excimer laser (Lambda Physik, LEXtra 200, Germany) producing pulse energies of about 350 mJ at a wavelength of 248 nm and a frequency of 10 Hz. The duration of every excimer laser pulse was 34 ns. The laser energy was transmitted onto the target in a high-vacuum chamber through an ultraviolet (UV)-grade fused silica window using an UV-grade fused silica lens. During the experiment, the target was rotating at a rate of 15 rpm to avoid drilling. The fluence was set at 5 J/cm^2 per pulse, corresponding to a total of approximately 1.5×10^5 laser pulses. The growth rate was estimated to be about 3×10^{-1} nm/s (or about 1 μm/h). The ablated substance was collected on a clean glass substrate, which was mounted on a substrate holder 2.5 cm away from the target. The high vacuum in the deposition chamber was achieved by using a cryopump (Edwards Coolstar 800). The base pressure prior to laser ablation was about 1×10^{-6} mbar, and the working pressure during laser ablation was about 2×10^{-6} mbar.

The as-prepared thin films were annealed at various temperatures ranging from 50 to 550 °C with a step of 50 °C for a fixed time of 30 min at each temperature. The annealing vacuum was about 2×10^{-3} Pa. Structure of these films was determined by recording X-ray diffraction (XRD) patterns at room temperature using Philips X'pert diffractometer equipped with Cu K_α radiation (1.5406 Å) in reflection geometry. Proportional counter with an operating voltage of 40 kV and a current of 40 mA was used. XRD patterns were recorded at a scanning rate of 0.05 $°s^{-1}$ in the 2θ ranges from 20° to 60°. Scanning electron micrographs were obtained using a JEOL-JSM6335F scanning electron microscope (SEM). Transmission electron micrographs were obtained using a Philips CM20 transmission electron microscope (TEM) at an acceleration voltage of 200 kV. The mean RMS surface roughness values for various films taken at different sites of each film were obtained at atmospheric pressure and room temperature by atomic force microscope (AFM). AFM measurements were made by a Digital Instruments Nanoscope Scanning Probe Microscope in tapping mode using window size of 6 × 6 $μm^2$ with a resolution of 256 × 256 pixels. The measuring procedure is as follows: for each AFM imaging, we chose five sites at random to get an average value. The average sizes of RMS were obtained by averaging the values of RMS with different sites. Because the systematic error was affected by many factors, e.g., scan size, pixel resolution, tip-sample force, and tip radius etc. In our experiments, the systematic error for RMS is about ± 0.2 nm [76].

The study of microstructural properties of SnO_2 thin films is significant for the understanding of whole structure features and the fabrication of novel functional materials with favorable properties. The lattice parameters of tetragonal phase SnO_2 rutile structure are $a = 4.7382(4)$ Å, and $c = 3.1871(1)$ Å [77]. The thin films deposited on clean glass substrates were physically stable and had very good adhesion to the substrates. Figure 7 shows the XRD patterns of as-prepared and annealed SnO_2 thin films at different temperature for 30 min annealing. The XRD patterns showed that the thin film deposited at room temperature is dominantly amorphous as indicated by the broad diffraction peaks (Figure 7a). When the annealing temperature reached 100 °C, the amorphous phase almost disappeared. With annealing temperature further increased, the thin films became polycrystalline (Figure 7b).

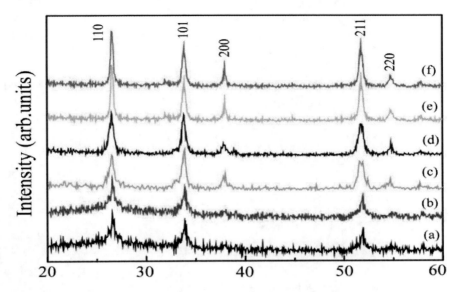

Figure 7. XRD patterns of as-prepared and annealed SnO$_2$ thin films at different annealing temperatures for 30 min. (a) as-prepared thin film; (b) 100 °C; (c) 200 °C; (d) 300 °C; (e) 400 °C; and (f) 500 °C.

It can be seen that all the thin films are polycrystalline SnO$_2$ tetragonal structure. No characteristic peak such as other tin oxides crystals or impurities is detected. The plane orientation was observed as (110), (101), (200), (211), and (220). On the other hand, the high intensity of peaks suggested that these thin films consist of mostly the crystalline phase. As annealing temperature increased, the crystallinity of the thin films improved as confirmed by the intensity and sharpness of the XRD peaks of the SnO$_2$ thin films annealed at 200, 300, 400, and 500 °C shown in Figures. 7c, d, e, and f, respectively. The temperature dependence can be interpreted mainly by the mobility of the atoms in thin films at different temperatures. At low temperature, the vapour species have a low surface mobility and will be located at different position on the surface. The low mobility of the species will prevent full crystallization of the thin films. However at high temperature such species with high enough mobility will arrange them at a suitable position in crystalline cell [30-32, 78]. The SnO$_2$ average grain sizes were calculated using the Scherrer formula: $D = K\lambda / \beta \cos\theta$, respectively, where D is the diameter of the nanoparticles, $K = 0.9$, λ (Cu K_α) = 1.5406 Å, and β is the full-width-half-maximum of the diffraction lines. Because each sample was tested for once, thus the inherent uncertainty is inevitable by XRD line broadening. The systematic errors can be determined by using Philips X'pert diffractometer. When the particle size is less than 100 nm, the systematic error for particle size is about ± 2 nm (~ 2 %). In order to further confirm the particle sizes, TEM observations were used. The average grain sizes for four thin films were estimated by measurement on TEM bright-field images shown as Figure9. The measuring procedure is as follows: for each TEM image, we chose twenty particles at random to get an average value. The average grain sizes for four thin films were obtained by averaging the values of TEM images with different orientations. Both methods indicate that TEM analysis is in agreement with those observed from XRD data. The results showed that the average grain sizes of the as-prepared and the annealing SnO$_2$ nanoparticles are in the range of 23.7-28.9 nm shown in Table 1. From Table 1, SnO$_2$ nanoparticle size first decreases from 27.6 nm (at room temperature) to smaller than 23.7 nm at 200 °C and then sharply increases to larger value of 27.1 nm at 300 °C and then gradually decreases to 25.7 nm at 400 °C, finally increases up to 28.9 nm for further annealing at higher temperatures.

Table 1. The average particle size versus annealing temperature for as-prepared (RT) and annealed SnO$_2$ thin films

Annealed temperature (°C)	RT	50	100	150	200	250
Average particle size (nm)	27.6	27.4	26.6	24.6	23.7	25.4
Annealed temperature (°C)	300	350	400	450	500	550
Average particle size (nm)	27.1	26.6	25.7	26.15	27.7	28.9

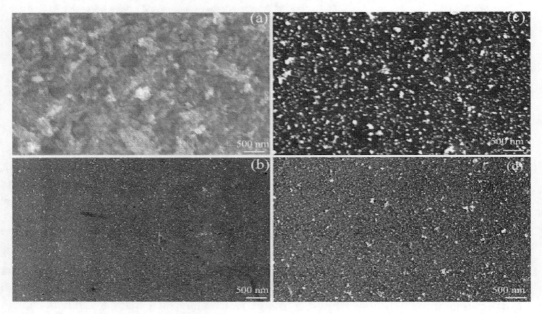

Figure 8. The typical SEM images of the as-prepared and annealed SnO$_2$ thin films. (a) as-prepared thin film; (b) 100 °C; (c) 200 °C; and (d) 300 °C.

Figure 8 shows the typical SEM images of the as-prepared and annealed SnO$_2$ thin films. At room temperature (RT), the formation of inhomogeneous SnO$_2$ particles can be seen from the image as shown in Figure 8a. Inspecting this image with more details, we can identify the presence of SnO$_2$ particles of nanometer size. However, bigger particles of the order of few microns formed by agglomeration of smaller particles are observed too. The surface of the SnO$_2$ thin film deposited at room temperature exhibits a rough surface. With increasing annealing temperature, the surface roughness of the thin films decreases. For the thin film annealed at 100 °C, the surface becomes relatively smooth and dense as shown in Figure 8b. When the SnO$_2$ thin film annealed at 200 °C, the SEM image indicated that the surface roughness increases as shown in Figure 8c, and decreases then in 300 °C annealing as shown in Figure 8d.

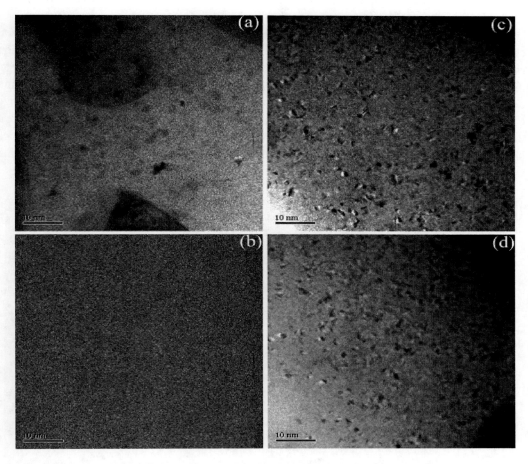

Figure 9. The TEM bright-field images of the as-prepared and annealed SnO$_2$ thin films. (a) as-prepared thin film; (b) 100 °C; (c) 200 °C; and (d) 300 °C.

The morphologies of the as-prepared and annealed SnO$_2$ thin films were evaluated in detailed using TEM as shown in Figure 9. Figure 9a shows the TEM bright-field image of the as-prepared thin film. We can see that the bigger particles are dispersed randomly in amorphous matrix due to the agglomeration of smaller particles. The size distribution width of particles was in the range of 12~36 nm according to the above sampling methodology (average in 27.6 nm). The selected area electron diffraction (SAED) patterns of the as-prepared thin film revealed that this observation may be simply indicative of a lower degree of crystallinity of the as-prepared thin films as they contain an amorphous phase. As seen in Figure 10a, we found that the amorphous matrix contains randomly in the thin film since the diffraction ring is a diffuse ring. Figure 9b displays the TEM bright-field image of the thin film annealed at 100 °C. We can see that the thin film is very smooth and dense. The size distribution width of particles was in the range of 24~29 nm (average in 26.6 nm). The SAED patterns as shown in Figure 10b demonstrated that the crystallinity of the thin films is improved since the diffuse diffraction ring weakens and the polycrystalline diffraction rings become bright. Figure 9c presents the TEM bright-field image of the thin film annealed at 200 °C. It was found that the particle sizes increased, and the smoothness and denseness decreased. The size distribution width of particles was in the range of 17~26 nm (average in 23.7 nm).

The SAED patterns as shown in Figure 10c testified that the microstructural characteristics of the typical tetragonal SnO$_2$ thin films relate to the polycrystalline diffraction rings d_{200} = 2.37Å, d_{210} = 2.12Å, d_{310} = 1.50Å, and d_{202} = 1.32 Å. Figure 9d exhibits the TEM bright-field image of the thin film annealed at 300 °C. We found that the smoothness and denseness of the thin film decrease. The size distribution width of particles was in the range of 25~32 nm (average in 27.1 nm). The SAED patterns as shown in Figure 10d proved that the crystallinity of the thin films is further improved since the polycrystalline diffraction rings become sharpness and bright. Above experimental results indicate that the annealing temperature is very important parameter to determining the thin film quality.

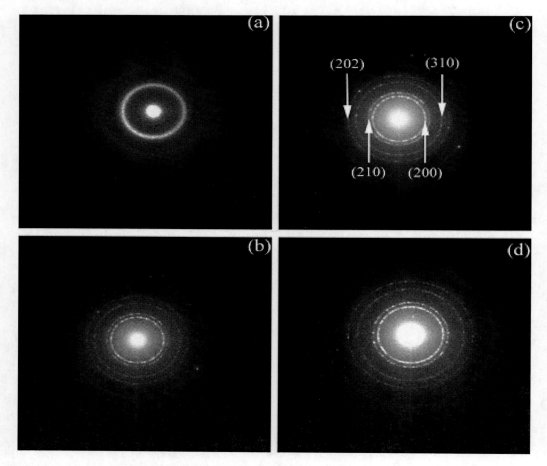

Figure 10. The SAED patterns of the as-prepared and annealed SnO$_2$ thin films. (a) as-prepared thin film; (b) 100 °C; (c) 200 °C; and (d) 300 °C.

Semiconducting oxides are a class of materials that are vitally important for developing new materials with functionality and smartness. The unique properties of these materials are related to the microstructural and morphological properties of the thin films. The micro-roughness of the thin films plays a vital role for developing optical coatings especially in the ultraviolet (UV) region for applications such as lithographic uses [79, 80]. To characterize an optical surface, the root-mean-square (RMS) surface roughness is normally used. The RMS roughness not only describes the light scattering but also gives an idea about the quality of the surface under investigation. Thus annealing process is a good parameter that could affect the RMS surface roughness behavior. The plot of RMS surface roughness values computed for as-prepared and annealed SnO_2 thin films is portrayed in Figure 11 as a function of annealing temperature in comparison with the microstructures and morphologies. It can be seen that the surface properties depend strongly on the annealing temperature, which correlates with the result of the microstructure analysis. At room temperature, the surface is characterized by the presence of very large particles aggregated in amorphous shape. The surface roughness is very high with average RMS value of about 25.6 nm. The average RMS surface roughness value for the deposited thin films decreased sharply with increasing annealing temperature. At 100 °C, the grains obviously started to form as confirmed by the diffuse diffraction ring weakened and the polycrystalline diffraction rings formed, which indicated that the crystallisation of the thin film is occurred. When the annealing temperature increased up to 200 °C, the smaller grains can be observed in the thin films. The RMS surface roughness trends to increase may be because of the larger grains size or the strain between the thin films and the substrates. Moreover, with increasing annealing temperature, a smoother surface might be expected due to an improvement of surface mobility of the species. At the higher annealing temperature of 300 °C, the grains grow with preferred orientation are formed obviously. While the RMS surface roughness of the thin film decreases, the surface smoothness increases and defects and strain variations may be produced. Lindström and co-authors also presented by power spectral density (PSD) function calculations in sputtered SnO_2 thin films that as the thin film surface smoothens, it gives rise to defects and strain variations for thicker films (> 300 nm) [81].

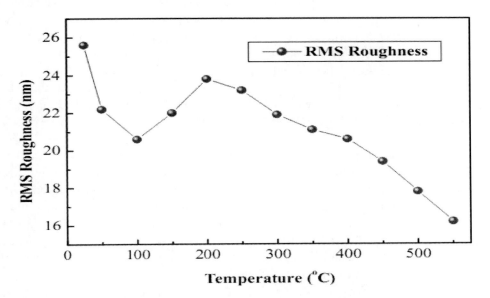

Figure 11. The plot of RMS roughness versus annealing temperature for the as-prepared and annealed SnO_2 thin films.

2.3. Defect Evolution of Tin Dioxide Thin Films

As a key functional material, nanocrystalline SnO$_2$ has been reported to show some unique characteristics different from the bulk crystals, and much attention has been focused on the synthesis of SnO$_2$ nanowires [12], nanotubes [12], nanorods [82, 83], nanobelts [9, 84, 85], and exploration of their novel properties. Nanocrystalline SnO$_2$ thin films have also aroused much attention since the higher-quality synthesis of SnO$_2$ thin films was achieved. This achievement is mainly due to the recognition of the bulk-quantity growth mechanism of SnO$_2$ thin films. The success in higher-quality synthesis of SnO$_2$ thin films has meant that the research is not limited to theoretical area, but it can be extended to the experimental area. A variety of methods, such as sol-gel [86, 87], chemical vapor deposition [88, 89], magnetron sputtering [90], sonochemical [91], and thermal evaporation [92], have been employed to prepare SnO$_2$ thin films or nanoparticles. To date, numerous experimental results on SnO$_2$ thin films have been reported [53, 54, 75, 93], including those from X-ray diffraction (XRD), transmission electron microscopy (TEM), electronical transport, Raman spectroscopy. These results have helped to speed up the study of its potential applications, since SnO$_2$ thin film is one of the promising materials for future transparent nanoelectrodes and solid-state gas nanosensors. However, the as-grown SnO$_2$ thin films typically possess a high density of defects, which would degrade their properties [94, 95]. Therefore, the synthesis of defect-free SnO$_2$ thin films is of great interest.

The diversity of thin films grown using pulsed laser deposition (PLD) is enormous and perhaps recommends its flexibility more persuasively than anything else. The rationale for using PLD in preference to other deposition techniques lies primarily in its pulsed nature, the possibility of carrying out surface chemistry far from thermal equilibrium, and under favorable conditions, the ability to reproduce in thin films the same elemental ratios of even highly chemically complex bulk ablation targets [73]. Recently, we have obtained nanocrystalline SnO$_2$ thin films by using a PLD approach [10, 31, 55, 78, 96]. However, these SnO$_2$ thin films contained many defects and amorphous components, since the presence of such an amorphous component and defect in the thin films greatly limits their practical use as gas-sensing devices. These defects will degrade the properties of the SnO$_2$ thin films. As we reported recently, bulk-quantities of higher purity SnO$_2$ thin films can be achieved with PLD growth [11, 19]. Therefore, the properties measured from this type of pure SnO$_2$ thin films are less affected by impurities. However, a considerable number of defects still remain inside the SnO$_2$ thin films, as revealed by high-resolution transmission electron microscopy (HRTEM) [97].

In this section, we report in detail that the microstructural defects of nanocrystalline SnO$_2$ thin films, prepared by pulsed laser deposition, have been investigated using transmission electron microscopy, high-resolution transmission electron microscopy and Raman spectroscopy. Experimental results indicate the defects inside nanocrystalline SnO$_2$ thin films could be significantly reduced by in-situ annealing SnO$_2$ thin films at 300 °C for 2 h. High-resolution transmission electron microscopy showed that the stacking faults and twins were annihilated upon in-situ annealing. In particular, the inside of the SnO$_2$ nanoparticles demonstrated perfect lattices free of defects after in-situ annealing. Raman spectra also confirmed that the in-situ annealed specimen was almost defect-free. By using in-situ annealing, the defect-free nanocrystalline SnO$_2$ thin films can be prepared in a simple and practical way, which holds promise for applications as transparent electrodes and solid-state gas sensors.

PLD experimental conditions are similar to the section 2.2. After PLD, the samples were cut into small circular wafers with a radius of 1.5 mm by the ultrasonic disk cutter. Then the wafers were polished to about 30 μm thickness with the disk grinder. The dimple grinder was used to thin the wafer thickness less than 5 μm. Finally, the wafers were milled by the ion polishing system. During the processes mentioned above only the sides of the glass substrates were polished, dimpled, and milled, thus SnO_2 thin films remained. The TEM specimen was in-situ annealed at 300 °C for 2 h in high-resolution transmission electron microscope equipped with a heating stage. Raman scattering measurements were obtained by backscattering geometry with a SPEX-1403 laser Raman spectrometer. The excitation source was an argon-ion laser operated at a wavelength of 514.5 nm in the backscattering configuration and a low incident power to avoid thermal effects. A Philips CM 20 transmission electron microscope operating at an acceleration voltage of 200 kV was used to determine the grain size distribution of the thin films. HRTEM images of nanocrystalline SnO_2 were obtained with a JEOL-2010 high-resolution transmission electron microscope with a point-to-point resolution 1.94 Å operating at 200 kV. The lattice images and the agglomerate states of the nanocrystalline SnO_2 were further carefully analyzed.

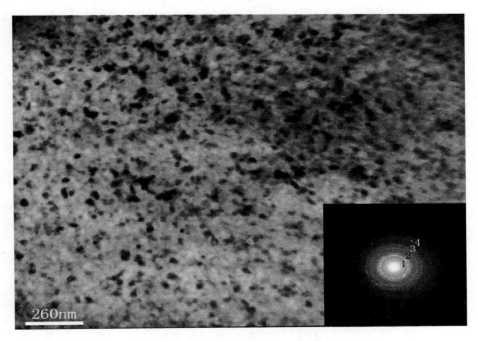

Figure 12. TEM morphology of as-grown SnO_2 thin films. The inset is a SAED pattern.

A typical TEM bright-field image and the corresponding selected area electron diffraction (SAED) pattern (inset) of the morphology of the as-grown nanocrystalline SnO_2 thin films is shown in Figure 12. As seen in the TEM bright-field image, there are many small particles of roughly spherical shape. The contrast of the particles in different regions of the TEM image indicates different density, which may be related to the grain sizes. The polycrystalline diffraction rings of the SAED pattern (inset in Figure 12) also demonstrate the microstructural characteristics of the typical tetragonal SnO$_2$ thin films (ring 1: d$_{200}$ = 2.37 Å, ring 2: d$_{210}$ = 2.12 Å, ring 3: d$_{310}$ = 1.50 Å, and ring 4: d$_{202}$ = 1.32 Å). The crystallites close to Bragg orientations were recognizable by their dark contrast. Analysis using energy dispersive X-ray spectroscopy (EDS) attached to the TEM confirmed that the thin film is composed of 38.3 at.% tin and 61.7 at.% oxygen, which Sn:O = 1:1.611 is departure to that of bulk SnO$_2$ (Sn:O = 1:2). The results indicate that the oxygen vacancies and nonstoichiometric SnO$_x$ (x < 2) relating to the as-grown nanocrystalline SnO$_2$ thin films in the surface layer cause a large number of oxygen defects in the surface region.

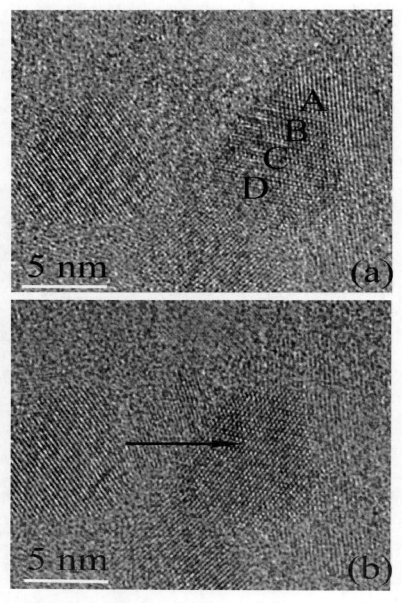

Figure 13. Typical HRTEM images of SnO$_2$ thin films: (a) an as-grown SnO$_2$ thin film with defects and (b) in-situ annealed SnO$_2$ thin film without defects.

However, the density of defects inside the in-situ annealed thin films decreased significantly. Figure 13a shows the HRTEM image of a typical as-grown thin film. The contrast of the SnO$_2$ thin film revealed a complicated feature, indicating the presence of many defects. The stacking faults (marked *A*, *B*, *C* and *D*) in the nanocrystalline SnO$_2$ thin films can be clearly observed. After in-situ annealing, the defect density in the nanocrystalline SnO$_2$ thin films was much lower, as shown in Figure 13b. The lattice fringes of the thin films (indicated by the arrow) can be seen clearly. It should be pointed out that the investigation was carried out on a statistical sampling of nanocrystalline SnO$_2$ thin films, so that the HRTEM images shown in Figures. 13 and 14 (below) of the in-situ annealed nanocrystalline SnO$_2$ thin films are representative morphologies.

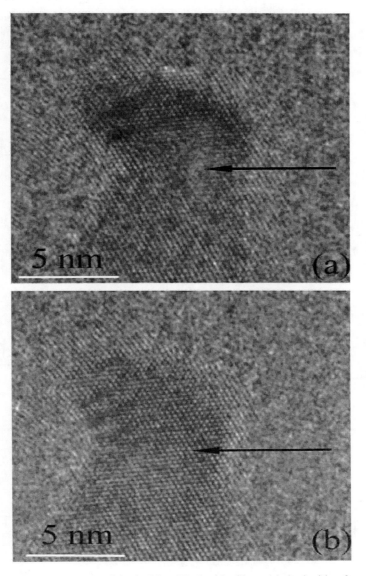

Figure 14. Typical HRTEM images of the inside of SnO$_2$ thin films: (a) the inside of an as-grown SnO$_2$ thin film with defects and (b) the inside of in-situ annealed SnO$_2$ thin film with a perfect lattice.

Defects in the inside of SnO$_2$ nanoparticles were also annihilated. The inside of an as-grown nanoparticle is shown in the HRTEM image in Figure 14a. The inside of the as-grown nanoparticle is generally cotton-like in shape and covered by a relatively thick amorphous layer (indicated by the arrow). The contrast of the amorphous layer was quite uniform and only the SnO$_2$ crystalline structure was observed. Similar to the microstructure in the body of the SnO$_2$ thin films, the SnO$_2$ crystal core in the inside also show a high density of defects. In our observation, this is a common phenomenon in the inside. The presence of these defects in the inside areas is considered to be responsible for the fast growth of SnO$_2$ thin films, since it is well known that dislocations can accelerate crystal growth. In other words, the defects in the inside of SnO$_2$ nanoparticles are necessary for the growth of SnO$_2$ thin films. However, the inside of the in-situ annealed SnO$_2$ nanoparticles demonstrate perfect lattices because the defects in the inside have been annihilated, as shown in Figure 14b (indicated by the arrow). Since the phenomenon of gas sensitivity is related to electronical transport, the geometry and nature of the SnO$_2$ particles is especially important.

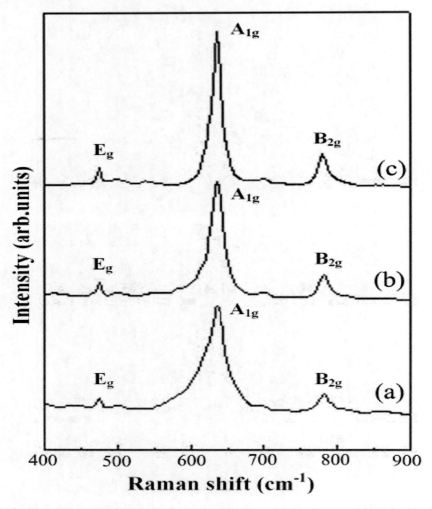

Figure 15. The Raman spectra of (a) as-grown SnO$_2$ thin films; (b) in-situ annealed SnO$_2$ thin films, and (c) the commercial SnO$_2$ bulks.

Defects in the bulk samples were further studied with Raman spectroscopy. Figure 15 shows the Raman spectra of the as-grown SnO₂ thin films (Figure 15a), the in-situ annealed SnO₂ thin films (Figure 15b) and the reference the commercial SnO₂ bulks (Figure 15c). The as-grown SnO₂ thin film shows a Raman peak with high asymmetry while the in-situ annealed SnO₂ thin film gives a more symmetric peak. The asymmetry of the Raman peak is contributed to by two factors: nanoscale size and defects [98, 99]. In the present experiment, the size of the nanoparticle does not change upon in-situ annealing. Moreover, the shifts of spectra (a) and (b) due to the size effect are almost the same. Thus, we conclude that the decrease in peak asymmetry is attributed to the decreasing density of defects. It should be noticed that the signal-to-noise ratio in spectrum (b) is much less than that in spectrum (c). As a result, the small ratio may have increased the asymmetry in spectrum (b). Therefore, the defects in the SnO₂ thin films (Figure 15b) are much less than what appears in the spectrum, and may even be comparable to the commercial SnO₂ bulks (Figure 15c).

2. 4. Optical Properties of Tin Dioxide Thin Films

Tin dioxide (SnO_2), an *n*-type semiconductor with a wide band gap of 3.6 eV, is one of the most important strategic materials and is used for very diverse technological applications such as optical and nanoelectronics [44, 45, 100], gas sensing [101], energy storage and conversion [102, 103], catalyst supports etc [104]. In recent years, the use of nanocrystalline particles with a high surface-to-volume ratio made it possible to improve the properties of devices such as gas sensing and optoelectronic devices, which are mainly influenced by the surface region and grain size of the particles. From an application point of view, it is important to obtain nanoparticles with specific size and microstructure. Thus, control of size, shape, and microstructure of the nanoparticles are amongst the most important issues in the synthesis of nanomaterials. SnO_2 is widely used under various forms in a broad range of important applications, ranging from solid-state gas sensors to liquid crystal displays, transparent conducting electrodes, photovoltaic cells etc [10, 46, 48]. SnO_2 gas-sensing material is a promising candidate for such applications due to its high sensitivity to various gases, high transparency in the visible wavelength range, high stability and low cost [49-51]. There is, in particular, an increasing interest in the fabrication of nanocrystalline thin films because the properties of thin films strongly depend on its microstructure, composition, crystal defects, and post-processing.

Recently, SnO_2 thin films have attracted much attention due to their unique microstructures and potential technological applications [53, 54]. SnO_2 thin film is one of the most extensively studied members of metal oxide family due to wide ranging applications such as gas sensors, transparent conductive oxides, catalysts, and far-infrared dichromatic mirrors. The immense technological importance keeps SnO_2 thin films under active focus of the research community, and new techniques to modify the properties and behaviour are being continually investigated. There are many different techniques used for depositing SnO_2 thin films, for example, RF-sputtering [56], DC-magnetron sputtering [57], thermal evaporation [58, 59], ion beam deposition [60], chemical vapour deposition [62-65], spray pyrolysis [66], successive ionic layer deposition (SILD) [67], and other chemical methods [51, 68]. Sberveglieri has presented a review of the techniques used in tin oxide thin films deposition [69]. As reported there, all available methods require high substrate temperature or post deposition annealing for the fabrication of good quality polycrystalline films. The influence of annealing temperature on material properties is especially remarkable. High temperature, however, damages the surface of thin films and increases the interface thickness, which has negative effects on optical properties, particularly on waveguide properties.

PLD offers many advantages of reduced contamination due to the use of laser light, control of the composition of deposited structure and in-situ doping. It is a versatile and powerful tool for the production of nanoparticles with desired size and composition under appropriate control of the experimental deposition conditions [70]. PLD techniques have been successfully applied for growing quality tin oxide thin films. They were produced by ablation of either a Sn target in an oxidizing oxygen atmosphere or a SnO_2 target [10, 75]. In order to provide guidance for the search of better sensor materials with suitable optical properties, it is necessary to investigate the temperature effects of the semiconductor oxide gas sensors. In this section, the microstructural properties of SnO_2 thin films deposited on glass substrates by pulsed laser deposition techniques were investigated before and after thermal annealing in the temperature range from 50 °C to 550 °C at 50 °C intervals. X-ray diffraction results confirmed that the various SnO_2 thin films consisted of nanoparticles with average grain size in the range of 23.7-28.9 nm as shown in Figure 7. The transmittance measurements indicated that present SnO_2 thin films can be used as a window material for solar cell application due to their > 75.3% transmittance. Various optical parameters such as optical band gas energy, refractive index and optical conductivity were calculated from the optical transmittance and reflectance data recorded in the wavelength range 300-2500 nm. We found that the optical band gas energy and refractive index of SnO_2 thin films annealed at various temperatures presented approximately linear behaviour, and SnO_2 thin films exhibited high optical conductivity. These findings revealed that present SnO_2 thin films annealed up to 400 °C are a good window material for solar cell application.

The information of optical transmittance is important in evaluating the optical performance of semiconducting oxide thin films. Figure 16 shows the optical transmittance spectra for (a) the as-prepared thin film, and those annealed for 30 min at the following temperatures: (b) 100 °C, (c) 200 °C, (d) 300 °C, (e) 400 °C, and (f) 500 °C. It can be seen that the transmittance behaviour of all thin films was similar in the wavelength range 300-2500 nm. However, the values of transmittance for the as-prepared thin film and the thin films annealed up to 400 °C are larger (>75.3%) than that for the thin film annealed at 500 °C (< 48.7%) in the wavelength range 300-750 nm. This wavelength range almost covers the entire solar spectrum. Thus, the thin films annealed up to 400 °C can be adopted as a window material for solar cell devices due to its high transmittance. The large and slow decrease of transmittance in the infrared (IR) region can be attributed to free carrier absorption. Similarly optical characterization also was found by Bhatti and co-authors in their SnO_2 thin films by using different method: rf-magnetron sputtered technique [38]. Our experimental results indicated that SnO_2 thin films with the high optical quality could be synthesized by PLD techniques. Figure 17 shows the optical reflectance spectra of (a) the as-prepared SnO_2 thin film and those annealed for 30 min at the following temperatures: (b) 100 °C, (c) 200 °C, (d) 300 °C, (e) 400 °C, and (f) 500 °C. The reflectance of the as-prepared and annealed thin films shows very similar behaviour in the entire wavelength range. When the thin film was annealed at 500 °C, the transmittance of the thin film decreased. We speculate that any variation in transmittance and reflectance caused by the annealing process might be related to the rearrangement of atoms and/or grain growth etc. to remove residual stresses/defects formed during thin film deposition [105].

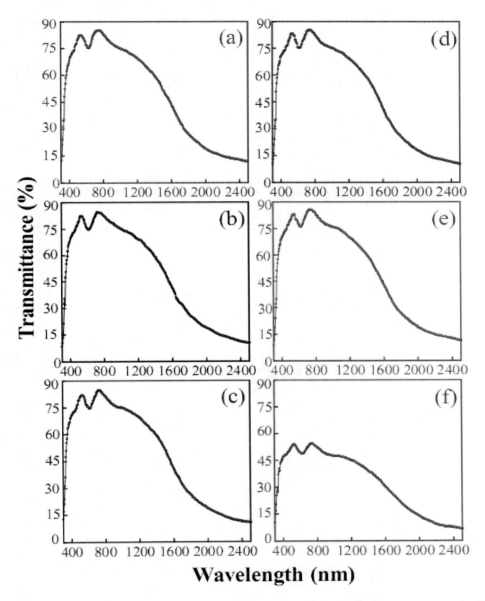

Figure 16. Optical transmittance spectra of SnO$_2$ thin films annealed at different temperatures for 30 min. (a) as-prepared thin film; (b) 100 °C; (c) 200 °C; (d) 300 °C; (e) 400 °C; (f) 500 °C.

To calculate the band gap energy (E_g) of all the thin films, the following relation was applied: $(\alpha h\nu)^2 = B^2(h\nu - E_g)$, in which the absorption coefficient α was determined from the transmittance data using the formula: $T = e^{-\alpha L}$, where B is a constant, T is the transmittance, and L is thickness of the thin films. The graphs of $(\alpha h\nu)^2$ versus photon energy ($h\nu$) are drawn for the as-prepared and annealed thin films.

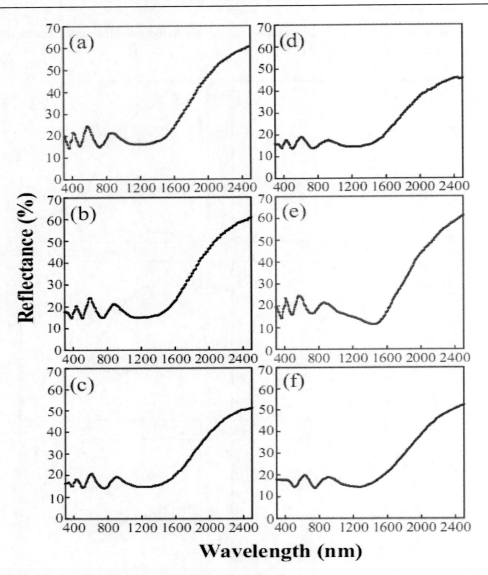

Figure 17. Optical reflectance spectra of SnO$_2$ thin films annealed at different temperatures for 30 min. (a) as-prepared thin film; (b) 100 °C; (c) 200 °C; (d) 300 °C; (e) 400 °C; (f) 500 °C.

The values of E_g obtained from the extrapolation of the linear portion of these curves to $(\alpha h\nu)^2 = 0$ are plotted in Figure 18a as a function of annealing temperature. Figure 18a shows a decrease in band gap energy as the annealing temperature is increased. The annealing may lead to a direct renormalization of band gap energy due to the temperature dependence of electron phonon interactions [106]. The optical band gap of the nanocrystalline particle depends on the particle radius due to quantum confinement. The approach to quantitative determination of the size dependence of the band gap energy is based on the effective mass approximation. The increase in optical band gap (ΔE_g) of a nanocrystalline semiconductor may be represented as: $\Delta E_g = E_g^{nano} - E_g^{bulk} = \dfrac{h^2}{8\mu R^2} - \dfrac{1.8e^2}{\varepsilon R}$, where E_g^{nano} is the band gap energy of the nanocrystalline material, E_g^{bulk} is the band gap energy of the material in bulk form, μ is the electron-hole effective mass, and ε is the static dielectric constant. The first term in this formula represents the particle-in-a-box quantum localization energy and has simple R^{-2} dependence, where R is the particle radius. The second term represents the Coulomb energy with R^{-1} dependence. The radii of the SnO$_2$ nanocrystals calculated from this formula match reasonably well with the previously calculated values [107]. Therefore, it can be assumed that during the annealing process atoms can rearrange themselves into more energetic and suitable positions in the valence band and cause the mean free path of electrons to increase, thus less energy is required by an electron to jump from the valence to the conduction band. Figure 18b is a graph of refractive index (n) versus annealing temperature of SnO$_2$ thin films at wavelength λ = 589.3 nm, which indicates a decreasing trend with increasing annealing temperature.

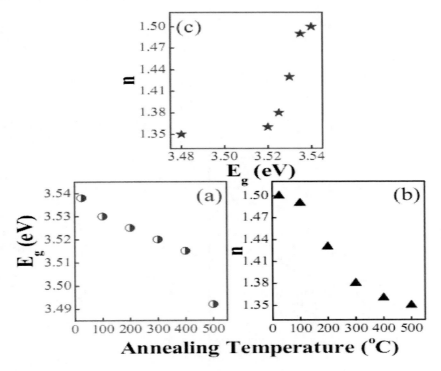

Figure 18. (a) Graph of optical band gap (E_g) versus annealing temperature for SnO$_2$ thin films. (b) Graph of refractive index (n) versus annealing temperature for SnO$_2$ thin films at wavelength λ = 589.3 nm. (c) Graph of optical band gap (E_g) against refractive index (n) for SnO$_2$ thin films.

The refractive index values measured at λ = 589.3 nm are in the range 1.49 to 1.36. In order to establish a relationship between the refractive index (n) and the optical band energy (E_g), a graph was plotted as shown in Figure 18c. An approximately linear relationship is found with $n = 2.1667E_g - 6.1801$. Figure 19 shows the graph of optical conductivity (σ) versus wavelength for SnO$_2$ thin films annealed at different temperatures for 30 min: (a) the as-prepared thin film, (b) 100 °C, (c) 200 °C, (d) 300 °C, (e) 400 °C, and (f) 500 °C. This figure shows a rapid rise of optical conductivity (σ) in the IR region. The result can be explained by an enhancement of crystallinity, and an increase of the grain size with annealing temperature. The value of optical conductivity (σ) was found to be very high, which may be applied to photovoltaic cells. To sum up, from the present investigation it can be concluded that the SnO$_2$ thin films annealed up to 400 °C can be used as a good window material for solar cell application due to their high transmittance (> 75.3%) and 400 °C may be the optimum annealing temperature. The optical band gap energy and refractive index of the present thin films annealed at various temperatures show approximately linear behaviour, and the SnO$_2$ thin films display high optical conductivity. These findings indicate that these SnO$_2$ thin films have potential technological applications in the fabrication of solar cell devices and gas sensors.

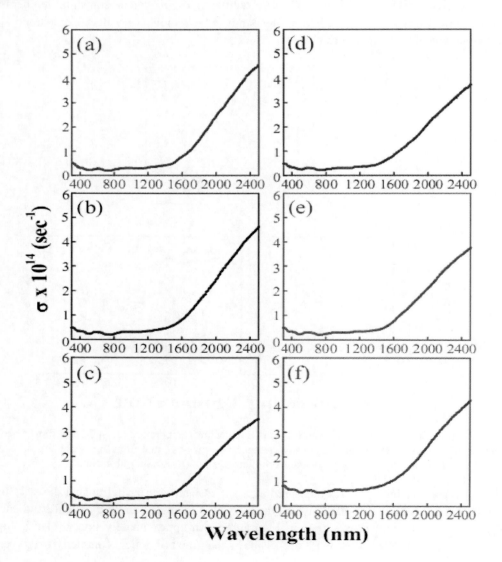

Figure 19. Graph of optical conductivity (σ) versus wavelength of SnO$_2$ thin films annealed at different temperatures for 30 min. (a) as-prepared thin film; (b) 100 °C; (c) 200 °C; (d) 300 °C; (e) 400 °C; (f) 500 °C.

2.5. Electrical Properties of Tin Dioxide Thin Films

Tin dioxide (SnO$_2$) is an *n*-type wide-band-gas semiconductor material (E$_g$ = 3.6 eV at 300 K) where inherent oxygen vacancies act as an *n*-type dopant [47, 48]. Research on SnO$_2$ attracts a lot of interest because it has been widely used in many applications, such as transparent electrodes, far-infrared detectors and high-efficiency solar cells [108-112]. In recent years, nanocrystalline SnO$_2$ has been reported to have some different characteristics from the bulk crystals, and much attention has been focused on the synthesis of SnO$_2$ nanowires [12], nanotubes [12], nanorods [82, 83], nanobelts [9, 84, 85], and exploration of their novel properties. Nanocrystalline SnO$_2$ thin films have also aroused much attention since the higher-quality synthesis of SnO$_2$ thin films was achieved. This achievement is mainly due to the recognition of the bulk-quantity growth mechanism of SnO$_2$ thin films. The success in higher-quality synthesis of SnO$_2$ thin films has meant that research is not limited to theoretical area, but can be extended to the experimental area.

A variety of methods, such as sol-gel [86, 87], chemical vapor deposition [88, 89], magnetron sputtering [90], sonochemical [91], and thermal evaporation [92], have been employed to prepare SnO$_2$ thin films or nanoparticles. To date, numerous experimental results on SnO$_2$ thin films have been reported [53, 54, 75, 93], including those from X-ray diffraction (XRD), transmission electron microscopy (TEM), electron transport, and Raman spectroscopy. These results have helped to speed up the study of its potential applications, since SnO$_2$ thin film is one of the promising materials for future transparent nanoelectrodes and solid-state gas nanosensors.

However, the diversity of thin films grown using pulsed laser deposition (PLD) is enormous and perhaps recommends its flexibility more persuasively than anything else. The rationale for using PLD in preference to other deposition techniques lies primarily in its pulsed nature, the possibility of carrying out surface chemistry far from thermal equilibrium, and under favorable conditions, the ability to reproduce in thin films the same elemental ratios of even highly chemically complex bulk ablation targets [73].

Being an *n*-type semiconductor, the resistivity and dielectric properties of SnO$_2$ thin films are two important factors for its characterization. The resistivity of these thin films is found to depend on oxygen vacancies [113] whose concentration can not be easily controlled. The thermal annealing processes are usually performed to reduce the intrinsic stress, to improve the lattice mismatch and to create longer mean paths for the free electrons to get better electrical conductivity. Other factors that alter the properties of SnO$_2$ thin films are dependent on their deviation from stoichiometry, on the nature and amount of impurities and on the microstructure [114]. Therefore, understanding the relationship between microstructure and electrical properties of SnO$_2$ thin films is essential in their applications [115].

In this section, the structural and electrical properties of nanocrystalline SnO$_2$ thin films deposited on glass substrates by pulsed laser deposition techniques have been investigated before and after thermal annealing in the temperature range from 50 to 550 °C with a step of 50 °C, respectively. This section attempts to explain the changes in the crystallinity, root-mean-square (RMS) surface roughness, and resistivity of nanocrystalline SnO$_2$ thin films as a result of thermal annealing. The results indicate that nearly opposite actions to RMS surface roughness and electrical resistivity make a unique performance with thermal annealing temperature.

Figure 20 shows the distribution of nanoparticle size versus thermal annealing temperature for as-prepared and thermal annealing thin films. SnO_2 nanoparticle size first decreases from 27.6 nm (at room temperature) to smaller than 23.7 nm at 200 °C and then sharply increases to larger value of 27.1 nm at 300 °C and then gradually decreases to 25.7 nm at 400 °C, finally increases up to 28.9 nm for further annealing at higher temperatures. Figure 21 shows the graph of resistivity versus annealing temperature for as-prepared and annealed SnO_2 thin films. Resistivity first rises from 1.24 mW-cm (at room temperature) to higher than 1.66 mW-cm at 100 °C and then sharply decreases to minimum value of 1.13 mW-cm at 200 °C and then gradually increases up to 1.45 mW-cm for further annealing at higher temperatures. The initial rise in resistivity caused by annealing at 100 °C might be attributed to the smaller mean particle size as depicted by XRD analysis. This rise of resistivity might also be explained on the basis of Lee's results as follows: there might be a decrease in oxygen vacancies taken up by excess metal ions during annealing in air (oxidizing environment). As a consequence, the carrier concentration might decrease and hence the mobility, resulting in an increase of resistivity [117]. The sudden fall of resistivity for annealing the films at 200 °C could be related to the rearrangement of atoms on valence band which could increase the mean free path of electrons and hence a rise in conductivity or fall in resistivity. The slow increase of resistivity for annealing above 200 °C could be associated with the oxygen vacancies and the appearance of nonstoichiometric SnO_x ($1 \leq x < 2$, for example, SnO phase) relating to nanocrystalline SnO_2 thin films in the surface layer cause a large number of oxygen defects in the surface region. Hall mobility measurements on SnO_2 films have already shown a decrease in mobility and consequently a rise in resistivity by the formation of SnO phase [117]. Furthermore, an improvement in the crystallinity of thermal annealed films can also be a source of increase of resistivity [118].

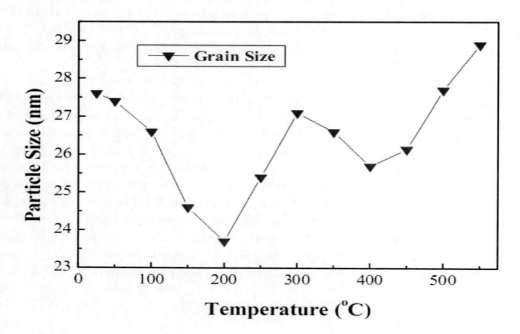

Figure 20. The distribution of nanoparticle size versus thermal annealing temperature for as-prepared and thermal annealing SnO$_2$ thin films.

Semiconducting oxides are a class of materials that are vitally important for developing new materials with functionality and smartness. The unique properties of these materials are related to the mobility of transmitted electrons. Micro-roughness of thin films plays a vital role for developing optical coatings especially in the ultraviolet (UV) region for applications such as lithographic uses [79, 80]. To characterize an optical surface (coatings), the root-mean-square (RMS) surface roughness is normally used. The RMS roughness not only describes the light scattering but also gives an idea about the quality of the surface under investigation. Thus annealing process is a good parameter that could affect the surface RMS roughness behavior. The plot of RMS surface roughness values computed for as-prepared and annealed SnO$_2$ thin films is portrayed in Figure 11 as a function of annealing temperature in comparison with resistivity and refractive index data shown in Figure 22. It can be seen that RMS roughness (Figure 11) and resistivity (Figure 21) show almost opposite trend while the refractive index (Figure 22) shows the opposite behavior with surface roughness up to 200 °C but beyond this annealing temperature, similar behavior is noted. Lindström and co-authors have shown by power spectral density (PSD) function calculations in sputtered SnO$_2$ films that for thicker films (> 300 nm), as the film surface smoothens it gives rise to defects and strain variations [81]. Therefore, if the RMS roughness of the film decreases, surface smoothness increases and defects and strain variations are produced which can hinder the flow of electrons and so the resistivity rises. With regard to refractive index variations in comparison with RMS roughness values, it is noted that where roughness is high, refractive index is low (up to 200 °C). Our results are similar to the behavior of Gd$_2$O$_3$ films [38, 80] which were reported by Senthilkumar and co-authors.

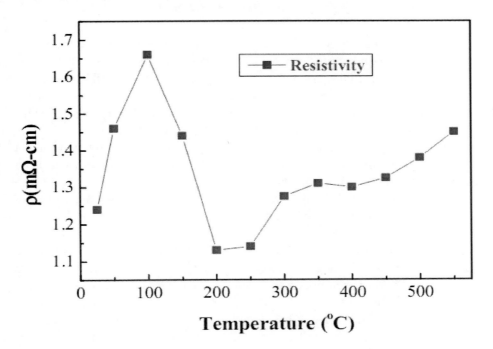

Figure 21. The graph of resistivity versus thermal annealing temperature for as-prepared and thermal annealing SnO$_2$ thin films.

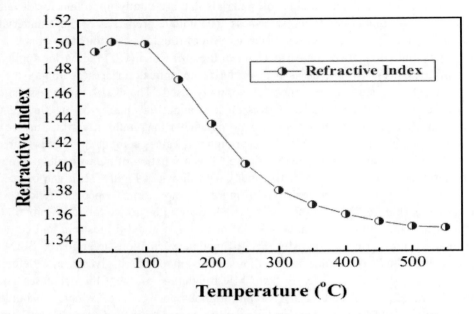

Figure 22. The outline of refractive index versus thermal annealing temperature for as-prepared and thermal annealing SnO$_2$ thin films.

3. TIN DIOXIDE NANOWIRES

One-dimensional nanostructures have stimulated increasing interest among materials scientists due to their peculiar properties and potential applications [119-124]. Nanostructured materials, such as nanoparticles, nanotubes, nanowires, nanorods, nanoribbons (or nanobelts), and nanorings, etc. have attracted extensive attention in the last few years due to their properties with important and unique applications in constructing nanoscaled electronic and optoelectronic devices, gas sensors, catalysts, and growth of thin films [10, 125-129]. Semiconducting oxides are fundamental to the development of smart and functional materials, devices, and systems. Among semiconducting oxide nanowires, tin dioxide (SnO$_2$) nanowires have been studied intensely for applications as lithium-ion batteries, varistors, gas sensors, and transparent conducting electrodes [9, 130-132], which employ the unique characteristics of SnO$_2$ nanowires including a wide-band-gap (E_g = 3.6 eV at 300 K) and large surface-to-volume ratios.

The synthesis of nanostructures for functional oxides, with a controlled structure and morphology, is critical for scientific and technological applications. Recently, the synthesis of nanostructures based on metal oxides was demonstrated extensively [12, 84, 133-135]. The early synthesis techniques of one-dimensional nanomaterials include the use of photolithography [136] and scanning tunneling microscopy [137]. While these methods have been providing nanomaterials for fundamental study, they are obviously not suitable for industrial applications. Screw-dislocations [138] and vapor-liquid-solid (VLS) [139] are two established mechanisms by which materials can grow one-dimensionally into wires, whiskers, or rods. A precondition for many SnO₂ nanowires based on the applications is an easy and large-scale production of SnO₂ nanowires. To date, however, the growth behavior of the SnO₂ nanowires prepared by pulsed laser deposition method is not fully understood yet.

In this section, we report in detail that SnO₂ nanowires have been synthesized direct by a pulsed laser deposition process based on a sintered cassiterite SnO₂ target, being deposited on Si (100) substrates at room temperature. Transmission electron microscopy shows that the nanowires are structurally perfect and uniform, and diameters range from 10 nm to 30 nm, and lengths of several hundreds nanometers to a few micrometers. X-ray diffraction indicates that the nanowires show the tetragonal rutile structure in the form of SnO₂. Selected area electron diffraction and high-resolution transmission electron microscopy reveal that the nanowires grow along the [110] growth direction. Electric properties are investigated by connecting a single SnO₂ nanowire in field-effect transistor configuration. The SnO₂ nanowires based on field-effect transistor devices exhibit that the SnO₂ nanowires prepared by our method hold better electrical properties. Our findings indicate that other one-dimension nanostructural materials may be manipulated by using pulsed laser deposition techniques, and might provide insight into the new opportunities to the applications in constructing nanoscaled electronic and optoelectronic devices.

SnO₂ nanowires were prepared direct by pulsed laser deposition (PLD) method [55]. The fabrication can be described in detail as follows: the sintered cassiterite SnO₂ bulk was used as the target. The circular target was consisted of high-purity commercial SnO₂ (99.8 %) disc. The size of the target was about ϕ15 mm × 4 mm, and the target was cleaned with methanol in an ultrasonic cleaner before installation to minimize contamination. The laser was a KrF excimer laser (Lambda Physik, LEXtra 200, Germany) producing pulse energies of 150 mJ at a wavelength of 248 nm, and the duration of every excimer laser pulse was 34 ns. During the experiment, the laser was operated at a repetition rate of 5 Hz at an incident angle of 45° to the polished sintered cassiterite SnO₂ target, which was rotating at a rate of 15 rpm to avoid drilling. The time of pulsed laser deposition was about 6 h. The ablated substance was collected on a Si (100) slide, which was mounted on a substrate holder 4 cm away from the target. The base oxygen pressure in the deposition chamber was about 3×10^{-4} mbar. All deposition processes were carried out at ambient temperature.

After above experiment, the SnO₂ nanowires were transferred from silicon substrates and a number of device structures were fabricated and tested. The SnO₂ nanowire field-effect transistor (FET) devices were fabricated by thermally growing SiO₂ film with 500 nm thickness on the top of *n*-type Si wafers, followed by deposition of an Au metal with sputtering and standard optical lithography. Parallel Au-electrode pairs, with a length of 10 μm and 80 nm in thickness, were patterned. The parallel Au electrode with a distance of 2 μm will be used as source and drain electrodes, and an *n*-type silicon layer serves as the back gate. The as-prepared SnO₂ nanowires were cleaned with ethanol by the ultrasonic into a suspension and then dispersed onto an *n*-type silicon chip coated with 500 nm of SiO₂ with predefined gold electrodes. The SnO₂-FETs samples were rapidly transferred to a stove to anneal in Ar at 500 °C for 30 min to improve the quality of Au–SnO₂ contacts. This single SnO₂ nanowire FET-device was constructed to characterize the electrical properties of the as-prepared nanowire.

X-ray diffraction (XRD) was performed with a Philips X'pert diffractometer using Cu K$_\alpha$ radiation (1.5406Å) in reflection geometry. Proportional counter with an operating voltage of 40 kV and a current of 40 mA was used. XRD patterns were recorded at a scanning rate of 0.05 °s^{-1} in the 2θ ranges from 20 to 60°. The quality of SnO$_2$ nanowires was characterized using high-resolution transmission electron microscopy (HRTEM) observations performed using a JEOL-2010 high-resolution transmission electron microscope with a point-to-point resolution 1.94 Å, operated at 200 kV.

Figure 23 shows the XRD patterns of as-prepared SnO$_2$ nanowires by pulsed delivery. From Figure 23, all diffraction peaks can be perfectly indexed to the rutile SnO$_2$ structure not only in peak position but also in their relative intensity. The SnO$_2$ lattice constants obtained by refinement of XRD data for as-prepared nanowires are a = 4.737 Å and c = 3.185 Å, which are consistent with those of bulk rutile SnO$_2$ tetragonal structure [140]. The crystallographic planes identified on the XRD patterns presented in Figure 23 are indeed typical of a pure crystalline SnO$_2$ phase, in terms of peak positions and relative peak intensities. In addition, no characteristic peaks of impurities, such as other forms of tin oxides, were detected, it demonstrated that the nanowires are pure phase of rutile SnO$_2$.

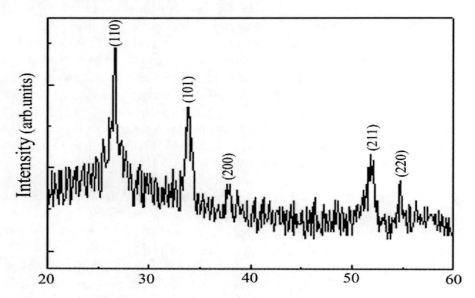

Figure 23. X-ray diffraction patterns of the as-prepared SnO$_2$ nanowires prepared by pulsed laser deposition method.

Figure 24a shows that the morphology of as-prepared nanowires was observed by transmission electron microscopy (TEM). It can be seen that the SnO$_2$ crystals display wire-like shape with diameters of 10–30 nm, and lengths of up to several hundreds nanometers to a few micrometers. We found that there are no spherical droplets at tops of these nanowires. This observation suggested that the nanowires may not grow by vapor-liquid-solid or solution-liquid-solid mechanism proposed for the nanowires growth by a catalytic-assisted technique, in which a metal liquid droplet was located at the growth front of the wire and acted as the catalytic active site. Figure 24b shows a typical TEM bright-field image of a single SnO$_2$ nanowire with about 15 nm in width and larger than 380 nm in length. It can be seen that the SnO$_2$ nanowire is very smooth, straight and uniform, and that the geometrical shape is structurally perfect. The inset at the upper left-hand corner shows the selected area electron diffraction (SAED) pattern of a single nanowire, taken along the [$\bar{1}$11] direction, which can be indexed on the d-spacing of (110) and (211) crystal planes. It can be determined a tetragonal cell with lattice parameters of a = 4.737 Å and c = 3.185 Å, consistent with the XRD results. The SAED pattern also confirms that the nanowire is a single rutile SnO_2. Figure 24c shows an HRTEM image of a single nanowire shown in Figure 24b, which confirmed that the nanowire grew along the [110] growth direction (indicated with an arrow, it is parallel to the long axis of a nanowire). HRTEM image proved that the nanowire is structurally uniform, and the clear lattice fringes illustrated that the nanowire is a single crystalline. The interplanar spacing is about 0.334 nm, which corresponds to the {110} planes of rutile crystalline SnO_2. This result revealed that the growth plane of the nanowires is along the {110} planes. Above experimental results indicated that the nanowires are a single pure crystallized SnO_2 rutile phase.

Figure 24. (a) the typical TEM bright-field image of SnO$_2$ nanowires. (b) the typical TEM bright-field image of a single SnO$_2$ nanowire; the inset at the upper left-hand corner shows the SAED patterns of the single SnO$_2$ nanowire; (c) the HRTEM image of the single SnO$_2$ nanowire shown in Figure (b).

Since the crystallized nanowires are deposited under ambient temperature, this might be caused by the high energy of the laser-vaporized clusters. Furthermore, as the vacuum pressure is about 3×10^{-4} mbar in our experiment, the mean free path of the nanoclusters can be estimated to be about 0.3 m, which is near the dimension of the deposition chamber. Therefore, the target-substrate distance and the power density on the target are very important for this process. In order to investigate the growth mechanism of one-dimensional oxide nanostructures, and depending on the presence or absence of metal catalysts in the synthesis processes, our TEM observation revealed that the growth behavior of SnO$_2$ nanowires may not be dominated by the vapor-liquid-solid process proposed for one-dimensional nanostructure, in which a catalytic metal particle was located at the growth front and acted as the energetically favorable site. Since no catalytic droplets were observed on any end of SnO$_2$ nanowires, which is the most remarkable sign of the VLS mechanism. Therefore, we can reasonably speculate that the SnO$_2$ nanowire's formation process undergoes a vapor-solid growth process. During pulsed delivery, the SnO$_2$ target was etched by using high-temperature laser plumes. The SnO$_2$ vapor originates from the starting materials in a higher temperature zone directly deposits in a lower temperature region, and the SnO$_2$ vapor deposits in the form of nanoclusters through aggregation of SnO$_2$ molecules on the deposition chamber near the target (due to the absence of a carrier gas in the chamber). As the ablation proceeds, more SnO$_2$ vapor is generated and more SnO$_2$ nanoclusters are formed. The as-formed SnO$_2$ nanoclusters are energetically favorable sites for rapid adhesion of additional SnO$_2$ molecules, which resulted in the formation of SnO$_2$ nanowires.

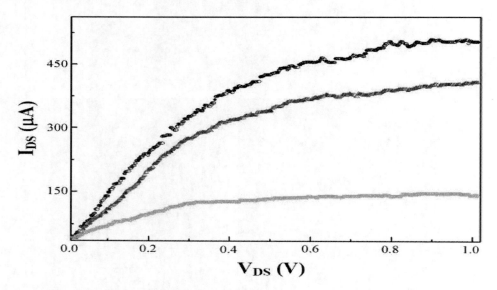

Figure 25. The current-voltage (I_{DS}-V_{DS}) curves of field-effect transistor devices for the back-gated single SnO$_2$ nanowire with V_{GS} = 5 V to -1 V in -2 V steps from top to bottom.

Figure 25 shows the current-voltage (I_{DS}-V_{DS}) curves of field-effect transistor (FET) device. The gate voltages were applied from -1 to 5 V and the drain voltages were from 0 to

1.0 V for room temperature samples. It was found that the higher conductance is obtained by increasing the back gate voltage, indicating *n*-type semiconductor characteristics of the SnO_2 nanowires. The possible reason might be due to oxygen vacancies and extra tin interstitial atoms in the lattice during the pulsed delivery process. An on/off current ratio (I_{on}/I_{off}) as high as 10^5 has been achieved at V_G from -1 to 5 V with V_{DS} = 15 mV.

Ideal field-effect transistor equations were used to determine the mobility and sub-threshold voltages. From the measurements, the total charge on the nanowires is $Q = CV_{g,t}$, where C is the nanowire capacitance and the threshold voltage $V_{g,t}$ necessary to completely deplete the wire. The capacitance of the nanowire is given by [119]

$$C = \frac{2\pi\varepsilon\varepsilon_0 L}{\ln(\frac{2h}{r})}, \qquad (1)$$

here, $\varepsilon = 3.9$, $\varepsilon_0 = 1$ and $h = 500$ nm are the relative dielectric constant, the vacuum dielectric constant, and the thickness of the silicon dioxide layer, respectively. The $r = 15$ nm is the nanowire radius and $L = 1.5$ μm is the channel length. Figure 26 shows the I_{DS} - V_{GS} curve of field-effect transistor device for the single SnO_2 nanowire in log scale at V_{DS} = 15 mV. The sub-threshold slope was estimated to be about 0.12 V/decade and independent of the bias voltage.

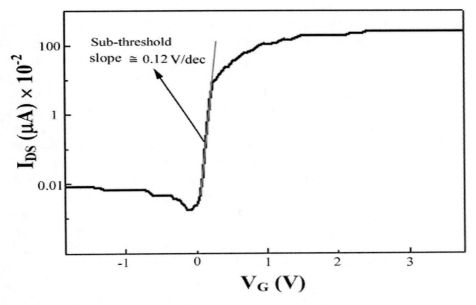

Figure 26. The I_{DS} - V_{GS} curve of field-effect transistor device for the single SnO_2 nanowire in log scale at V_{DS} = 15 mV.

From Figure 26, an on/off current ratio as high as 10^5 was achieved when V_{DS} = 15 mV. This value is consistent with those obtained in planar single-crystalline SnO_2 nanowires and other mental oxide nanowires [141, 142]. We get a one-dimensional electron density of n =

$Q/eL \approx 3.86 \times 10^5$ cm^{-1}. The mobility of the electrons can also be deduced from the trans-conductance of the FET. In the linear region, the trans-conductance g_m is expressed as

$$g_m = \frac{dI_{DS}}{dV_{DS}}, \tag{2}$$

And

$$g_m = \mu_e \left(\frac{C}{L^2}\right) V_{DS} \tag{3}$$

From Figure 26, the FET device was found that the mobility of the channel stoichiometry is about 191 cm^{2}V^{-1}s^{-1}. Transistor composed of the SnO$_2$ nanowires exhibited better electrical properties which affected both the on/off ratio and threshold voltage of the devices. Our findings indicate that other one-dimension nanostructural materials may be manipulated by using pulsed laser deposition techniques, and might provide insight into the new opportunities to the applications in constructing nanoscaled electronic and optoelectronic devices.

4. TIN DIOXIDE NANORODS

Semiconductor oxides are essential constituents for the development of smart and functional materials, devices, and systems [8, 9]. These oxides have two unique structural features: mixed cation valences and an adjustable oxygen deficiency. Such features are the bases for creating and tuning many novel materials' properties, from chemical to physical [12, 13]. The preparation of functional oxides, with different microstructural features, is critical for scientific and technological applications [10, 11, 55]. The usefulness of these materials lies not just in the direct miniaturization of micrometer-scale devices. There has also been widespread interest in understanding the influence of the nanometer size scale, the surface, the interface, and their nucleation and growth on the materials' chemical and physical properties. Research on tin dioxide (SnO$_2$) attracts a lot of interest because it has been widely used in many applications, such as transparent electrodes, far-infrared detectors, and high-efficiency solar cells [108-112]. Many studies of SnO$_2$ have been focused on two-dimensional thin films and zero-dimensional nanoparticles, which can be readily synthesized with various well-established techniques, such as sputtering (for thin films) and sol-gel (for particles). In contrast, the investigations of SnO$_2$ nanorods, nanowires, nanotubes, and nanobelts are often cumbersome because of the difficulty in obtaining such nanostructures. This topic has attracted extensive interest over the past decade due to its potential in addressing some basic issues about dimensionality and space-confined transport phenomena as well as novel applications [143]. One-dimensional nanostructured systems have recently attracted much attention due to their novel properties and potential applications in numerous areas such as nanoscale electronics and photonics [133, 144-146]. In recent years, one-dimensional nanostructured SnO$_2$ has been reported to have some different characteristics from the bulk crystals, and much attention has been focused on the synthesis of SnO$_2$ nanowires [12], nanorods [82, 83], nanotubes [12], nanobelts [9, 84, 85], and exploration of their novel properties. A series of binary semiconducting oxides, such as In_2O_3, SnO_2, ZnO, and CdS, have distinctive properties and are widely used as transparent conducting oxide materials and gas sensors. For example, fluorine-doped SnO_2 thin film is used in architectural glass applications because of its low emissivity for thermal infrared heat [147, 148].

In this section, we report in detail a simple and novel method for the preparation of SnO_2 nanorods by annealing precursor powders in which NaCl, Na_2CO_3, and $SnCl_4$ were homogeneously mixed. The precursors were synthesized in an inverse micro-emulsion (IµE) system. The synthesis route is distinguished by the simplicity of the apparatus used and the high efficiency of crystal growth. X-ray diffraction and energy dispersive X-ray spectroscopy as well as Raman analysis indicated that these nanorods have the same crystal structure and chemical composition found in the tetragonal rutile form of SnO_2. Transmission electron microscopy showed that the nanorods are structurally perfect and uniform, with widths of 10-25 nm, and lengths of several hundred nanometers to a few micrometers. Selected area electron diffraction and high-resolution transmission electron microscopy revealed that the nanorods grow along the [110] crystal direction. Electric properties are investigated by connecting a single SnO_2 nanorod in field-effect transistor configuration. The results show that the SnO_2 nanorods prepared by our method based on field-effect transistor devices exhibit good electrical properties. The electrical measurements confirmed that the SnO_2 nanorods were *n*-type semiconductors. The findings indicate that other one-dimension nanostructured materials may be manipulated by using this simple technique. This work might provide insight into new opportunities for the applications as building blocks for nanoelectronics and active sensing materials.

The method used is somewhat similar to the molten salt synthesis (MSS) mechanism [82, 149], which was reported to be one of the simplest techniques for preparing ceramic powders with whisker-like [150], needle-like [151], and plate-like [152-154] morphologies. In order to obtain the micro-emulsion (µE), the cyclohexane was used as the oil phase and a mixture of poly (oxyethylene)-5-nonyl phenolether (NP5) and poly (oxyethylene)-9-nonyl phenolether (NP9) as well as *p*-octyl-polyethylene glycol phenylether (OP) (all analytical grade) with weight ratio 1:1:1 was used as non-ionic surfactant. The mixture was stirred until it became transparent (referred to as µE). Thereafter 8 mL of 1 M Na_2CO_3, 32 mL of 2 M NaCl, and 8 mL of 0.5 M $SnCl_4$ (all analytical grade) aqueous solutions were, respectively, added to 60, 100, and 60 mL of the above µEs with vigorous stirring. The solution containing Na_2CO_3 (µE$_1$) was mixed with that containing NaCl (µE$_2$), followed by adding the solution containing $SnCl_4$ (µE$_3$) to the system (µE$_1$+µE$_2$). All mixings were performed under vigorous stirring conditions. The solution µE$_4$ (µE$_1$+µE$_2$+µE$_3$) containing Na_2CO_3, NaCl, and $SnCl_4$ was washed with acetone, and then dried to form the precursor powders. The dried precursor powders were calcined in a tube furnace at 800 °C for 2 h in oxygen partial pressure of 3×10^{-2} Pa atmosphere. After heat treatment, the mixtures were cooled naturally to room temperaturer. Subsequently, the salts were removed from the resulting mixtures by washing with distilled water. In partial O_2 atmosphere, since the oxygen species (O_2^- and O^-) continuously exchange electrons with the products and the neutral oxygen atom may contribute to the lattice of the products via the faster diffusion of the grain boundary, which results in the reduction of the oxygen vacancies and hence the SnO_2 nanorods grew perfectly.

Structure of these nanorods was determined by recording X-ray diffraction (XRD) patterns at room temperature using Philips X'pert diffractometer equipped with Cu K$_\alpha$ radiation (1.5406 Å) in reflection geometry. Proportional counter with an operating voltage of 40 kV and a current of 40 mA was used. XRD patterns were recorded at a scanning rate of 0.05°s^{-1} in the 2θ range from 15° to 65°. The SnO$_2$ nanorods were characterized using transmission electron microscopy (TEM) and high-resolution transmission electron microscopy (HRTEM). Such observations were performed using a Philips CM20 transmission electron microscope operating at an acceleration voltage of 200 kV, and a JEOL-2010 high-resolution transmission electron microscope with a point-to-point resolution of 1.94 Å, operating at 200 kV. These instruments were also equipped with energy dispersive X-ray spectroscopy (EDS) capabilities. Raman scattering measurements were obtained by using a SPEX-1403 laser Raman spectrometer in backscattering geometry. The excitation source was an argon-ion laser operated at a wavelength of 514.5 nm in the backscattering configuration and a low incident power to avoid thermal effects.

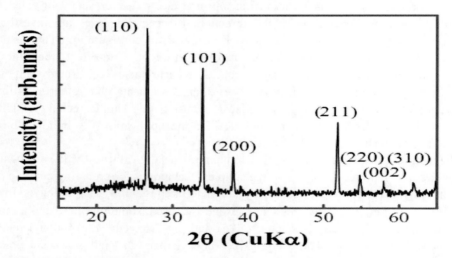

Figure 27. A typical X-ray diffraction (XRD) pattern of the SnO$_2$ nanorods obtained using Cu K$_\alpha$ radiation (1.5406 Å).

After the above experiment, the SnO$_2$ nanorods were transferred from the silicon substrates and a number of device structures were fabricated and tested. A single SnO$_2$ nanorod field-effect transistor (FET) was constructed to characterize the electrical properties of the nanorods. Devices were fabricated by thermally growing SiO$_2$ films with 500 nm thickness on top of *n*-type Si wafers, followed by deposition of Au metal using sputtering and standard optical lithography technique. Parallel Au-electrode pairs, with a length of 15 μm and 90 nm in thickness, were patterned. The parallel Au electrode with a distance of 3 μm was used as source and drain electrodes, with an *n*-type silicon layer served as the back gate. The SnO$_2$ nanorods were ultrasonically agitated with ethanol into a suspension and then dispersed onto an *n*-type silicon chip coated with 500 nm of SiO$_2$ with predefined gold electrodes. The SnO$_2$-FETs samples were rapidly transferred to a stove to anneal in Ar at 500 °C for 10 min to improve the quality of Au–SnO$_2$ contacts since the Ar is a protection gas to further prevent the oxidation of the SnO$_2$ nanorods. Figure 27 shows the XRD pattern of

SnO$_2$ sample synthesized at 800 °C for 2 h in oxygen partial pressure of 3 × 10^{-2} Pa atmosphere. The XRD analysis revealed the overall crystal structure and phase purity of the products. All the diffraction peaks can be indexed to the tetragonal rutile structure of SnO$_2$ with lattice constants a = 4.738 Å and c = 3.187 Å, which are consistent with the standard values for bulk SnO$_2$ [30]. No characteristic peaks of impurities, such as other forms of tin oxides, were detected. Figure 28 shows that the energy dispersive x-ray spectrum (EDS) recorded on the SnO$_2$ nanorods. EDS analysis of the sample revealed that the nanorods were composed of 33.5 at.% Sn and 66.5 at.% O, and the ratio Sn:O = 1:1.985 was in good agreement with that of bulk SnO$_2$ (Sn:O = 1:2). The Cu peaks came from the Cu grids. The above chemical composition results indicate that the nanorods mainly consisted of the SnO$_2$ form.

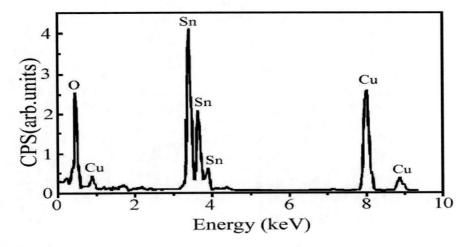

Figure 28. The energy-dispersive X-ray spectrum (EDS) recorded on the SnO$_2$ nanorods.

Figure 29 shows a lower-magnification TEM bright-field image. This is a typical distribution in morphology for the products obtained by the precursor powders calcined at 800 °C for 2 h in oxygen partial pressure of 3 × 10^{-2} Pa atmosphere. It is evident that the products consist of one-dimensional nanorods in bulk-quantity with lengths from several hundred nanometers to a few micrometers, and diameters range from 10-25 nm. Figure 30a shows a typical TEM bright-field image of a single SnO$_2$ nanorod of about 20 nm in width and larger than 550 nm in length. It can be seen that the SnO$_2$ nanorod is very smooth, straight and uniform, and that the geometrical shape is structurally perfect. The chemical composition of the nanorods was also determined by EDS to be close to SnO$_2$. The inset at the upper left-hand corner shows the end of a single SnO$_2$ nanorod. The profile of the fringes implies that the geometrical shape of this nanostructure is likely to be a nanorod. The growth directions of the nanorods were determined from selected area electron diffraction (SAED) patterns and HRTEM images of individual nanorods (in our experiment, we examined more than 10 individual nanorods). The inset at the bottom right-hand corner is the SAED pattern of the single SnO$_2$ nanorod shown in this figure. From the SAED patterns, it can be determined that the d-spacing of (110) crystal plane is 3.35Å, (200): 2.37Å, (101): 2.64Å, and (310): 1.50Å. A more-detailed analysis can be made based on the highly magnified HRTEM image shown in Figure 30b, which confirmed that the nanorod grew along the [110] direction (indicated with an arrow parallel to the long axis of the nanorod). In the HRTEM observation, the field of view mainly consisted of a single-domain crystallite. On closer inspection, recurrent values of separation distance between lattice layers were found (in particular, 0.33 nm, shown in Figure 30b corresponding to the lattice parameters of the rutile structure of the SnO_2 cassiterite phase (arising from the [110] reflections). These results were also confirmed by the analysis of the SAED patterns collected by the TEM electron probe. All the above data indicate that the nanorod segregates as a single phase belonging to the SnO_2 cassiterite structure, and it was a single crystal with a growth direction of [110]. In our experiments, HRTEM observations also indicated that the SnO_2 nanorods were structurally uniform and no dislocations or other planar defects were detected in the examined area of the SnO_2 nanorods.

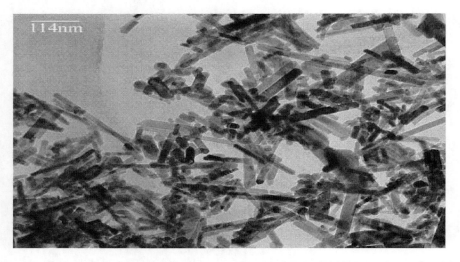

Figure 29. A typical transmission electron microscopy (TEM) bright-field image of clusters of the SnO_2 nanorods.

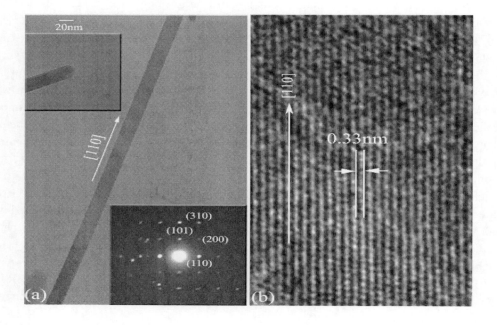

Figure 30. (a) A typical TEM bright-field image of a single SnO$_2$ nanorod with [110] growth direction. Its growth direction is shown by the arrow. The inset at the upper left-hand corner shows the end of a single SnO$_2$ nanorod. The inset at the bottom right-hand corner shows the SAED pattern of the single SnO$_2$ nanorod. (b) An HRTEM image of the single SnO$_2$ nanorod shown in Figure a, where the arrow shows its [110] growth direction.

Figure 31 shows the Raman spectrum of the SnO$_2$ nanorods. It can be seen that there are three fundamental Raman scattering peaks at 477.8, 638.6, and 774.8 cm^{-1} respectively, which are in good agreement with those of a rutile SnO$_2$ single crystal. The peak at 477.8 cm^{-1} can be assigned to the E_g mode, the peak at 638.6 cm^{-1} can be identified to the A_{1g} mode, and the peak at 774.8 cm^{-1} can be indexed to the B_{2g} mode. Thus, the three Raman peaks show the typical features of the rutile phase of synthesized SnO$_2$ nanorods [44, 155].

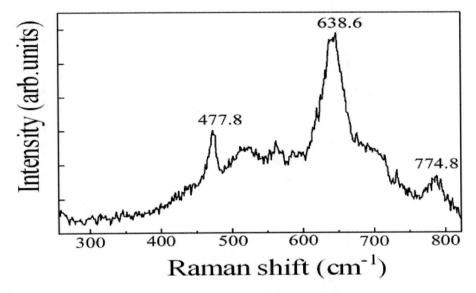

Figure 31. Room temperature Raman spectrum of the SnO$_2$ nanorods.

In the formation of SnO$_2$ nanorods, the nucleation and growth of SnO$_2$ nanorods are attributed to the key effect of the NaCl and surfactant. We believed that NaCl may significantly decrease the viscosity of the melt and thus enhance the mobility of components in the flux, i.e., provides a favorable environment for the nucleation and growth of nanorods. The surfactant (NP9/NP5/OP) is favorable to the formation of fine particles by making a surrounding shell to prevent them from aggregating into larger particles during the formation of the precursor. Concurrently, during the formation of SnO$_2$ nanorods, the surfactant was thought to be able to act as a template, with the template action resulting in the expitaxial growth of the products. This understanding of the formation process of SnO$_2$ nanorods is scientifically interesting and important. By choosing the appropriate surfactant, it is reasonable to expect that the present route can be extended to producing other nanostructured semiconducting metal oxides.

Figure 32 shows the current-voltage (I_{DS}-V_{DS}) curves of the field-effect transistor (FET) device. The gate voltages were applied from -1 to 5 V and the drain voltages were from 0 to 1.0 V for the room temperature samples. It was found that higher conductance was obtained by increasing the back gate voltage, indicating *n*-type semiconductor characteristics of the SnO$_2$ nanorods. The possible reason might be due to oxygen vacancies and extra tin interstitial atoms in the lattice during the annealing of precursor powders. An on/off current ratio (I_{on}/I_{off}) as high as 10^7 has been achieved at V_G from -1 to 5 V with V_{DS} = 15 mV. Ideal field-effect transistor equations were used to determine the mobility and sub-threshold voltages. From the measurements, the total charge on the nanorods is $Q = CV_{g,t}$, where C is the nanorod capacitance and the threshold voltage $V_{g,t}$ necessary to completely deplete the rod.

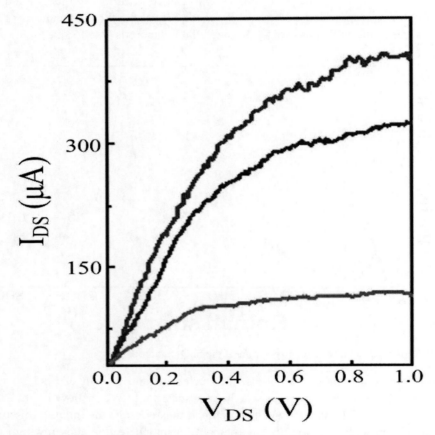

Figure 32. The current-voltage (I_{DS}-V_{DS}) curves of field-effect transistor devices for the back-gated single SnO$_2$ nanorod with V_{GS} = 5 V to -1 V in -2 V steps from top to bottom.

The capacitance of the nanorod is given by as the formula (1), where, ε = 3.9, ε_0 = 1 and h = 500 nm are the relative dielectric constant, the vacuum dielectric constant, and the thickness of the silicon dioxide layer respectively. r = 10 nm is the nanorod radius and L = 3 μm is the channel length. Figure 33 shows the I_{DS} - V_{GS} curve of the field-effect transistor device for the single SnO$_2$ nanorod in log scale at V_{DS} = 15 mV. The sub-threshold slope was estimated to be about 0.06 V/decade and independent of the bias voltage. From Figure 33, an on/off current ratio as high as 10^7 was achieved at V_{DS} = 15 mV. This value is consistent with those obtained in planar single-crystalline SnO$_2$ nanorods and other mental oxide nanorods [34, 35]. We obtained a one-dimensional electron density of $n = Q/eL \approx 1.17 \times 10^7$ cm^{-1}. The mobility of the electrons can also be deduced from the trans-conductance of the field-effect transistor. In the linear region, the trans-conductance g_m is expressed as the formulae (2) and (3). From Figure 33, in the field-effect transistor device, the mobility of the channel stoichiometry was found to be about 82 $cm^2V^{-1}s^{-1}$. Transistors composed of the SnO_2 nanorods exhibited good electrical properties which affected both the on/off ratio and threshold voltage of the devices. Electrical measurements confirmed that the SnO_2 nanorods were *n*-type semiconductors. Our findings indicate that other one-dimension nanostructural materials may be manipulated by using this simple technique. This might provide insight into new opportunities for the applications in building blocks for nanoelectronics and active sensing materials.

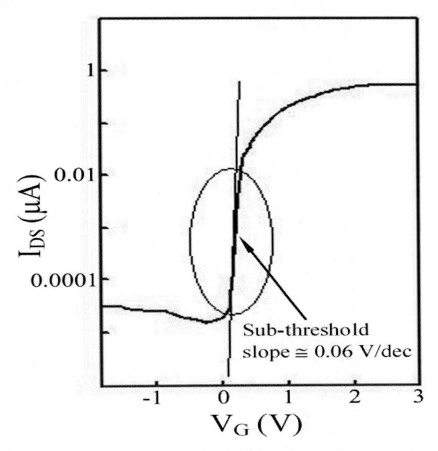

Figure 33. The I_{DS} - V_{GS} curve of field-effect transistor device for the single SnO_2 nanorod in log scale at V_{DS} = 15 mV.

5. CONCLUSIONS

In summary, we summarized in detail the synthesis, characterization and applications of SnO_2 thin films and nanowires by using pulsed laser deposition techniques as well as the SnO_2 nanorods by using micro-emulsion method. The following conclusions can be drawn from the present work:

(1) We have successfully prepared SnO$_2$ thin films at different substrate temperatures with interesting fractal features. The experimental evidence indicated that the fractal clusters with various sizes, densities, and fractal dimensions were affected by different substrate temperatures. The formation process of significant fractal features could be reasonably described by a novel model: (i) operation of the KrF excimer laser; (ii) production of the SnO$_2$ plasma; (iii) cooling of the SnO$_2$ plumes; (iv) deposition of the SnO$_2$ plume on the Si substrate; (v) grain rotation; (vi) formation of coherent boundary between grains followed by coalescence; (vii) growth of SnO$_2$ nanocrystals along preferred crystallographic directions; and (viii) formation of the fractal structure. CO gas sensitivity measurement confirmed that the gas sensing behavior is sensitively dependent on fractal dimensions, fractal densities, and average sizes of the fractal clusters. It was found that the sensitivity increases with increasing CO concentration and decreasing fractal dimension. We have shown that fractal methodology can be applied to the evaluation of SnO$_2$ thin films. This gas sensing behavior could be explained by the Random Tunneling Junction Network (RTJN) mechanism.

(2) The microstructural and morphological analyses of SnO$_2$ thin films demonstrated that the annealing temperature is very important parameter to determining the thin film quality. It affects the phase formation, crystalline microstructure and preferred orientation of the thin films. It was found that the amorphous microstructure almost transformed into a polycrystalline SnO$_2$ phase and preferred orientation related to the (110), (101) and (211) crystal planes with increasing annealing temperature. SnO$_2$ thin films prepared PLD techniques are mostly nanocrystalline in nature and crystallinity improves with increasing annealing temperature. The thin film deposited at 200 °C demonstrated the best crystalline properties, viz. optimum growth conditions. However, the thin film annealed at 100 °C revealed the minimum average root-mean-square roughness of 20.6 nm with average grain size of 26.6 nm.

(3) The defects inside SnO$_2$ thin films were significantly decreased after SnO$_2$ thin films were annealed at 300 °C for 2 h by in-situ observation. HRTEM investigation of microstructures of SnO$_2$ thin films after in-situ annealing indicated that the stacking faults were annihilated in the as-grown SnO$_2$ thin films. Raman studies confirmed that the in-situ annealed sample is consisted of low defect density. The SnO$_2$ thin films with low defect density are promising materials for future transparent nano-electrodes and solid-state gas nano-sensors.

(4) The microstructural and optical properties of SnO$_2$ thin films deposited on glass substrates by PLD techniques were investigated before and after thermal annealing in the temperature range from 50 °C to 550 °C at 50 °C intervals. XRD confirmed that the various SnO$_2$ thin films consisted of nanoparticles with average grain size in the range of 23.7-28.9 nm. The microstructural analysis demonstrated that the annealing temperature is a very important parameter. It affects crystalline microstructure and optical properties of thin films. The transmittance measurements indicated that present SnO$_2$ thin films can be used as a window material for solar cell devices due to their > 75.3 % transmittance. Various optical parameters such as optical band gas energy, refractive index and optical conductivity were calculated from the optical transmittance and reflectance data recorded in the wavelength range 300-2500 nm. We found that the optical band gas energy and refractive index of SnO$_2$ thin films

annealed at various temperatures presented approximately linear behaviour, and SnO$_2$ thin films exhibited high optical conductivity. These findings revealed that SnO$_2$ thin film annealed up to 400 °C is a good window material for solar cell devices. Our experimental results indicated that SnO$_2$ thin films with the high optical quality could be synthesized by PLD techniques.

(5) The post annealing shows greater tendency to affect the structural and electrical properties of SnO$_2$ thin films fabricated by PLD method which composed of nanoparticles. Thermal annealing decreases the RMS roughness values. Electrical resistivity of present thin films shows an oscillatory behavior. We found that nearly opposite actions to RMS surface roughness and electrical resistivity make a unique performance with thermal annealing.

(6) PLD techniques can be successfully developed to synthesize single crystalline rutile SnO$_2$ nanowires. SnO$_2$ nanowires with diameters of about 10-30 nm and lengths of several hundreds nanometers to a few micrometers have been successfully grown along the [110] growth direction. The as-prepared SnO$_2$ nanowires were structurally uniform, single crystalline. The growth behavior of SnO$_2$ nanowires was most likely controlled by the VS growth mechanism. The SnO$_2$ nanowires based on field-effect transistor devices exhibited that the SnO$_2$ nanowires hold better electrical properties, which might provide insight into the new opportunities to the applications in constructing nanoscaled electronic and optoelectronic devices.

(7) One-dimensional nanostructures of SnO$_2$ nanorods have been successfully synthesized in bulk-quantity via a calcining process based on annealing precursor powders. The SnO$_2$ nanorods showed the single-crystal features and had preferred [110] growth direction. Particularly, the surfactant (NP9/NP5/OP) is of critical importance in the formation of SnO$_2$ nanorods. The SnO$_2$ nanorods based field-effect transistor devices prepared by our method exhibit good electrical properties. The electrical measurements confirmed that the SnO$_2$ nanorods are *n*-type semiconductors. We believe these nanorods could be used as active sensing materials and building blocks for nanoelectronics. The findings indicate that other nanorods or nanowires may be manipulated by using this simple technique.

This review article is an interdisciplinary work that integrates the areas of physics, chemistry and materials science. Some challenges are still remained in the motivation to clarify the intricate aspects and potential applications of SnO$_2$ thin films and nanostructured materials. The results may enable novel SnO$_2$ functional materials with appropriate microstructures to be tailor made for a large number of applications and provide new opportunities for future study of SnO$_2$ architectures with the goal of optimizing functional material properties for specific applications.

ACKNOWLEDGMENTS

The work described in this review article was financially supported by the Shanghai Pujiang Program (10PJ1404100), China, Innovation Funds of Shanghai Municipal Education Commission (10ZZ64) and Shanghai University, National Natural Science Foundation of

China (Program 20871081 and Key Program 40830744), Science and Technology Commission of Shanghai Municipality (10JC1405400, 09530501200 and 09XD1401800), and Shanghai Leading Academic Discipline Project (S30109). This work also was supported by a grant from the CityU (Project No. 7002295).

REFERENCES

[1] G. Ansari, P. Boroojerdian, S. R. Sainkar, R. N. Karekar, R. C. Alyer and S. K. Kulkarni, *Thin Solid Films* 295 (1997) 271.
[2] O. K. Varghese and L. K. Malhotra, *Sens. Actuat. B* 53 (1998) 19.
[3] Y. S. He, J. C. Campbell, R. C. Murphy, M. F. Arendt and J. S. Swinnea, *J. Mater. Res.* 8 (1993) 3131.
[4] D. Z. Wang, S. L. Wen, J. Chen, S. Y. Zhang and F. Q. Li, *Phys. Rev. B* 49 (1994) 14282.
[5] Z. W. Chen, D. Y. Pan, B. Zhao, G. J. Ding, Z. Jiao, M. H. Wu, C. H. Shek, C. M. L. Wu and J. K. L. Lai, *ACS Nano* 4 (2010) 1202.
[6] Z. W. Chen, H. J. Zhang, Z. Li, Z. Jiao, M. H. Wu, C. H. Shek, C. M. L. Wu and J. K. L. Lai, *Acta Materialia* 57 (2009) 5078.
[7] Z. W. Chen, G. Liu, H. J. Zhang, G. J. Ding, Z. Jiao, M. H. Wu, C. H. Shek, C. M. L. Wu and J. K. L. Lai, *J. Non-Cryst. Solids* 355 (2009) 2647.
[8] Z. L. Wang and Z. C. Kang, *Functional and Smart Materials-Structural Evolution and Structure Analysis,* (Plenum Press, New York 1998).
[9] Z. W. Pan, Z. R. Dai and Z. L. Wang, *Science* 291 (2001) 1947.
[10] Z. W.Chen, J. K. L. Lai and C. H. Shek, *Phys. Rev. B* 70 (2004) 165314.
[11] Z. W. Chen, J. K. L. Lai and C. H. Shek, *Appl. Phys. Lett.* 88 (2006) 033115.
[12] Z. R. Dai, J. L. Gole, J. D. Stout and Z. L.Wang, *J. Phys. Chem. B* 106 (2002) 1274.
[13] Z. L. Wang, *Adv. Mater.* 15 (2003) 432.
[14] H. T. Ng, J. Li, M. K. Smith, P. Nguyen, A. Han, J. Cassell and M. Meyyappan, *Science* 300 (2003) 1249.
[15] B. Cheng, J. M. Russell, W. S. Shi, L. Zhang and E. T. Samulski, *J. Am. Chem. Soc.* 126 (2004) 5972.
[16] U. Mikko, L. Hanna, V. Heli, N. Lauri, R. Resch and F. Gernot, *Mikrochim. Acta* 133 (2000) 119.
[17] P. Hidalgo Falla, H. E. Peres, M. D. Gouvêa and F. J. Ramirez-Fernandez, *Materials Science Forum* 636 (2005) 498.
[18] T. Oyabu, *J. Appl. Phys* 53 (1982) 2785.
[19] Z. W. Chen, J. K. L. Lai and C. H. Shek, *Appl. Phys. Lett.* 89 (2006) 231902.
[20] V. Demarne and A. Grisel, *Sens. Actuat.* 13 (1988) 301.
[21] O. D. Santos, M. L. Weiller, D. Q. Junior and A. N. Medina, *Sens. Actuta. B* 75 (2001) 83.
[22] S. Shukla, S. Seal, L. Ludwing and C. Parish, *Sens. Actuta. B* 97 (2004) 256.
[23] S. Harbeck, A. Szatvanyi, N. Barsan, U. Weimar and V. Hoffmann, *Thin Solid Films* 436 (2003) 76.
[24] H. Windischmann and P. Mark, *J. Electrochem. Soc.* 126 (1979) 627.

[25] V. Lantto and P. Romppainen, *Surf. Sci.* 192 (1987) 243.
[26] P. Romppainen and V. Lantto, *J. Appl. Phys.* 63 (1988) 5159.
[27] C. H. Shek, J. K. L. Lai and G. M. Lin, *J. Phys. Chem. Solids* 60 (1999) 189.
[28] Z. W. Chen, S. Y. Zhang, S. Tan, J. G. Hou and Y. H. Zhang, *Thin Solid Films* 322 (1998) 194.
[29] J. Feder, *Fractal,* (Plenum Press, New York 1988).
[30] H. R. Fallah, M. Ghasemi and A. Hassanzadeh, *Phys. E* 39 (2007) 23.
[31] Z. W. Chen, J. K. L. Lai, C. H. Shek and H. D Chen, *Appl. Phys. A* 81 (2005) 959.
[32] Y. J. Kim, Y. T. Kim, H. K. Yang, J. C. Park, J. I. Han, Y. E. Lee and H. J. Kim, *J. Vac. Sci. Tech. A* 15 (1997) 1103.
[33] S. Zhu, Y. F. Lu, M. H. Hong and X. Y. Chen, *J. Appl. Phys.* 89 (2001) 2400.
[34] D. Kim and H. Lee, *J. Appl. Phys.* 89 (2001) 5703.
[35] S. Zhu, Y. F. Lu and M. H. Hong, *Appl. Phys. Lett.* 79 (2001) 1396.
[36] L. Berthe, R. Fabbro, P. Peyre, L. Tollier and E. Bartnicki, *J. Appl. Phys.* 82 (1997) 2826.
[37] R. L. Penn and J. F. Banfield, *Geochim. Cosmochim. Acta* 63 (1999) 1549.
[38] R. L. Penn and J. F. Banfield, *Science* 281 (1998) 969.
[39] J. G. Hou and Z. Q. Wu, *Phys. Rev. B* 40 (1989) 1008.
[40] Z. W. Chen, S. Y. Zhang, S. Tan, J. G. Hou and Y. H. Zhang, *J. Vac. Sci. Technol. A* 16 (1998) 2292.
[41] R. B. Cooper, G. N. Advani and A. G. Jordan, *J. Electron. Mater.* 10 (1981) 455.
[42] Cicera, A. Dieguez, R. Diaz, A. Cornet and J. R. Morante, *Sens. Actuat. B* 58 (1999) 360.
[43] K. L. Chopra, S. Major and D. K. Pandya, *Thin Solid Films* 102 (1983) 1.
[44] L. Abello, B. Bochu, A. Gaskov, S. Koudryavtseva, G. Lucazeau and M. Roumyantseva, *J. Solid State Chem.* 135 (1998) 78.
[45] S. G. Ansari, P. Boroojerdian, S. R. Sainkar, R. N. Karekar, R. C. Aiyer and S. K. Kulkarni, *Thin Solid Films* 295 (1997) 271.
[46] Z. W. Chen, C. M. L. Wu, C. H. Shek, J. K. L. Lai, Z. Jiao and M. H. Wu, *Critical Reviews in Solid State and Materials Sciences* 33 (2008) 197.
[47] S. W. Lee, Y. W. Kim and H. D. Chen, *Appl. Phys. Lett.* 78 (2001) 350.
[48] M. G. Mason, L. S. Hung, C. W. Tang, S. T. Lee, K. W. Wong and M. Wang, *J. Appl. Phys.* 86 (1999) 1688.
[49] W. Göpel and K. D. Schierbaum, *Sens. Actuat. B* 26 (1995) 1.
[50] M. J. Madou and S. R. Morrison, *Chemical sensing with solid state devices,* (Academic Press, New York, 1989).
[51] J. H. Sung, Y. S. Lee, J. W. Lim, Y. H. Hong and D. D. Lee, *Sens. Actuat. B* 66 (2000) 149.
[52] C. K. Kim, S. M. Choi, I. H. Noh, J. H. Lee, C. Hong, H. B. Chae, G. E. Jang and H. D. Park, *Sens. Actuat. B* 77 (2001) 463.
[53] M. A. El Khakani, R. Dolbec, A. M. Serventi, M. C. Horrillo, M. Trudeau, R. G. Saint-Jacques, D. G. Rickerby and I. Sayago, *Sens. Actuat. B* 77 (2001) 383.
[54] R. Dolbec, M. A. El Khakani, A. M. Serventi, M. Trudeau and R. G. Saint-Jacques, *Thin Solid Films* 419 (2002) 230.
[55] Z. W. Chen, J. K. L. Lai, C. H. Shek and H. D. Chen, *J. Mater. Res.* 18 (2003) 1289.

[56] T. W. Kim, D. U. Lee, J. H. Lee, D. C. Choo, M. Jung and Y. S. Yoon, *J. Appl. Phys.* 90 (2001) 175.
[57] G. G. Mandayo, E. Castano, F. J. Gracia, A. Cirera, A. Cornet and J. R. Morante, *Sens. Actuat. B* 95 (2003) 90.
[58] H. Ogawa, M. Nishikawa and A. Abe, *J. Appl. Phys.* 53 (1982) 4448.
[59] V. R. Katti, A. K. Debnath, K. P. Muthe, M. Kaur, A. K. Dua, S. C. Gadkari, S. K. Gupta and V. C. Sahni, *Sens. Actuat. B* 96 (2003) 245.
[60] B. K. Min and S. D. Choi, *Sens. Actuat. B* 98 (2004) 239.
[61] G. Korotcenkov, V. Brinzari, J. Schwank and A. Cerneavschi, *Mater. Sci. Eng. C* 19 (2002) 73.
[62] C. Alfonso, A. Charai, A. Armigliato and D. Narducci, *Appl. Phys. Lett.* 68 (1996) 1207.
[63] S. Shukla, S. Patil, S. C. Kuiry, Z. Rahman, T. Du, L. Ludwig, C. Parish and S. Seal, *Sens. Actuat. B* 96 (2003) 343.
[64] D. Kotsikau, M. Ivanovskaya, D. Orlik and M. Falasconi, *Sens. Actuat. B* 101 (2004) 199.
[65] G. Zhang and M. L. Liu, *Sens. Actuat. B* 69 (2000) 144.
[66] G. Korotcenkov, V. Brinzari, V. Golovanov and Y. Blinov, *Sens. Actuat. B* 98 (2004) 41.
[67] G. Korotcenkov, V. Macsanov, V. Tolstoy, V. Brinzari, J. Schwank and D. Faglia, *Sens. Actuat. B* 96 (2003) 602.
[68] C. N. Xu, J. Tamaki, N. Miura and N. Yamazoe, *Sens. Actuat. B* 3 (1991) 147.
[69] G. Sberveglieri, *Sens. Actuat. B* 6 (1992) 239.
[70] D. B. Chrisey and G. K. Hubler, *Pulsed laser deposition of thin films,* (Wiley New York, 1994).
[71] P. R. Willmott and J. R. Huber, *Rev. Mod. Phys.* 72 (2000) 315.
[72] O. Auciello and J. Engemann, *Multicomponent and multilayered thin films for advanced microtechnologies: techniques, fundamentals and devices,* (Kluwer, Netherlands, 1993).
[73] D. Bäuerle, *Laser processing and chemistry,* (Springer, New York, 1996).
[74] M. Von Allmen, A. Blatter, *Laser-beam interactions with materials,* (Springer, New York, 1995).
[75] R. Dolbec, M. A. El Khakani, A. M. Serventi and R. G. Saint-Jacques, *Sens. Actuat. B* 93 (2003) 566.
[76] P. Petrik, L. P. Biró, M. Fried, T. Lohner, R. Berger, C. Schneider, J. Gyulai and H. Ryssel, *Thin Solid Films* 315 (1998) 186.
[77] G. McCarthy and J. Welton, *J. Powder Diffraction* 4 (1989) 156.
[78] Z. W. Chen, J. K. L. Lai, C. H. Shek and H. D. Chen, *Appl. Phys. A* 81 (2005) 1073.
[79] Duparre, *Handbook of optical properties,* (CRC, Press, Vol.1, 1995).
[80] M. Senthilkumar, N. K. Sahoo, S. Thakur and R. B. Tokas, *Appl. Sur. Sci.* 245 (2005) 114.
[81] T. Lindström, J. Isidorsson and G. A. Niklasson, *Thin Solid Films* 401 (2001) 165.
[82] Y. K. Liu, C. L. Zheng, W. Z. Wang, C. R. Yin and G. H. Wang, *Adv. Mater.* 13 (2001) 1883.
[83] C. K. Xu, G. D. Xu, Y. K. Liu, X. L. Zhao and G. H. Wang, *Scripta Mater.* 46 (2002) 789.

[84] Z. R. Dai, Z. W. Pan and Z. L. Wang, *Solid State Commun.* 118 (2001) 351.
[85] J. Q. Hu, X. L. Ma, N. G. Shang, Z. Y. Xie, N. B .Wong, C. S. Lee and S. T. Lee. *J. Phys. Chem. B* 106 (2002) 3823.
[86] A, Maddalena, R. D Maschio, S. Dire and A. J. Raccanelli, *J. Non-Cryst. Solids* 121 (1990) 365.
[87] C. H. Shek, J. K. L. Lai and G. M. Lin, *NanoStuct. Mater.* 11 (1999) 887.
[88] R. N. Ghostagore, *J. Electrochem. Soc.* 125 (1978) 110.
[89] R. D. Tarey and T. A. Raju, *Thin Solid Films* 128 (1995) 181.
[90] T. Minami, H. Nanto and S. J. Takata, *J. Appl. Phys.* 27 (1988) L287.
[91] J. J. Zhu, Z. H. Lu, S. T. Aruna, D. Aurbach and A. Gedanken, *Chem. Mater.* 12 (2000) 2557.
[92] V. Schosser and G. Wind, *Proceedings of the 8th EC Photovoltaic Solar Energy Conference,* (Florence, Italy, 1998).
[93] M. Serventi, R. Dolbec, M. A. El Khakani, R. G. Saint-Jacques and D. G. Rickerby *J. Phys. Chem. Solids* 64 (2003) 2097.
[94] P. Serrini, V. Briois, M. C. Horrillo, A. Traverse and L. Manes, *Thin Solid Films* 304 (1997) 113.
[95] C. Xu, J. Tamaki, N. Miura and N. Yamazoe, *J. Mater. Sci. Lett.* 8 (1989) 1092.
[96] Z. W. Chen, J. K. L. Lai and C. H. Shek, *Chem. Phys. Lett.* 422 (2006) 1.
[97] Z. W. Chen, J. K. L. Lai and C. H. Shek, *Solid State Chem.* 178 (2005) 892.
[98] G. Nolsson and G. Nelin, *Phys. Rev. B* 6 (1972) 3777.
[99] S. L. Zhang, B. F. Zhu, F. M. Huang, Y. Yan, E. Y. Shang, S. S. Fan and W. G. Han, *Solid State Commun.* 111 (1999) 647.
[100] R. E. Presley, C. L. Munsee, C. H. Park, D. Hong, J. F. Wager and D. A. Keszler, *J. Phys. D* 37 (2004) 2810.
[101] M. Law, H. Kind, B. Messer, F. Kim and P. D. Yang, *Angew. Chem. Int. Ed.* 41 (2002) 2405.
[102] C. Nayral, E. Viala, P. Fau, F. Senocq, J. C. Jumas, A. Maisonnat and B. Chaudret, *Chem.-Eur. J.* 6 (2000) 4082.
[103] S. de Monredon, A. Cellot, F. Ribot, C. Sanchez, L. Armelao, L. Gueneau and L. Delattre, *J. Mater. Chem.* 12 (2002) 2396.
[104] Y. A. Cao, X. T. Zhang, W. S. Yang, H. Du, Y. B. Bai, T. J. Li and J. N. Yao, *Chem. Mater.* 12 (2000) 3445.
[105] M. T. Bhatti, A. M. Rana and A. F. Khan, *Mater. Chem. Phys.* 84 (2004) 126.
[106] D. Olguín, M. Cardona and A. Cantarero, *Solid State Commun.* 122 (2002) 575.
[107] S. Das, S. Kar and S. Chaudhuri, *J. Appl. Phys.* 99 (2006) 114303.
[108] R. Peaker and B. Horsley, *Rev. Sci. Instrument* 42 (1971) 1825.
[109] M. Von Ortenberg, J. Link and R. Helbig, *J. Opt. Soc. Am.* 67 (1977) 968.
[110] E. Bucher, *Appl. Phys.* 17 (1978) 1.
[111] K. Ghosh, C. Fishman and T. Feng, *J. Appl. Phys.* 49 (1978) 3490.
[112] J. Watson, *Sens. Actuat.* 5 (1984) 29.
[113] X. Feng, J. Ma, F. Yang, F. Ji, F. Zong, C. Luan and H. Ma, *Mater. Lett.* 62 (2008) 1779.
[114] N. Mukashev, S. Z. Tokmoldin, N. B. Beisenkhanov, S. M. Kikkarin, I. V. Valitova, V. B. Glazman, A. B. Aimagambetov, E. A. Dmitrieva and B. M. Veremenithev, *Mater. Sci. Eng. B* 118 (2005) 164.

[115] Q. H. Wu, J. Song, J. Kang, Q. F. Dong, S. T. Wu and S. G. Sun, *Mater. Lett.* 61 (2007) 3679.
[116] Z. W. Chen, S. Y. Zhang, S. Tan, J. G. Hou, Y. H. Zhang and H. Sekine, *J. Appl. Phys.* 89 (2001) 783.
[117] J. Lee, *Thin Solid Films* 516 (2008) 1386.
[118] Y. Ku, I. H. Kim, I. Lee, K. S. Lee, T. S. Lee, J. H. Jeong, B. Cheong, Y. J. Baik and W. M. Kim, *Thin Solid Films* 515 (2006) 1364.
[119] P. V. Braun, P. Osenar and S. I. Stupp, *Nature* 380 (1996) 325.
[120] Y. Zhang, K. Suenaga, C. Colliex and S. Lijima, *Science* 281 (1998) 973.
[121] Y. Zhang, T. Ichihashi, E. Landree, F. Nihey and S. Lijima, *Science* 285 (1999) 1719.
[122] X. Peng, L. Manna, W. Yang, J. Wickham, E. Scher, A. Kadavanich and A. P. Alivisatos, *Nature* 404 (2000) 59.
[123] X. F. Duan, Y. Huang, Y. Cui, J. F. Wang and C. M. Lieber, *Nature* 409 (2001) 66.
[124] H. Y. Peng, Z. W. Pan, L. Xu, X. H. Fan, N. B. Wang, C. S. Lee and S. T. Lee, *Adv. Mater.* 13 (2001) 317.
[125] B. Murray, C. R. Cagan and M. G. Bawendi, *Science* 270 (1995) 1335.
[126] R. Leite, I. T. Weber, E. Longo and J. A. Varela, *Adv. Mater.* 12 (2000) 965.
[127] H. W. Postma, T. Teepen, Z. Yao, M. Grifoni and C. Deckker, *Science* 293 (2001) 76.
[128] R. G. Gordon, *MRS Bull.* 25 (2000) 52.
[129] C. Li, D. H. Zhang, S. Han, X. L. Liu, T. Tang and C. W. Zhou, *Adv. Mater.* 15 (2003) 143.
[130] Y. Idota, T. Kubota, A. Matsufuji, Y. Maekawa and T. Miyasaka, *Science* 276 (1997) 1395.
[131] Z. Q. Liu, D. H. Zhang, S. Han, C. Li, T. Tang, W. Jin, X. L. Liu, B. Lei and C. W. Zhou, *Adv. Mater.* 15 (2003) 1754.
[132] D. F. Zhang, L. D. Sun, J. L. Yin and C. H. Yan, *Adv. Mater.* 15 (2003) 1022.
[133] M. H. Huang, Y. Wu, H. Feick, N. Tran, E. Weber and P. Yang, *Adv. Mater.* 13 (2001) 113.
[134] R. Leite, J. W. Gomes, M. M. Oliveira, E. J. H. Lee, E. Longo, J. A. Varela, C. A. Paskocimas, T. M. Boschi, F. Lanciotti Jr, P. S. Pizani and P. C. Soares Jr, *Appl. Sci. Res.* 2 (2002) 125.
[135] M. Law, H. Kind, B. Messer, F. Kim and P. Yang, *Angew. Chem.* 114 (2002) 2511.
[136] I. Liu, N. I. Maluf and R. F. W. Pease, *J. Vac. Sci. Technol. B* 10 (1992) 2846.
[137] T. Ono, H. Saitoh and M. Esashi, *Appl. Phys. Lett.* 70 (1997) 1852.
[138] F. C. Frank, *Discovery Faraday* 5 (1949) 48.
[139] R. S. Wagner and W. C. Ellis, *Appl. Phys. Lett.* 4 (1964) 89.
[140] J. G. Zheng, X. Q. Pan, M. Schweizer, U. Weimar, W. Göpel and M. Rühle, *Philos. Mag. Lett.* 73 (1996) 93.
[141] Y. Cheng, P. Xiong, L. Fields, J. P. Zheng, R. S. Yang and Z. L. Wang, *Appl. Phys. Lett.* 89 (2006) 093114.
[142] M. G. McDowell, R. J. Sanderson and I. G. Hill, *Appl. Phys. Lett.* 92 (2008) 013502.
[143] C. Dekker, *Phys. Today* 52 (1999) 22.
[144] M. Morales and C. M. Lieber, *Science* 279 (1998) 208.
[145] X. Duan and C. M. Lieber, *J. Am. Chem. Soc.* 122 (2000) 188.
[146] Y. Wu and P. Yang, *Chem. Mater.* 12 (2000) 605.
[147] D. S. Ginley and C. Bright, *Mater. Res. Soc. Bull.* 25 (2000) 15.

[148] N. Yamazoe, *Sens. Actuat. B* 5 (1991) 7.
[149] K. H. Yoon, Y. S. Cho and D. H. Kang, *J. Mater. Sci.* 33 (1998) 2977.
[150] Wada, K. Sakane and T. Kitamura, *J. Mater. Sci. Lett.* 10 (1991) 1076.
[151] S. Hashimoto and A. Yamaguchi, *J. Eur. Ceram. Soc.* 20 (2000) 397.
[152] M. Kajiwara, *J. Mater. Sci.* 22 (1987) 1223.
[153] C. Tas, *J. Am. Ceram. Soc.* 84 (2001) 295.
[154] Y. Hayashi, T.Kimura and T. Yamaguchi, *J. Mater. Sci.* 21 (1986) 757.
[155] M. Ocana and C. Serna, *J. Spectrochim. Acta, Part A* 47 (1991) 765.

Chapter 3

CURRENT METHODS OF DENTAL CERAMIC AND METAL SURFACE TREATMENT FOR BONDING

*Boonlert Kukiattrakoon**

*Associate Professor, Department of Conservative Dentistry,
Faculty of Dentistry, Prince of Songkla University, Thailand*

ABSTRACT

This chapter focuses on the current methods of dental ceramic and metal surface treatment intraorally for the purpose of repairing or bonding to ceramic restorations. Fractured ceramics in ceramic restorations cause esthetic problems, functional problems, loss of time and economic cost to patients for fabricating a new one. Repairing the fractured restoration is an alternative choice for the patients. However, successful repair requires the bonding of repair materials to appropriately prepared surfaces of ceramics or metals. Ceramic surface treatments can be treated by both mechanical and chemical methods. Mechanical methods include grinding with coarse diamond burs, air abrasion, etching with hydrofluoric acid or acidulated phosphate fluoride gel, and laser irradiation, whilst chemical bonding involves silica coated aluminium oxide, and silane treatment. Similarly, metal surface preparations incorporate mechanical procedures such as grinding with carborundum discs or diamond burs, air abrasion, and chemical etching or electrolytic etching. Chemical bonding to metal has also been treated by tin plating, silica coated aluminium oxide, and primers or silane application. Furthermore, this chapter has also included the method of choice in clinical applications for current ceramic and metal surface treatment.

INTRODUCTION

Dental ceramics are increasingly utilized for both anterior and posterior restorations including inlays, onlays, laminate veneers, metal-ceramic and all-ceramic crowns or fixed partial dentures [1]. Their uses have been increased substantially due to their excellent

---

* E-mail address: boonlert.k@psu.ac.th.

esthetic properties, biocompatibility, resistance to wear, and high resistance to compression. They also have low thermal and electrical conductance, and a coefficient of thermal expansion that is similar to enamel and dentin, resulting in minimal marginal leakage [2,3]. In addition, dental ceramics are considered chemically inert restorative materials.

However, dental ceramics are brittle and highly susceptible to fracture propagation caused by inherent flaws (cracks or pores) or defects on the surface and body, which are created during fabrication (Figure 1).

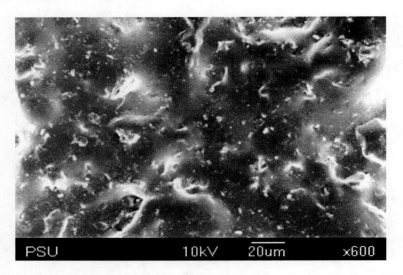

Figure 1. Illustration of surface flaws of a feldspathic ceramic (original magnification ×600).

These flaws behave as sharp notches whose tips may be as narrow as the spacing between several atoms in the dental ceramics. When the induced mechanical stress surpasses the actual strength of the dental ceramics, the bonds at the notch tip smash and form a crack [2]. Tensile or bending stresses then open and widen inherent flaws in the ceramic, and subsequently, failure occurs. These characteristics impair physical properties of dental ceramics such as surface roughness, surface hardness, strength [4,5], and influence the clinical success and failure of ceramic restorations [6]. On the contrary, compressive stresses close these flaws and then improve strength of the ceramics (Figure 2).

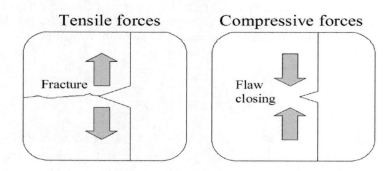

Figure 2. Illustration of the effect of tensile and compressive forces on ceramics (modified from O'Brien [9]).

Therefore, to overcome this problem, many methods have been developed to improve mechanical properties of dental ceramics and expand their clinical applications by "closing" these flaws, for example, bonding ceramic to metal (forming metal-ceramic restorations) [7,8], residual surface stressing including ion exchange and thermal tempering, adding crystalline materials in glass (dispersion strengthening) as in some types of all-ceramic restorations, surface treatments including polishing and glazing to eliminate surface flaws, or transformation toughening (as in zirconia ceramics) [2,9].

Even though these attempts have been utilized to improve strength of the dental ceramics, fractured ceramics may occur as a result of various factors such as extensive occlusal forces, impact and fatigue load, poor substructure design, flexural fatigue of the metal substructure, insufficient porcelain thickness, an accident, or long-term use [10]. Hence, ceramic fractures have been reported to be the second greatest cause for restoration replacement after dental caries [10]. Fractured ceramics in ceramic or metal-ceramic restorations cause both esthetic and functional problems. Restoration replacement may be required, which results in issues related to the time required at the clinic and laboratory, and additional cost for the patients especially in multiple-unit fixed partial dentures [11]. Therefore, patients may prefer to have fractured ceramic repairing as an alternative choice to help maintain their esthetics, functions, and definitely reduce their financial cost. Patients also demand rapid case resolution and may occasionally delay the replacement of a fractured ceramic or metal-ceramic restoration [3]. Dentists usually search for various methods to lengthen the service life time of existing restorations.

The repair of fractured ceramic or metal-ceramic restorations has been reported to produce acceptable esthetics and functional successes depending on the materials, methods, and techniques [12-16]. For successful bonding, some necessities are required. The first is adequate micromechanical retention when clinical adhesion is negligible. The second is adequate wetting of the ceramic surface by the bonding agent. The third and final necessity is the need for resistance to fatigue, stress, erosion, and stress relief [17,18]. From these requirements, one of the important factors for success is sufficient micromechanical retention, achieved by ceramic or metal surface treatments for bonding to repaired materials, even for the purpose of cementation. Therefore, the aim of this chapter was to focus on the current methods of dental ceramic and metal surface treatment intraorally for the purpose of repairing or bonding all-ceramic or metal-ceramic restorations.

CLASSIFICATION OF BOND FAILURE

Prior to describing all methods, a classification of bond failure in metal-ceramic restorations was mentioned so that dentists might know the failures which could occur and eliminate the causes. In 1977, O'Brien [19] classified the type of failures in metal-ceramic restorations into 6 types (Figure 3).

Type I, failure occurred between the metal and ceramic. Type II, failure happened between the metal oxide layer and ceramic. Type III, failure raised inside the ceramic (cohesive failure within the ceramic). Type IV, failure occurred between the metal oxide layer and metal. Type V, failure took place inside the metal oxide layer (cohesive failure within the oxide layer). Type VI, failure arose inside the metal (cohesive failure within the metal).

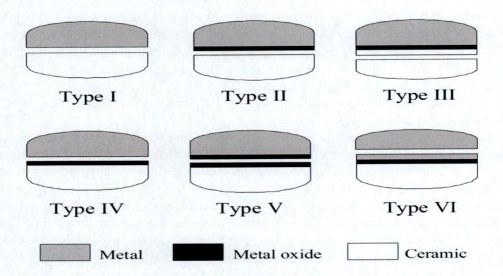

Figure 3. Illustration of classification of bond failure (modified from O'Brien [19]).

According to this classification, simply speaking, the two remaining surfaces after fractures have occurred are ceramic or metal, or more likely a combination of both, as illustrated in Figure 4.

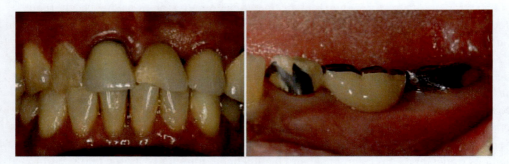

Figure 4. Illustrations of fractured restorations: (left) a fractured ceramic on the left maxillary central incisor (pontic), and (right) a combination of ceramic and metal exposed surfaces on the left mandibular second premolar.

Hence, appropriate surface treatments of ceramic or metal are the most important factors in successful bonding for repairing or cementation.

CERAMIC SURFACE TREATMENT

The surface treatment methods of ceramics have been divided into 2 methods, namely, mechanical and chemical methods. Mechanical methods include grinding with coarse diamond burs, air abrasion (or sandblasting), etching with hydrofluoric (HF) acid or acidulated phosphate fluoride (APF) gel, and laser irradiation, whilst the chemical method involves silica coated aluminium oxide and silane treatment.

Mechanical Methods of Ceramic Surface Treatment

The purposes of this method are to develop rough, groove, or undercut archetype surfaces of ceramic and increase the surface area for adhesive cement to flow into this prepared surface, forming a mechanical interlocking. Mechanical methods of ceramic surface treatment involve grinding with coarse diamond burs, sandblasting, etching with HF acid or APF gel, and laser irradiation.

Grinding with Diamond Burs

In 1973, Jochen [20] was the first person who reported a surface treatment of ceramic with a green stone bur for repairing a porcelain tooth with a resin composite. He provided the reason for grinding the ceramic with a green stone bur was that it created a rough and an irregular surface necessary for improving resin composite bonding. Subsequently, Jochen and Caputo [21] further investigated this by comparing the shear bond strength between a resin composite and a ceramic treated surface with a coarse diamond bur, a heatless stone bur, a green stone bur, and a carborundum disc. The results presented that ceramic surface treatment with a coarse diamond bur showed significantly the highest shear bond strength. The reason was that a coarse diamond bur had a rough and high surface hardness, therefore, it caused a rougher surface for the ceramic which improved adhering to resin composite more than other burs. A scanning electron microscope (SEM) photomicrograph of a grinded surface of a ceramic is presented in Figure 5.

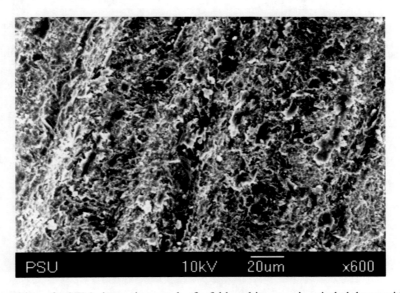

Figure 5. Illustration of a SEM photomicrograph of a feldspathic ceramic grinded down with a coarse diamond bur (original magnification ×600).

Air Abrasion or Sandblasting

In 1989, Bertolotti and co-workers [22] first introduced ceramic surface treatment by sandblasting with 50 microns of aluminium oxide for roughening. Consequent investigation by Wolf and co-workers [23] in 1993 confirmed that sandblasting with 50 microns of aluminium oxide provided a higher shear bond strength than other sizes of aluminium oxide particles. A SEM photomicrograph of a sandblasted ceramic surface is illustrated in Figure 6. Presently, several intraoral sandblasting devices have been launched in the market, for instance, the MicroEtcher IIA (Danville Materials, San Ramon, CA, USA), the Micro-Etcher II (DynaFlex, St.Ann, MO, USA), the CoJet Prep (3M ESPE, St. Paul, MN, USA), or the MiniBlaster (Deldent Ltd., Petach Tikva, Isael). Figure 7 is an example of an intraoral sandblasting device.

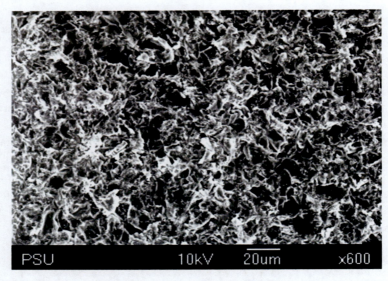

Figure 6. Illustration of a SEM photomicrograph of a feldspathic ceramic sandblasted with 50 microns of aluminium oxide (original magnification ×600).

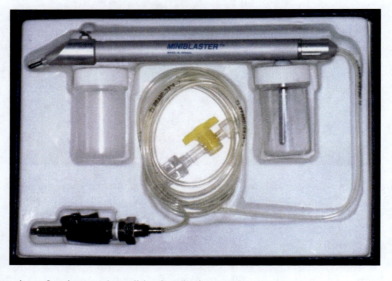

Figure 7. Illustration of an intraoral sandblasting device.

Etching with HF Acid

HF acid is an organic acid which is used extensively for etching glass or ceramic surfaces. The starting point for HF application in dentistry began in 1983 by Simonsen and Calamia [24], who first advocated a ceramic surface treatment with 7.5% HF substitute acid compared with a non-etch surface. The results demonstrated that etching with 7.5% HF substitute acid significantly improved the tensile bond strength with resin composite. Subsequently, in 1987, Stangel and co-workers [25] also used 20% HF acid to etch a ceramic surface for 2.5 minutes in comparison with a non-etched surface. The result was similar to a study by Simonsen and Calamia [24] which showed significant high shear bond strength. Since then, HF acid has been widely used for dental ceramic surface treatment.

In principle, the effect of HF etching on feldspathic ceramic can be explained by the chemical nature of the etching process. HF acid, at low concentrations, first attacks the leucite crystals, followed by undissolved feldspar silica. The next composition to react is the glass matrix or silica phase, followed by alumina oxide, if present. HF acid reacts selectively with the silica phase to form hexafluorosilicates at the rate of 0.44 microns/minute [26]. As a result, the surface of the ceramic becomes honeycomb-like, which is expected for microretentions [27], as demonstrated in Figure 8.

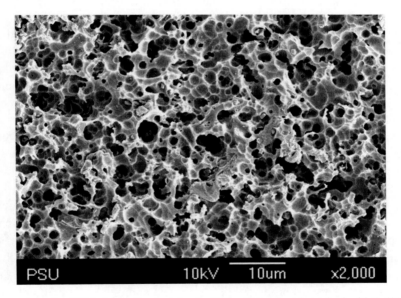

Figure 8. Illustration of a SEM photomicrograph of a feldspathic ceramic etched with 9.6% HF acid for 2 minutes (original magnification ×2000).

The forming equations of hexafluorosilicates are described as follows [28]: First, a volatile silicon tetrafluoride (tetrafluorosilane) is formed (equation 1):

$$2(HF)_2(l) + SiO_2(S) \rightarrow SiF_4(l) + 2H_2O(l) \quad (1)$$

Secondly, silicon tetrafluoride then forms with the HF soluble complex ion, hexafluorosilicate (equation 2):

$$(HF)_2(l) + SiF_4(l) \rightarrow (SiF_6)^{2-}(aq) + 2H^+ \quad (2)$$

Finally, hexafluorosilicate can further react with the protons to form tetrafluorosilicic acid that can be rinsed off with water (equation 3):

$$(SiF_6)^{2-}(aq) + 2H^+(aq) \rightarrow H_2SiF_6(l) \qquad (3)$$

Ceramic etched with HF acid demonstrates a microstructure that appears most conducive to the development of high strength as a function of the number of large porosities with its amorphous surface. This causes physical alteration to promote adhesion of resin composite to the porous surface of the fractured ceramic and produces greater roughness on the ceramic surfaces than other acids. Numerous previous studies have documented concentration used and time required of HF acid varying between 5 through 10% and 1 through 20 minutes [23,29-34], respectively. Effective etching depends on the concentration and time applied, and the type of ceramics being utilized, which will be briefly described below.

Currently, there are several different types of ceramics used in dentistry. Feldspathic ceramics, which are produced from a mixture of potassium feldspar ($K_2O \cdot Al_2O_3 \cdot 6SiO_2$) and a silica ($SiO_2$) network, form a 19 weight percentage (wt%) of leucite crystals ($K_2O \cdot Al_2O_3 \cdot 4SiO_2$) after incongruent melting [35]. Incongruent melting is the process by which one material melts to form a liquid plus a different crystalline material [2]. Feldspathic ceramics are not resistant to tension and shear, therefore, it is being employed as a veneering material for metal-ceramic restorations. Aluminous ceramics are composed of mixtures similar to that of feldspathic ceramics, but with increased amounts of 40 to 50 wt% aluminium oxide crystals [36]. Aluminous ceramics may be used as inlays, onlays, laminate veneers, and as a covering material for all-ceramic restorations. High leucite content ceramics (injection molded or pressable ceramics) are available as prefabricated ceramic ingots. The lost-wax technique is used to fabricate a mold for this ceramic in which they are melted at high temperatures and pressed into a mold. IPS Empress Esthetic (Ivoclar Vivadent AG, Schaan, Liechtenstein) is an example of this ceramic, which originates from IPS Empress Original. They contain 40 to 50 wt% leucite crystals [37]. Recently, the new all ceramic systems, IPS e.max (Ivoclar Vivadent AG, Schaan, Liechtenstein), have been launched into the market. This ceramic system has various types of ceramics including IPS e.max Press and ZirPress for press techniques, IPS e.max ZirCAD and CAD for CAD/CAM (computer-aided design/computer-aided manufacturing) technology, and IPS e.max Ceram as a veneering material for this system. IPS ZirPress and CAD are lithium disilicate ($Li_2Si_2O_5$) glass ceramics which have needle-like crystal structures that offer excellent strength and durability as well as outstanding optical properties. They contain approximately a 70 volume percentage (vol%) of lithium disilicate crystals with an approximate crystal size of 1.5 microns. IPS e.max Ceram and Press belong to the feldspathic-based ceramics group, namely, fluorapatite ceramics. These glass ceramics consist of dispersed fluorapatite crystals ($Ca_{10}(PO_4)_6F_2$) in a feldspathic glassy matrix and have a microstructure unlike that of any other commercially available dental ceramics [38]. IPS e.max ZirCAD is a presintered yttrium stabilized zirconium oxide block (Y-TZP) for CAD/CAM technology. After the restoration has been milled into shape with CAD/CAM technology, the material is sintered to densify the microstructure. The final restoration is densely sintered and consists of tetragonal grains. The density is approximately 99.5% of the theoretical density. Strength and toughness have now reached the high values desired [39]. Fluorapatite-leucite glass ceramics (IPS d.SIGN, Ivoclar Vivadent AG, Schaan, Liechtenstein) are another type of ceramic in this system and are used as a veneering for metal-ceramic restorations. They contain dispersed fluorapatites incorporated with leucite crystals in a feldspathic glassy matrix [40]. By uniting these two types of crystals in one glass ceramic, very diverse properties can be combined and increase the overall strength of this ceramic.

The hazards of HF acid are well documented. Even though it is effective, HF acid can cause tissue rash and burn [41,42]. The severity of the burn is dependent upon the concentration of the acid and the duration of the exposure. The severity of tissue damage may range from swelling and erythema to blistering and tissue coagulation. Tissue destruction following severe burns may slowly progress to ulceration with extensive tissue loss [41,42]. Therefore, in case of inadequate rubber dam isolation, dentists should be careful. For this reason, an HF substitute agent which can reduce risk and serve as a safe and effective substitute for etching ceramic surfaces to bond, has been investigated. The best agent found is 1.23% APF gel.

Etching with APF Gel

APF gel, widely used for in-office fluoride application, consists of sodium fluoride, phosphoric acid, and hydrofluoric acid [43]. It is safe for oral tissue, unlike hydrofluoric acid. Hence, APF gel is proposed to be an alternative for ceramic surface etching before bonding with resin composite [44]. In 1973, Gau and Kerause [44] first compared the effect of 1.23% APF gel (pH 4.0 to 4.7) and stannous fluoride (SnF_2) (pH 7.0) in etching ceramic surfaces. The results showed that only 1.23% APF gel, which is acidic, could etch the ceramic. From microscopic observation of the etched feldspathic ceramic surfaces by APF gel, an interesting phenomena was revealed. Initially, as in the etching process, small crystal-like structures protruded from the etched surfaces. Then as etching advanced, these initial crystals become surrounded with large irregular areas, almost as though the initial crystals were surrounded by snowflake-like structures (Figure 9).

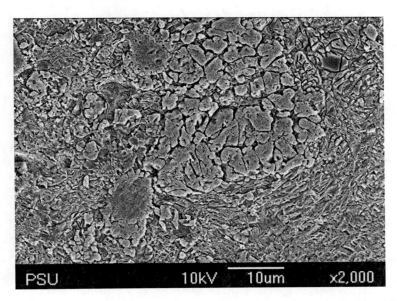

Figure 9. Illustration of a SEM photomicrograph of a feldspathic ceramic etched with 1.23% APF gel for 10 minutes (original magnification ×2000).

APF etching appeared to have a relatively superficial effect and was possibly a combination of surface dissolution and precipitation. The action of acid fluoride solutions may at first build up surface deposits, preferentially on the leucite crystal phase. Secondly, it may be a simultaneous nucleophilic-electrophilic attack on the silica network as well as an electrophilic attack on non-bridging oxygen atoms. As a result, the glass matrix dissolved more readily than the leucite one such that the leucite crystals projected from the surface [44,45]. Accordingly, numerous studies have confirmed this and documented that the bond strength of resin composite to silanized ceramic after being etched by 1.23% APF gel was comparable to that of HF acid etching [46-50].

Previous studies have reported that surface treatment of silanized feldspathic ceramics with 1.23% APF gel for 10 minutes was comparable to treatment with 9.5% HF acid for 4 [29] to 5 [31] minutes for bonding with resin composites. Another study by Brentel and co-workers [51] demonstrated that etching unsilanized feldspathic ceramic with 10% HF acid was comparable to etching with 1.23% APF gel for 5 minutes. However, these studies have not revealed the reason why these etching times with APF gel were used and besides, 5 or 10 minutes of etching with APF gel is relatively time consuming compared with HF acid etching. It would be more beneficial to the patient if the etching time of APF gel, which serves as a safe and effective substitute for etching ceramic surfaces, could be shortened. Accordingly, a recent study by Kukiattrakoon and Thammasitboon [52] investigated and revealed that a seven- to ten-minute application of 1.23% APF gel on leucite containing ceramics produced a shear bond strength to resin composites comparable to a four-minute etch with 9.6% HF acid. A SEM photomicrograph of an etched surface of a leucite containing ceramic is presented in Figure 10. Also, a study by the same authors [53] reported that a six- to ten-minute etching with 1.23% APF gel on feldspathic ceramics yielded a shear bond strength to resin composites comparable to a two-minute etch with 9.6% HF acid.

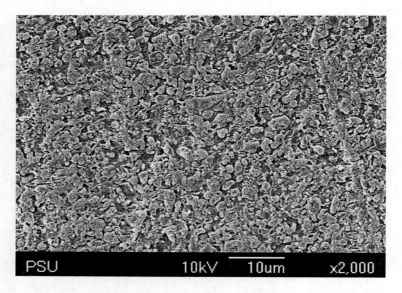

Figure 10. Illustration of a SEM photomicrograph of a leucite containing ceramic etched with 1.23% APF gel for 7 minutes (original magnification ×2000). Reproduced from [52] with permission from Elsevier.

However, remarkable differences in the etched ceramic surface micromorphology between etching with HF acid and APF gel in visual comparisons were observed [54]. Etching with 1.23% APF gel was found to create smooth, homogenous surfaces on the exposed ceramic surfaces, resulting in very shallow etching patterns when compared to HF acid, which produced porous irregular surfaces with numerous microundercut, channels and voids. However, al Edris and co-workers [54] have concluded that there was no correlation between micromorphologic characteristics of etched surfaces and actual bond strength, which corresponded with the results of previous studies as mentioned above.

Laser Irradiation

A new alternative method for surface treatment of ceramics was proposed in 2000 by Li and co-workers [55]. They reported adequate bond strength with the application of a neodymium: yttrium-aluminium-garnet (Nd:YAG) laser (15 Hz, 60 mJ, and 0.9 W or 15 Hz, 80 mJ, and 1.2 W) on feldspathic ceramics, attaining the same effectiveness as HF acid etching. Their SEM showed that ceramic surfaces irradiated by a Nd:YAG laser demonstrated rough, umbilicate, lava crater-like structures, producing mechanical retention between resin composite and ceramic. The Nd:YAG laser roughened the ceramic surface by removing the glassy phase of the ceramic [56], by surface melting and random crystallization [57]. Subsequently, in 2005, da Silveira and co-workers [58] examined microtensile bond strength of aluminous ceramic (In-Ceram alumina blocks, Vita Zahnfabrik, Bad Säckingen, Germany) treated with 50 microns of aluminium oxide sandblasting, sandblasting associated with silica coating (Rocatec Plus, 3M ESPE, St.Paul, MN, USA), or sandblasting associated with a Nd:YAG laser (20 Hz, 100 mJ, and 2 W), and then bonded to resin cement after silane application. The results revealed that sandblasting associated with a Nd:YAG laser significantly yielded the highest microtensile bond strength.

Among the different types of lasers used, the erbium: yttrium-aluminium-garnet (Er:YAG) laser is one of the most often recommended types to be used on the dental surface. The reasons for this are that its wavelength (2.94 µm) concurs with the main absorption band of water (approximately 3.0 µm) and that it is also well absorbed by OH– groups in hydroxyapatite [59]. Therefore, laser irradiations of both Nd:YAG and Er:YAG lasers were further investigated in surface treatment of feldspathic ceramics [56,59,60], zirconia ceramic [57], and lithium disilicate ceramic [61]. The results revealed that laser treatment, either Nd:YAG or Er:YAG laser alone, was not adequately effective in roughening feldspathic ceramic surfaces to create a high-strength bond with resin composites [56,59,60]. However, the bond strength for feldspathic ceramics after laser irradiation could be increased by HF acid etching or alumina blasting [56,59,60]. Whereas for zirconia ceramic, alumina blasting associated with a Nd:YAG laser (20 Hz, 100 mJ, and 2 W) provided the best results as compared to alumina blasting alone [57]. An Er:YAG laser (20 Hz, 300 mJ, and 4 to 10 W) reported a superior bond strength for lithium disilicate ceramic compared with 9.5% HF etching for 30 seconds [61].

Presently, different laser types are available for use in dental application including Nd:YAG laser, Er:YAG laser, holmium (Ho):YAG laser, erbium: yttrium-scandium-gallium-garnet (Er:YSGG) laser, or carbon dioxide (CO_2) laser. Therefore, further studies are required to enhance this laser irradiation technique with various types of lasers and with other types of dental ceramics.

Chemical Methods of Ceramic Surface Treatment

These methods have been categorized by means of promoting chemical bonding between prepared restoration surfaces and repaired resin composite. Two methods have been described; silica coated aluminium oxide and silane application.

Silica Coated Aluminium Oxide

This technique was first proposed with the initial objective to treat metal surfaces in 1984 by Musil and Tiller [62]. They introduced the silicoater technique (coating metal with silica and aluminium oxide) for increased bond strength between resin veneer and metal, and reducing the marginal gap. This technique, in principle, begun by air abrasion with 50 microns of aluminium oxide to clean, roughens, and improves adhering capability. The metal surface was then coated with silica coated aluminium oxide at 30 pound/inch² for 15 seconds to achieve a 0.1 to 1.0 microns silicon oxide layer (SiO_x–C). Subsequently silane, which can bond to –OH groups of silicon oxide, was applied. Silane can also further bond to resin cement [63].

Currently, the basic principle of the silicoater technique is developed for intraoral application and recommended for treating metal, ceramic, and resin composite surfaces — namely, the CoJet System (3M ESPE AG, Seefeld, Germany) [10,11,64-67]. The CoJet System includes CoJet Sand, the silica blast-coating medium (approximate particle size 30 µm) for cold silicatisation of the restoration surface; ESPE Sil, the silane coupling agent for silicated surfaces; Sinfony Opaquer for masking exposed metal surfaces; and finally, Visio-Bond, the bonding agent for bonding between the prepared restoration surfaces and resin composite. This system has provided a simplified method that does not require pretreatment with alumina blasting. Silica coating was simply achieved by blasting in one step. The coating step is performed by sandblasting with silicatized Cojet Sand. Blasting procedure causes the ceramic coating to be tribochemically anchored. Tribochemistry means the creation of a chemical bond by the use of mechanical energy. This energy can be supplied by rubbing, grinding, or blasting. During this process, components of the blasting silica are incorporated onto the restoration surface, to a depth of 15 µm. Since this effect is limited to microscopically small areas of the surface, no temperature increase over the entire metal substructure can be observed. The surfaces modified in this way are then conditioned with silane. Silanization allows a chemical bond between the ceramic bonding agent layer and the Opaquer or any other commercial methacrylated monomer system (these details are described later in Silane application). In order to obtain an optimum, microgap-free bond to repair composites that are usually highly filled and therefore viscous, the next step requires the application of Visio-Bond to the previously silanized surface [67].

Several research studies have reported that silica coating increased bond strength with a resin composite of zirconia and aluminous ceramics, as compared to sandblasting or HF acid etching [68-72]. Meanwhile for feldspathic ceramics, the results showed a comparable bond strength [73], as well as a lower bond strength when compared with HF acid etching [74].

Silane Application

Silane coupling agents (also known as silane, or ceramic primer in product name) are silicon-based chemicals that contain two types of reactivity, inorganic and organic, in the same molecules [75,76]. A general formula of silane shows two classes of moieties to the silicon (Si) atom as follows (equation 4):

$$X-CH_2CH_2CH_2-Si-(OR)_3 \qquad (4)$$

where X represents organofunctional groups such as vinyl ($-CH=CH_2$), allyl ($-CH_2CH=CH_2$), amino ($-NH_2$), or isocyanato ($-N=C=O$), and OR is hydrolysable or alkoxy groups, for example, methoxy ($-O-CH_3$) or ethoxy ($-O-CH_2CH_3$). The alkoxy groups can react with an inorganic substrate such as ceramic, some oxidized metals, and glass filler in resin-based composites; and the organofunctional groups can polymerize with an organic matrix, for example, resin-based materials. In both groups, covalent bonds are formed between the matrices. Generally, silane may or may not contain reactive groups, for instance, chloride ($-Cl$). There can also be a propylene link ($-CH_2CH_2CH_2-$) between Si and the organofunctional groups.

Numerous different silane compounds exist and they are used extensively in industry and manufacturing. In dentistry, Bowen and Rodriguez [77] first introduced silane as a silane coupling agent in resin composite in 1962. Subsequently, in 1967, Paffenbarger and co-workers [78] used silane to bond resin composite and ceramic and found that silane significantly increased bond strength as compared with no silane treatment. Lacy and co-workers [33] also found that silane could increase 25% of the bond strength, which corresponded with a study by Della Bona and co-workers [79] who revealed that even only with a silane application, the bond strength increased. Therefore, it has recently been shown that silane coupling treatment significantly increases the bond strength between resin composite and ceramics, and therefore recommends the use of silane agents with etched ceramics [29,80-84], even though there is controversy regarding the efficacy of silane coupling treatment in long-term adhesion between resin composite and ceramic [32,85,86].

The reaction of silane involves four steps [62,87,88]. Firstly, silane must be hydrolyzed (activated) in the ratio of silane : H_2O = 1 : 3, and secondly, condensed. In an adequate solution, silane alkoxy groups react with water to form reactive, hydrophilic, acidic silanol groups (Si$-$OH) and release free alcohols as side products. The acidity of the silanol groups depends upon the organofunctional group of the silane. The silanol can be formed by the reaction in equation 5:

$$X-Si-(OR)_3 + 3H_2O \rightarrow X-Si-(OH)_3 + 3R-OH$$
$$\text{(silanol)} \qquad (5)$$

This reaction is the hydronium ion (H_3O^+) catalyzed, for example, in acidic solution. At about pH 4 (for organotrialkoxysilane), the rate of condensation between silanol groups of monomeric silane molecules to larger oligomers is at the minimum, and the silane solutions have the highest stability. Acetic acid is often used for the pH adjustment and activates or hydrolyzes silane by reacting with three alkoxy groups located at one end of the silane molecule. The hydrolysis time varies depending on the silane concentration, solution, and the temperature, but usually 0.5 to 2.0 hours is sufficient. It is at this first stage that many of the commercially available organosilanes differ. The silanes can be purchased in either a prehydrolyzed or nonhydrolyzed form. The major difference between the two is that the prehydrolyzed silane is easier to use but generally has a shorter shelf life and is less stable in its container [89]. Furthermore, both silanes have limited shelf lives and require proper handling and storage according to the manufacturer's recommendations. Research has also shown that operator activated nonhydrolyzed silane was associated with lower bond strengths at 6 months [90].

During the condensation reaction in the second step, silane molecules react with each other, forming dimers according to equation 6:

$$X-Si-(OH)_3 + X-Si-(OH)_3 \rightarrow X-Si-(OH)_2-O-Si-X-(OH)_2 + H_2O \qquad (6)$$

Dimers then condense to form siloxane oligomers. Also, in the third step, hydrogen bonding between the siloxane monomers and oligomers occurs in the solution. Silane oligomers react with each other, forming branched hydrophobic siloxane bonds (O−Si−O) and with an organic matrix (for example, silica in ceramic, metal oxides that contain hydroxyl (−OH) groups), they can form −Si−O−M− bonds (M=metal) in the fourth step. This process is slow and usually takes anywhere from 2 to 24 hours for bonds to develop and stabilize. In air, the metal surfaces are oxidized and become covered by hydroxyl groups. The acidic silanol groups then can react with the OH− groups on the metal. On the inorganic substrate (metal), siloxane bonds of both types will thus be formed, −Si−O−M− and −Si−O−Si. If the substrate is silica (quartz, SiO_2) or silicate, only a siloxane layer (−Si−O−Si) will be formed. A simplified equation for the reaction of silanols with the metal surface reaction would be written as follows (equation 7):

$$….X-Si-(OH)_2-O-Si-X(OH)-…. + 2OH-M \rightarrow$$

$$\begin{array}{cc} X & X \\ | & | \\ -X-Si-O-Si-O-…+ H_2O \\ | & | \\ O & O \\ | & | \\ M & M \end{array}$$

(siloxane bond) \qquad (7)

The methacrylate group on the other end of the silane molecule can now react, through free radical addition polymerization, with methacrylate groups in subsequently placed adhesives and methacrylate-based materials. This process is called silanization or silanation. A conclusion of silane bonding to organic and inorganic materials is schematically presented in Figure 11 [91].

The assumption of the role of silane in adhesion is that it increases the bond strength by performing two functions [88,92]. Firstly, silane provides a chemical link between dental ceramic and resin composite as already described above. Secondly, the organic portion of the silane molecule increases the wettability of the ceramic surface and thus, enhances the flow of the resin cement into the complicated archetype of the microundercuts of the ceramic surface, resulting in a more intimate micromechanical bond. This also effectively reduces the size and numbers of the surface flaws, and then strengthens the ceramic.

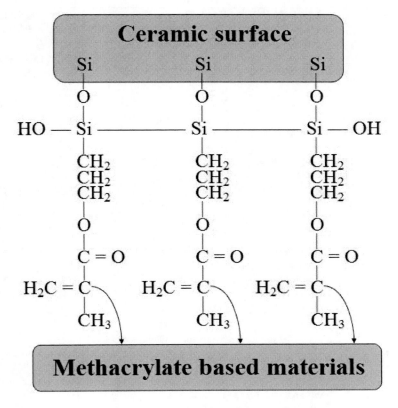

Figure 11. Illustration of silane molecules which have covalently bonded to the ceramic surface on the one end, and to the methacrylate group on the other end (modified from Alex [91]).

In prosthetic and restorative dentistry, the silane most commonly used is a monofunctional γ-methacryloxypropyl trimethoxysilane or 3-trimethoxysilylpropyl methacrylate (MPS) [87]. The formula of this silane is as follows (equation 8):

$$H_2C=C(CH_3)-C(=O)-O-CH_2CH_2CH_2-Si(O-CH_3)_3 \tag{8}$$

This silane is usually diluted, being often by less than 2 wt% in water-ethanol solution, with a pH of 4 to 5, adjusted with acetic acid, by being prehydrolyzed.

The silanes used in dentistry are also usually in 90% to 95% ethanol or isopropanol solutions, but more diluted alcohol solutions, about 20% or even 40% to 50%, are also used. An acetone-ethanol mixture is also known to be used. Several silane products available for professional dental use are indicated in Table 1 [87,93-97].

Table 1. Product of silanes used in dentistry [87,93-97]

Product	Effective silane	pH	Solution	Manufacturer
Monobond-S	MPS 1.0%	4	Ethanol 52%, distilled water 47%	Ivoclar Vivadent AG, Schaan, Liechtenstein
Vectris Wetting Agent	MPS 1.0%	4	Ethanol 50%-52%	Ivoclar Vivadent AG, Schaan, Liechtenstein
ESPE Sil	A silane (%NA)	NA	Ethanol > 90%	3M ESPE, St.Paul, MN, USA
RelyX Ceramic Primer	A silane < 1%	NA	Ethanol 70%-80%, water 20%-30%	3M ESPE, St.Paul, MN, USA
Porcelain Repair Primer	A silane 15%-20%	NA	Ethanol 80%-85%	Kerr Corp., Orange, CA, USA
Pulpdent Silane Bond Enhancer	A silane (%NA)	NA	Ethanol 92.6%, acetone 7.4%	Pulpdent Corp., Watertown, MA, USA
Silicoup A and B (two-bottle system)	MPS (%NA)	NA	Ethanol 25%-50%, ethylacetate 25%-50%, acetic acid 5%-10%	Heraeus Kulzer GmbH and Co., Wehrheim, Germany
Ultradent	MPS (%NA)	NA	Isopropanol 92%	Ultradent Inc., South Jordan, UT, USA
Bisco Porcelain Primer	Silane with methacrylate	NA	Alcohol	Bisco Inc., Schaumburg, IL, USA
Clearfil Porcelain Bond Activator	MPS (%NA)	NA	Bisphenol-a-polyethoxy-dimethacrylate	Kuraray Co. Ltd., Tokyo, Japan
Clearfil Ceramic Primer	MPS<5%	NA	Ethanol>80%, 10-MDP	Kuraray Co.Ltd., Tokyo, Japan
Quadrant Porcelain Coupling Agent (Liquid A)	Based on 4-META 10%; no silane	NA	MMA	Cavex Holland BV, Haarlem, Holland
Quadrant Porcelain Coupling Agent (Liquid B)	MPS 5%	NA	MMA, water	Cavex Holland BV, Haarlem, Holland
Bifix DC	MPS (%NA)	NA	Alcohol, water	Voco GmbH, Cuxhaven, Germany
Versa-Link Porcelain Bonding	MPS > 10%	NA	Methanol	Sultan Healthcare, Englewood, NJ, USA
Ormco Porcelain Primer	A silane 15%-20%	NA	Ethyl alcohol 80%-85%	Ormco Corp., Glendora, CA, USA
Shofu Porcelain Primer	MPS (%NA)	NA	Ethanol 95%	Shofu Inc., Kyoto, Japan
AZ primer (for zirconia ceramics)	Based on 6-MHPA; no silane	NA	Acetone >99%	Shofu Inc., Kyoto, Japan

NA = Not available; MPS = γ-methacryloxypropyl trimethoxysilane; 10-MDP = 10-methacryloyloxydecyl dihydrogen phosphate; 4-META = 4-methacryloxyethyl trimellitate anhydride; MMA = Methyl methacrylate; 6-MHPA = 6-methacryloxyhexylphosphono acetate.

In conclusion, the surface treatment protocols of the dental ceramics are designated according to Table 2 [71,72,98-106].

Table 2. Types of ceramics used and surface treatment protocols [71,72,98-106]

Ceramic	Surface treatment protocols
Feldspathic ceramics	9.5% HF acid for 2 to 2.5 minutes, 1 minute washing, and silane application
Leucite-reinforced ceramics	9.5% HF acid for 60 seconds, 1 minute washing, and silane application
Lithium dilsilicate ceramics	9.5% HF acid for 20 seconds, 1 minute washing, and silane application; or air abrasion with 50 microns aluminium oxide particles for 5 seconds associated with 4% HF acid etching for 5 minutes
Aluminous ceramics	9.6% HF acid for 1 to 2 minutes, 1 minute washing, and silane application
Fluorapatite ceramics	4.9% HF acid for 20 seconds and silane application
Fluorapatite-leucite ceramics	4.9% HF acid for 90 seconds and silane application
Glass-infiltrated aluminium oxide ceramics	Air abrasion with 50 microns aluminium oxide particles, and 1 minute washing; or using silica coated aluminium oxide
Densely sintered aluminium oxide ceramics	Air abrasion with 50 microns aluminium oxide particles
Zirconia ceramics	Silica coated aluminium oxide alone, or associated with ceramic or metal primer*

* Details of metal primer are described later in Mechanical methods of metal surface treatment: Primer application).

METAL SURFACE TREATMENT

When metal-ceramic restorations have a fractured ceramic in combination with metal substructure exposed, metal surface treatment should be included in the treatment sequence. Metal surface treatment methods are similar to ceramic surface treatments which include both mechanical and chemical methods. Mechanical methods incorporate grinding with carborundum discs or diamond burs, air abrasion, and chemical etching or electrolytic etching. Chemical bonding to metal has also been successful with tin plating, silica coating aluminium oxide, and primers or silane applications.

Mechanical Methods of Metal Surface Treatment

Metal surface treatment methods have various applications which have the same objectives to that of ceramic surface treatment to produce rough, grooved, or complicated surface textures for adhesive resin penetration. These methods; grinding with carborundum discs or diamond burs, air abrasion, electrolytic etching, and chemical etching, are presented below.

Grinding with Carborundum Discs or Diamond Burs

The aim of this method is to roughen metal surfaces where it is appropriate for the cement flow to adhere to, similar to that of ceramic results [106-108].

Air Abrasion

The size of the aluminium oxide used is 50 microns. The purposes of air abrasion are to clean, roughen, and increase wettability of cement on the metal surface [109]. Additionally, it is to increase aluminium oxide particles on metal surfaces, in which Kern and Thompson reported a 37% increasing [110]. The exact role of aluminium oxides is unknown. It seems probable that aluminium oxides can bond to phosphate ester and methacrylate functional monomers. Air abrasion can also be applied in combination with other surface treatments to improve bond strength [111-113].

Electrolytic Etching

In 1982, Livaditis and Thompson [112] pioneered a surface treatment of base metal alloys for bonding with acid and electricity which was developed for resin bonded fixed partial dentures in laboratory use. Subsequently, in 1983, Simonsen and co-workers [114] reported electrolytic etching methods for cobalt-chromium, nickel-chromium, and nickel-chromium-beryllium alloys. For cobalt-chromium and nickel-chromium alloys, 0.5 N nitric acid at 250 mA/cm^2 was applied for 5 minutes. While for nickel-chromium-beryllium alloys, 10% sulfuric acid at 200 to 300 mA/cm^2 was used for 3 to 6 minutes. The effect of acid is mostly found in the etch interdendritic phase, while there is a minor etch effect on the intradendritic gamma phase [115]. As seen, these techniques have been developed for laboratory or extraoral use, but in 1985, Jackson and Healey [116] developed an electrolytic etcher (Jackson Electrolytic Etcher, Solid State Innovations Inc., Mt. Airy, NC, USA) for clinical application and suggested it could be applied for intraoral use. The technique was quick and produced a clean uniform etch without immersing the restoration in a beaker. The procedure was started with attaching an anode directly to the restoration by means of a small clip and a cathode clip was attached to a cotton applicator, which was saturated with diluted nitric or sulfuric acid depending on the alloy to be etched. The cathode applicator then attached to the specific area to be etched. A 6 to 12 volts direct current at 200 to 300 mA was used for 120 seconds. Then the restoration was rinsed and dried. Dark oxide layers were revealed and eliminated by attaching 10% hydrochloric acid soaked cotton on this oxide layer. An anode was attached on the restoration. A 6 volts alternating current at 300 to 450 mA was applied for 30 to 60 seconds. Dark oxide layers were then removed, retaining a clean, gray-white etched surface. Jackson and Healey [116] suggested that the Electrolytic Etcher should be used with a specially designed digital thermal probe (Electro Therm 99, Solid State Innovations Inc., Mt. Airy, NC, USA) to monitor a temperature not to exceed 60°C. However, there are currently no reports of use of the Jackson Electrolytic Etcher.

Chemical Etching

According to the limitation of the special equipment required for electrolytic etching, chemical agents which are simple and easy to manipulate have been developed. In 1985, Love and Breitman [117] reported a surface treatment of nickel-chromium-beryllium alloy for resin-bonded restorations by a chemical agent composed of 50% nitric acid, 25% hydrochloric acid, and 25% methanol for 5 minutes, then rinsed with 18% hydrochloric acid for 10 minutes. After that, Assure-Etch (Ivoclar/Williams USA, Amherst, NY, USA) and Met-Etch (Gresco Products Inc., Staffords, TX, USA) were introduced into the market. Assure-Etch was applied to a restoration for 60 minutes at 70°C, and then rinsed with distilled water for 15 minutes [118]. Met-Etch was used with 2 techniques, firstly, etching a restoration for 3 minutes at 65.5°C, then etching again for 7 to 10 minutes, rinsing with 18% hydrochloric acid for 10 minutes, and subsequently cleaned with distilled water [119]. Secondly, the restoration was etched for 20 to 25 minutes at room temperature. Even though these techniques have been used for resin-bonded restorations with no reports for intraoral uses, Met-Etch may be used intraorally because of being very simple to use chairside and etching is possible at room temperature. However, further studies are required.

Chemical Methods of Metal Surface Treatment

Similar to ceramic surface treatment, chemical bonding of prepared metal surfaces and repaired resin composite can be formed by either tin plating, silica coated aluminium oxide, primer application, or silane treatment. These methods are described below.

Tin Plating

In 1976, McLean and Sced [120] used the tin plating technique for fabricating an aluminous crown with the twin foil technique, and suggested that 0.2 to 2.0 microns of tin layer was appropriate. The tin plating technique is currently developed for intraoral use (MicroTin, Danville Materials, San Ramon, CA, USA (Figure 12), and Kura Ace Mini, J Morita, Tustin, CA, USA — Kura Ace Mini has not been currently supplied into the market) [22,110,121,122].

Figure 12. Illustration of an intraoral tin plating device.

This technique has been approached by air abrasion with 50 microns of aluminium oxide to clean, roughen, and improve tin adhering. A cathode tip was then attached to the restoration. An anode tip, wrapped with tin saturated cotton, was also attached to the restoration. A 4.5 volts direct current was used for 5 to 30 seconds. Patients may feel a little sensitivity or none at all [64,123]. The color of an applied surface on the restoration changed to yellow-brown from tin oxide, and then the restoration was rinsed with distilled water and dried. A crystal tin layer of 0.5 microns was coated on the restoration which was appropriate for cement penetration, producing micromechanical and chemical retention [110,124,125].

The tin plating technique is suitable for noble alloys [120,123,126,127]. However, for base metal alloys, Inoue and co-workers [128] have reported that tin plating increased bond strength, while Kiatsirirote and co-workers [122] documented that tin plating did not increase bond strength but did not cause any negative effect either.

Silica Coated Aluminium Oxide

As previously described, the principle of this technique for chemical methods of ceramic surface treatment is that the intraoral application of silica coating (CoJet System, 3M ESPE AG, Seefeld, Germany) can be used with metal, ceramic, and resin composite restorations. For metal surface treatment, this method can be used with all types of alloys but the varying bond strength achieved depends on the type of alloys. The contributing factors of this are the quantity and type of metal oxide which can bond to silicon oxide and the surface hardness of alloys which present different degrees of roughness from air abrasion [107,129-131].

Primer Application

Primer is a surface-adjusted solution of metal for bonding to resin cement. Currently, available metal primers can be classified into three types consistent with the functional monomers occupied (Table 3).

In the first category, primer is composed of organic acid including phosphate or carboxylic monomer. These primers are able to bond with base metal oxide layers [132-137], but for noble alloys, the oxide layer should be formed before bonding [133,137] or additional surface alteration such as tin plating or silica coating [139]. Consequently, a second type of primer which has an organic sulfur compound including thiol (−SH), thione (=S), and disulfide monomers, have been developed and provide relatively good bonding to noble alloys [140-144]. The mechanism is a changing of thione to thiol, and then thiol is further bonded to noble alloys. Additionally, it is able to promote resin cement penetration [145]. Recently, a third type, being dual functional monomers, is now available for use. These primers utilized sulfur-containing monomers associated with a phosphate monomer and were effective for both noble and base metal alloys [139,146].

Silane Application

The advantages of silane in ceramic surface treatment have already been explained. It can also improve bonding between metal and resin cement or resin composite. The mechanism of silane to metal is similar to silane and ceramic. Briefly, silane is hydrolyzed to silanol by acid or heat. Then one end of the silane is covalently adhered to the metal oxide and the other end is bonded to the methacrylate groups [147,148]. Additionally, silane acts as a wetting agent lowering the surface tension at the interface between the metal and composite, due to its polar and non-polar heads.

Anagnostopoulos and co-workers [148] have reported that silane improved bond strength between resin composite and nickel-chromium alloys more than between resin composite and gold-palladium alloys, and between resin composite and high palladium alloys. They proposed that the reason for the results was that the oxide of nickel-chromium alloys has more acidity than the oxide of noble alloys. Accordingly, this acidic oxide is able to accelerate the effect of silane [148].

Table 3. Product, composition, and manufacturer of metal primers classified according to functional monomer [132-146]

Product	Composition	Manufacturer
Phosphate functional monomer		
Cesead I and II Opaque Primer	MDP	Kuraray Co. Ltd., Tokyo, Japan
Epricord	MDP	Kuraray Co. Ltd., Tokyo, Japan
Targis Link	MDP	Ivoclar Vivadent AG, Schaan, Liechtenstein
Carboxylic functional monomer		
Metacolor Opaque Bonding Liner (Super Bond Liquid)	4-META	Sun Medical Co. Ltd., Moriyama, Japan
Acryl Bond (Solidex Metal Photo Primer)	4-AET	Shofu Inc., Kyoto, Japan
MR Bond	MAC-10	Tokuyama Corp., Tokyo, Japan
Thiol functional monomer		
Metal Primer I and II	MEPS	GC Corp., Tokyo, Japan
V-Primer (Infis Opaque Primer)	VBATDT	Sun Medical Co. Ltd., Moriyama, Japan
Luna-Wing Primer	PETP	Yamamoto Precious Metal Co. Ltd., Osaka, Japan
Metaltite	MTU-6	Tokuyama Corp., Tokyo, Japan
Thione-phosphate dual functional monomer		
Alloy Primer	VBATDT, MDP	Kuraray Co. Ltd., Tokyo, Japan
Metal Link Primer	10-MDDT, 6-MHPA	Shofu Inc., Kyoto, Japan

MDP = 10-methacryloyloxydecyl dihydrogen phosphate; 4-META = 4-methacryloyloxyethyl trimetallic anhydride; 4-AET = 4-acryloyloxyethyl trimetallic acid; MAC-10 = 11-methacryloyloxyl undecan-1,1-dicarboxylic acid; MEPS = Methacryloyloxyalkyl thiophosphate; PETP = Pentaerythritol tetrakis(3-mercaptopropionate); VBATDT = 6-(4-vinylbenzyl-n-propyl) amino-1,3,5-triazine-2,4-dithione tautomer; MTU-6 = 6-methacryloyloxyhexyl 2-thiouracil-5-carboxylate; 10-MDDT = 10-methacryloxydecyl-6,8-dithiooctanate; 6-MHPA = 6-methacryloxy hexyl phosphonoacetate.

COMPARISONS AMONG VARIOUS METHODS

Much research has been conducted to investigate the different bond strengths acquired among the various methods which have already been described. Such research will be grouped and explained in this chapter into two topics; research about ceramic surface treatment and research about metal surface treatment.

Ceramic Surface Treatment Comparisons

A number of previous studies about comparisons among different ceramic surface treatment methods have been reported. These indicated great variation in the results due to the differences in materials, methods, and techniques used. For comparisons between HF acid and APF gel surface treatment, Lacy and co-workers [29] reported no significant difference in the shear bond strength between resin composite (P-30, 3M ESPE, St.Paul, MN, USA) and feldspathic ceramic (Will-Ceram, Williams Gold and Refining, Buffalo, NY, USA) treated with 9.5% HF acid (Ceram-Etch, Gresco Dental Products, Stafford, TX, USA) for 4 minutes, or 1.23% APF gel (Sultan, Sultan Healthcare, Englewood, NJ, USA) for 10 minutes. The results were similar to the results of Tylka and Stewart [31] in which no significant difference in bond strengths was found between resin composite (Command Ultrafine, Kerr/Sybron Corp., Romulus, MI, USA) and 2 types of ceramics (Biobond, Dentsply Intl. Inc., York, PA, USA and Will-Ceram, Williams Gold and Refining, Buffalo, NY, USA) after treatment with either 9.5% HF acid (Ceram-Etch, Gresco Dental Products, Stafford, TX, USA) for 5 minutes or 1.23% APF gel (Sultan, Sultan Healthcare, Englewood, NJ, USA) for 10 minutes. A recent study by Kukiattrakoon and Thammasitboon [52] revealed that a seven- to ten-minute application of 1.23% APF gel on a leucite containing ceramic produced a shear bond strength to resin composite comparable to a four-minute etch with 9.6% HF acid, and also reported that a six- to ten-minute etching with 1.23% APF gel on feldspathic ceramics yielded a shear bond strength to resin composites comparable to a two-minute etch with 9.6% HF acid [53]. However, further studies are required for other types of dental ceramics.

For comparisons of air abrasion and other methods, Wolf and co-workers [32] investigated the tensile bond strength between resin composite (Herculite, Kerr Corp., Orange, CA, USA) and feldspathic ceramic (Ceramco, Ceramco Inc., Center Conway, NH, USA) treated with either etching with 9.5% HF acid (Ceram-Etch, Gresco Dental Products, Stafford, TX, USA) for 5 minutes, sandblasting with 50 microns of aluminium oxide at 0.48 MPa for 3 seconds, or grinding with coarse pear-shaped diamond burs at high speed for 60 seconds. The results showed that sandblasting significantly provided the best bond strength. Wolf and co-workers [32] then examined the tensile bond strength between resin composite (Herculite, Kerr Corp., Orange, CA, USA) and feldspathic ceramic (Ceramco II, Ceramco Inc., Center Conway, NH, USA) treated with 9.5% HF acid or air abrasion with aluminium oxide. The results demonstrated that 9.5% HF acid etching yielded more bond strength than sandblasting, which related to a study by Cochran and co-workers [149]. Chung and Hwang [150] found there to be no significant differences of shear bond strength between ceramic treated with 9.5% HF acid (Ultradent Inc., South Jordan, UT, USA) for 4 minutes, or sandblasting with 50 microns aluminium oxide, which was consistent with other studies [33,81,151]. However, Kupiec and co-workers [152] revealed that etching with 8% HF acid, in addition to 50 microns of aluminium oxide sandblasting, significantly achieved the best bond strength between resin composite (Prisma TPH, Dentsply Intl., York, PA, USA) and feldspathic ceramic (Ceramco II, Ceramco Inc., Center Conway, NH, USA), compared with only etching or sandblasting. The results were also similar to other studies [34,153,154].

In recent years, zirconia ceramics have been increasingly used in fabrication of all-ceramic restorations due to their excellent strength. However, surface treatment of this ceramic is complicated. Etching with HF acid is not adequate because of the highly crystalline phase. Therefore, alumina blasting and silica coating were proposed and achieved good results [68-72,75]. Recently, metal primers have been used to combine with sandblasting and provided the best bond strength as compared to blasting alone [155-157]. Primers containing a phosphonic acid monomer (AZ primer, Shofu Inc., Kyoto, Japan) or a phosphate ester monomer (Clearfil Ceramic Primer, Kuraray Co. Ltd., Tokyo, Japan) also improved resin bonding to zirconia ceramics [158]. However, there is still a controversial result [159], so further studies are required. Intraoral silica coating also presented the best results as compared to sandblasting or HF acid etching alone [68-72].

According to the results of previous studies described, there is still controversy about the results regarding the different methods of ceramic surface treatment. However, considering the failure that occurred, most studies have demonstrated cohesive fractures within resin composite or ceramic. This showed that the bond strength acquired is relatively high and also higher than the cohesive strength of ceramic or resin composite. Hence, from the results of previous studies, it can be concluded that all methods, which caused rough surface to the ceramic, can be applied and should be combined with silane application. In the author's opinion, the methods used should be simple and easy to use for clinical application. In regard to all the methods described, roughening with coarse diamond burs may break more ceramic surface due to vibration during grinding. Sandblasting may not be available in some clinics because of the high cost, and if it is not used with care, it may damage a large volume of the ceramic surface [99]. Inhalation of aluminium oxides may cause coughing or shortness of breath to the patient if there is inadequate rubber dam isolation. Silica coating or laser applications may also not be practical in clinics due to their cost. Therefore, in the author's opinion, application of HF acid or APF gel should be the method of choice (except densely sintered aluminium oxide and zirconia ceramics — should use a silica coating method) in clinical situations, due to their easy use and low cost. However, if HF acid is used, dentists should be careful not to cause tissue rash and burns to the patient, and should always use rubber dam protection, or alternatively use with 1.23% APF gel.

Metal Surface Treatment Comparisons

Much research has been done regarding the comparisons among various methods. Vallitu [160] compared the bond strength between resin composite (Prisma TPH, Dentsply Intl., York, PA, USA) and gold-palladium alloys (LM-Ceragold 3, LM Dental, Turku, Finland) using three methods; grinding with diamond burs, grinding with silicon carbide burs, and sandblasting with aluminium oxide particles. The results showed that sandblasting significantly provided the highest bond strength due to their rough effectiveness. Lynde and co-workers [161] then investigated the bond strength between resin composite and palladium-copper alloys treated with air abrasion and tin plating. They found that tin plating significantly showed more bond strength than air abrasion. The reason was that tin plating supported both mechanical interlocking from rough surfaces and chemical bonding from tin oxide. This result corresponded with a study by van der Veen and co-workers [162]. On the contrary, Breeding and Dixon [163] found there to be no significant differences of shear bond strength between sandblasted base metal alloys (Rexillium III, Jeneric Gold Co., Wallingford, CT, USA) and tin plated noble alloys (Olympia and Jelstar, J.F. Jelenko Co., Armonk, NY, USA) in bonding between alloys and enamel with resin cement (Panavia OP, Kuraray Co. Ltd., Osaka, Japan), which correlated to the results of other studies [164-166]. However, this study [163] did not investigate or compare it with tin plated base metal alloys.

For comparisons between metal primers and other methods, Naegeli and co-workers [167], and Moulin and co-workers [168] found that alloys that were surface treated with silica coated aluminium oxide achieved more bond strength than those treated with primer. Another study by Abreu and co-workers [169] found that a metal primer application (Alloy Primer, Kuraray Co. Ltd., Kurashiki, Japan) significantly enhanced the tensile bond strength to base and noble metal alloys as compared to 50 microns aluminium oxide sandblasting alone. Comparing metal primers and tin plating, the results showed both significant increasing [148] and no significant differences of bond strength [170]. While comparisons between tin plating and silica coating aluminium oxide demonstrated both significant increasing [171] and no significant differences of bond strength [172]. These results may conclude that the effectiveness of primers vary considerably according to the types of primers and alloys used.

Studies related to metal surface treatment methods revealed that, for base metal alloys, air abrasion with aluminium oxide should be appropriate for bonding, or in addition with silica coated aluminium oxide, to provide better results. For noble alloys, mechanical methods in combination with chemical methods such as tin plating and primer or silane application, should be suitable for bonding. However, in the author's opinion, further studies are also required for alloy treatment by chemical etching with acid because it is simple, easy to apply, and of course, low cost as compared to tin plating, sandblasting, or silica coating. It is a great benefit for the patients if the etching time could be shortened, and still achieve a high bond strength comparable to other methods.

However, regarding the type of failure, most results of the studies have demonstrated adhesive fracture between resin cement and alloys. This showed less bond strength than ceramic results. Therefore, when repairing fractured metal-ceramic restorations which have metal exposed, the service life time and durability may be less than repairing only fractured ceramic.

CONCLUSION

Surface treatment of all-ceramic or metal-ceramic restorations for repairing or bonding can be accomplished by both mechanical and chemical methods. The appropriate method should depend upon the clinical situation in which the dentist can complete repairs simply, as well as save time and cost for the patients. Etching with HF acid or APF gel in addition to the silane application should be the method of choice for treating a ceramic surface (except densely sintered aluminium oxide and zirconia ceramics should be treated with silica coated aluminium oxide). For base metal alloy surface treatment, sandblasting or silica coating should be appropriate; while for noble alloys, tin plating in combination with a metal primer or silane should be applied.

REFERENCES

[1]   Raptis, NV; Michalakis, KX; Hirayama, H. Optical behavior of current ceramic systems. *Int. J. Periodontics Restorative Dent.*, 2006, 26, 31-41.
[2]   Anusavice, KJ. Philips' science of dental materials. 11th ed. St. Louis: *Mosby Elsevier*; 2003.
[3]   Raposo, LH; Neiva, NA; da Silva, GR; Carlo, HL; da Mota, AS; do Prado, CJ; Soares, CJ. Ceramic restoration repair: report of two cases. *J. Appl. Oral Sci.*, 2009, 17, 140-4.

[4]   de Jager, N; Feilzer, AJ; Davidson, CL. The influence of surface roughness on porcelain strength. *Dent. Mater*, 2000, 16, 381-8.
[5]   Fischer, H; Schäfer, M; Marx, R. Effect of surface roughness on flexural strength of veneer ceramics. *J. Dent. Res.*, 2003, 82, 972-5.
[6]   Goodacre, CJ; Bernal, G; Rungcharassaeng, K; Kan, JY. Clinical complications in fixed prosthodontics. *J. Prosthet. Dent.*, 2003, 90, 31-41.
[7]   Weinstein, M; Weinstein, LK; Katz, S; Weinstein, AB. Fused porcelain-to-metal teeth. US Patent No. 3052982, Sep 11, 1962.
[8]   Weinstein, M; Weinstein, LK; Weinstein, AB. Porcelain covered metal-reinforced teeth. US Patent No. 3052983, Sep 11, 1962.
[9]   O'Brien, WJ. Strengthening mechanisms of current dental porcelains. *Compend. Contin. Educ. Dent.*, 2000, 21, 625-30.
[10] Latta, MA; Barkmeier, WW. Approaches for intraoral repair of ceramic restorations. *Compend. Contin. Educ. Dent.* 2000, 21, 635-46.
[11] Özcan, M. The use of chairside silica coating for different dental applications: a clinical report. *J. Prosthet. Dent.*, 2002, 87, 469-72.
[12] Margeas, RC. Salvaging a porcelain-fused-to-metal bridge with intraoral ceramic repair. *Compend. Contin. Educ. Dent.*, 2002, 23, 952-6.
[13] Denehy, G; Bouschlicher, M; Vargas, M. Intraoral repair of cosmetic restorations. *Dent. Clin. North Am.*, 1998, 42, 719-37.
[14] Cardoso, AC; Spinelli Filho, P. Clinical and laboratory techniques for repair of fractured porcelain in fixed prostheses: a case report. *Quintessence Int.*, 1994, 25, 835-8.
[15] Bertolotti, RL; Paganetti, C. Adhesion monomers utilized for fixed partial denture (porcelain/metal) repair. *Quintessence Int.*, 1990, 21, 579-82.
[16] Özcan, M; Niedermeier, W. Clinical study on the reasons for and location of failures of metal-ceramic restorations and survival of repairs. *Int. J. Prosthodont.*, 2002, 15, 299-302.
[17] Özcan, M; Pfeiffer, P; Nergiz, I. A brief history and current status of metal-and ceramic surface-conditioning concepts for resin bonding in dentistry. *Quintessence Int.*, 1998, 29, 713-24.
[18] Blatz, MB; Sadan, A; Kern, M. Resin-ceramic bonding: a review of the literature. *J. Prosthet. Dent.*, 2003, 89, 268-74.
[19] O'Brien, WJ. Dental materials and their election. 4th ed. Chicago: *Quintessence*; 2008.
[20] Jochen, DG. Repair of fractured porcelain denture teeth. *J. Prosthet. Dent.*, 1973, 29, 228-30.
[21] Jochen, DG; Caputo, AA. Composite resin repair of porcelain denture teeth. *J. Prosthet. Dent.*, 1977, 38, 673-9.
[22] Bertolotti, RL; Lacy, AM; Watanabe, LG. Adhesive monomers for porcelain repair. *Int. J. Prosthodont.*, 1989, 2, 483-9.
[23] Wolf, DM; Powers, JM; O'Keefe, KL. Bond strength of composite to etched and sandblasted porcelain. *Am. J. Dent.*, 1993, 6, 155-8.
[24] Simonsen, RJ; Calamia, JR. Tensile bond strength of etched porcelain. *J. Dent. Res.*, 1983, 62, 297.
[25] Stangel, I; Nathanson, D; Hsu, CS. Shear strength of the composite bond to etched porcelain. *J. Dent. Res.*, 1987, 66, 1460-5.

[26] Yen, TW; Blackman, RB; Baez, RJ. Effect of acid etching on the flexural strength of a feldspathic porcelain and a castable ceramics. *J. Prosthet. Dent.*, 1993, 70, 224-33.

[27] Chen, JH; Matsumura, H; Atsuta, M. Effect of different etching periods on the bond strength of a composite resin to a machinable porcelain. *J. Dent.*, 1998, 28, 53-8.

[28] Matinlinna, JP; Vallittu, PK. Bonding of resin composites to etchable ceramic surfaces – an insight review of the chemical aspects on surface conditioning. *J. Oral Rehabil.*, 2007, 34, 622-30.

[29] Lacy, AM; LaLuz, J; Watanabe, LG; Dellinges, M. Effect of porcelain surface treatment on the bond to composite. *J. Prosthet. Dent.*, 1988, 60, 288-91.

[30] Nelson, E; Barghi, N. Effect of APF etching time on resin bonded porcelain. *J. Dent. Res.*, 1989, 68, 27.

[31] Tylka, DF; Stewart, GP. Comparison of acidulated phosphate fluoride gel and hydrofluoric acid etchants for porcelain-composite repair. *J. Prosthet. Dent.*, 1994, 72, 121-7.

[32] Wolf, DM; Powers, JM; O'Keefe, KL. Bond strength of composite to porcelain treated with new porcelain repair agents. *Dent. Mater*, 1992, 8, 158-61.

[33] Suliman, AHA; Swift, EJ; Perdigao, J. Effects of surface treatment and bonding agents on bond strength of composite resin to porcelain. *J. Prosthet. Dent.*, 1993, 70, 118-20.

[34] Shahverdi, S; Canay, S; Sahin, E; Bilge, A. Effect of different surface treatment methods on the bond strength of composite resin to porcelain. *J. Oral Rehabil.*, 1998, 25, 699-705.

[35] Kelly, JR; Campbell, SD. Ceramics in dentistry: historical roots and current perspectives. *J. Prosthet. Dent.*, 1996, 75, 18-32.

[36] McLean, JW; Hughes, TH. The reinforced of dental porcelain with ceramic oxide. *Br. Dent. J.*, 1965, 119, 251-67.

[37] Höland, W; Rheinberger, V; Schweiger, M. Control of nucleation in glass ceramics. *Philos. Transact. A Math. Phys. Eng. Sci.*, 2003, 361, 575-89.

[38] Sinmazişik, G; Oveçoğlu, ML. Physical properties and microstructural characterization of dental porcelains mixed with distilled water and modeling liquid. *Dent. Mater*, 2006, 22, 735-45.

[39] IPS e.max ZirCAD: Scientific documentation. Schaan: Ivoclar Vivadent AG; 2005.

[40] Höland, W; Rheinberger, V; Apel, E; van 't Hoen, C; Höland, M; Dommann, A; Obrecht, M; Mauth, C; Graf-Hausner, U. Clinical applications of glass-ceramics in dentistry. *J. Mater Sci. Mater Med.*, 2006, 17, 1037-42.

[41] Moore, PA; Manor, RC. Hydrofluoric acid burns. *J. Prosthet. Dent.*, 1982, 47, 338-9.

[42] Barbosa, VL; Almeida, MA; Chevitarese, D; Keith, O. Direct bonding to porcelain. *Am. J. Orthod. Dentofacial. Orthop.*, 1995, 107, 159-64.

[43] Wellock, WP; Brudevold, F. A study of acidulated fluoride solution-II. *Arch. Oral Biol.*, 1963, 8, 179-82.

[44] Gau, DJ; Krause, EA. Etching effect of topical fluorides on dental porcelains: a preliminary study. *J. Can. Dent. Assoc.*, 1973, 39, 410-5.

[45] Della Bona, A; van Noort, R. Ceramic surface preparations for resin bonding. *Am. J. Dent.*, 1998, 11, 276-80.

[46] Sposetti, VJ; Shen, C; Levin, AC. The effect of topical fluoride application on porcelain restorations. *J. Prosthet.Dent.*, 1986, 55, 677-82.

[47] Wunderlich, RC; Yaman, P. In vitro effect of topical fluoride on dental porcelain. *J. Prosthet. Dent.,* 1986, 55, 385-8.

[48] Abbasi, J; Bertolotti, RL; Lacy, AM; Watanabe, LG. Bond strengths of porcelain repair monomers. *J. Dent. Res.,* 1988, 67, 223.

[49] Copps, DP; Lacy, AM; Curtis, T; Carman, JE. Effects of topical fluorides on five low-fusing dental porcelains. *J. Prosthet. Dent.,* 1984, 52, 340-3.

[50] Jones, DA. Effects of topical fluoride preparations on glazed porcelain surfaces. *J. Prosthet. Dent.,* 1985, 53, 483-4.

[51] Brentel, AS; Özcan, M; Valandro, LF; Alarca, LG; Amaral, R; Bottino, MA. Microtensile bond strength of a resin cement to feldspathic ceramic after different etching and silanization regimens in dry and aged conditions. *Dent. Mater,* 2007, 23, 1323-31.

[52] Kukiattrakoon, B; Thammasitboon, K. The effect of different etching times of acidulated phosphate fluoride gel on the shear bond strength of high-leucite ceramics bonded to composite resin. *J. Prosthet. Dent.,* 2007, 98, 17-23.

[53] Kukiattrakoon, B; Thammasitboon, K. Optimal acidulated phosphate fluoride gel etching time for surface treatment of feldspathic porcelain: on shear bond strength to resin composite. *Eur. J. Dent.,* 2011, In press.

[54] al Edris, A; al Jabr, A; Cooley, RL; Barghi, N. SEM evaluation of etch patterns by three etchants on three porcelains. *J. Prosthet. Dent.,* 1990, 64, 734-9.

[55] Li, R; Ren, Y; Han, J. Effects of pulsed Nd:YAG laser irradiation on shear bond strength of composite resin bonded to porcelain. *Hua Xi Kou Qiang Yi Xue Za Zhi,* 2000, 18, 377-9.

[56] Akyıl, MS; Yilmaz, A; Karaalioğlu, OF; Duymuş, ZY. Shear bond strength of repair composite resin to an acid-etched and a laser-irradiated feldspathic ceramic surface. *Photomed Laser Surg.,* 2010, 28, 539-45.

[57] Spohr, AM; Borges, GA; Júnior, LH; Mota, EG; Oshima, HM. Surface modification of In-Ceram Zirconia ceramic by Nd:YAG laser, Rocatec system, or aluminum oxide sandblasting and its bond strength to a resin cement. *Photomed Laser Surg.,* 2008, 26, 203-8.

[58] da Silveira, BL; Paglia, A; Burnett, LH; Shinkai, RS; Eduardo, C de P; Spohr, AM. Micro-tensile bond strength between a resin cement and an aluminous ceramic treated with Nd:YAG laser, Rocatec System, or aluminum oxide sandblasting. *Photomed Laser Surg.,* 2005, 23, 543-8.

[59] da Silva Ferreira, S; Hanashiro, FS; de Souza-Zaroni, WC; Turbino, ML; Youssef, MN. Influence of aluminum oxide sandblasting associated with Nd:YAG or Er:YAG lasers on shear bond strength of a feldspathic ceramic to resin cements. *Photomed Laser Surg.,* 2010, 28, 471-5.

[60] Shiu, P; De Souza-Zaroni, WC; Eduardo, C de P; Youssef, MN. Effect of feldspathic ceramic surface treatments on bond strength to resin cement. *Photomed Laser Surg.,* 2007, 25, 291-6.

[61] Gökçe, B; Ozpinar, B; Dündar, M; Cömlekoglu, E; Sen, BH; Güngör, MA. Bond strengths of all-ceramics: acid vs laser etching. *Oper. Dent.,* 2007, 32, 173-8.

[62] Musil, R; Tiller, HJ. The adhesion of dental resins to metal surfaces: the Kulzer silicoater technique. Wehrheim: *Kulzer and Co GmbH*; 1984.

[63] Haselton, DR; Diaz-Arnold, AM; Dunne, JT Jr. Shear bond strengths of 2 intraoral porcelain repair systems to porcelain or metal substrates. *J. Prosthet. Dent.*, 2001, 86, 526-31.

[64] Cobbs, DS; Vargas, MA; Fridrich, TA; Bouschlicher, MR. Metal surface treatment characterization and effect on composite-to-metal bond strength. *Oper. Dent.*, 2000, 85, 427-33.

[65] Sun, R; Suansuwan, N; Kilpatrick, N; Swain, M. Characterization of tribochemically assisted bonding of composite resin to porcelain and metal. *J. Dent.*, 2000, 28, 441-5.

[66] Frankenberger, R; Kramer, N; Sindel, J. Repair strength of etched VS silica-coated metal-ceramic and all-ceramic restorations. *Oper. Dent.*, 2000, 25, 209-15.

[67] CoJet Intraloral Adhesive Repair System: Instruction for use. Seefeld: 3M ESPE AG; 2004.

[68] Valandro, LF; Della Bona, A; Antonio Bottino, M; Neisser, MP. The effect of ceramic surface treatment on bonding to densely sintered alumina ceramic. *J. Prosthet. Dent.*, 2005, 93, 253-9.

[69] Della Bona, A; Borba, M; Benetti, P; Cecchetti, D. Effect of surface treatments on the bond strength of a zirconia-reinforced ceramic to composite resin. *Braz. Oral Res.*, 2007, 21, 10-5.

[70] Özcan, M; Valandro, LF; Amaral, R; Leite, F; Bottino, MA. Bond strength durability of a resin composite on a reinforced ceramic using various repair systems. *Dent. Mater,* 2009, 25, 1477-83.

[71] Abbas, MH; Mosleh, I; Badawi, MF. Micro tensile bond strength of a ceramic repair system to all-ceramic coping materials. *Egypt Dent. J.,* 2010, 56, 121-30.

[72] Attia, A. Influence of surface treatment and cyclic loading on the durability of repaired all-ceramic crowns. *J. Appl. Oral Sci.*, 2010, 18, 194-200.

[73] de Melo, RM; Valandro, LF; Bottino, MA. Microtensile bond strength of a repair composite to leucite-reinforced feldspathic ceramic. *Braz. Dent. J.*, 2007, 18, 314-9.

[74] Boscato, N; Della Bona, A; Del Bel Cury, AA. Influence of ceramic pre-treatments on tensile bond strength and mode of failure of resin bonded to ceramics. *Am. J. Dent.*, 2007, 20, 103-8.

[75] Witucki, GL. A silane primer: chemistry and applications of alkoxy silanes. *J. Coating Technol.*, 1993, 822, 57-60.

[76] Larson, TD. The uses of silane and surface treatment in bonding. *Northwest Dent.*, 2006, 85, 27-30.

[77] Bowen, RL; Rodriguez, MS. Tensile strength and modulus of elasticity of tooth structure and several restorative materials. *J. Am. Dent. Assoc.*, 1962, 64, 378-87.

[78] Paffenbarger, GC; Sweeney, WT; Bowen, RL. Bonding porcelain teeth to acrylic resin denture bases. *J. Am. Dent. Assoc.*, 1967, 74, 1018-23.

[79] Della Bona, A; Anusavice, KJ, Hood, JA. Effect of ceramic surface treatment on tensile bond strength to a resin cement. *Int. J. Prosthodont.*, 2002, 15, 248-53.

[80] Kanchanatawewat, K; Kukiattrakoon, B. Effect of surface treatments on shear bond strength on high leucite content porcelain and aluminous porcelain. *CU Dent. J.*, 2001, 24, 175-86.

[81] Kanchanatawewat, K; Kukiattrakoon, B. Effect of surface treatments on shear bond strength on high leucite content porcelain and feldspathic porcelain. *J. Dent. Assoc. Thai.*, 2002, 52, 165-76.

[82] James, WB; Rogers, LB; Feller, PR; Price, WR. Bonding agents for repairing porcelain and gold: an evaluation. *Oper. Dent.*, 1977, 2, 118-24.

[83] Filho, AM; Vieira, LC; Araujo, E; Monteiro Junior, S. Effect of different ceramic surface treatments on resin microtensile bond strength. *J. Prosthodont.*, 2004, 13, 28-35.

[84] Guler, AU; Yilmaz, F; Ural, C; Guler, E. Evaluation of 24-hour shear bond strength of resin composite to porcelain according to surface treatment. *Int. J. Prosthodont.*, 2005, 18, 156-60.

[85] Diaz-Arnold, AM; Aquilino, SA. An evaluation of the bond strengths of four organosilane materials in response to thermal stress. *J. Prosthet. Dent.*, 1989, 62, 257-60.

[86] Bailey, JH. Porcelain-to-composite bond strengths using four organisilane materials. *J. Prosthet. Dent.*, 1989, 61, 174-7.

[87] Matinlinna, JP; Lassila, LV; Ozcan, M; Yli-Urpo, A; Vallittu, PK. An introduction to silanes and their clinical applications in dentistry. *Int. J. Prosthodont.*, 2004, 17, 155-64.

[88] Major, PW; Koehler, JR; Manning, KE. 24-hour shear bond strength of metal orthodontic brackets bonded to porcelain using various adhesion promoters. *Am. J. Orthod. Dentofac. Orthop.*, 1995, 108, 322-9.

[89] Andreasen, G; Stieg, M. Bonding and debonding brackets to porcelain and gold. *Am. J. Orthod. Dentofac. Orthop,.* 1988, 93, 341-5.

[90] Stokes, A; Hood, J; Tidmarsh, B. Effect of six month water storage on silane treated resin/porcelain bonds. *J. Dent.*, 1988, 16, 294-6.

[91] Alex, G. Preparing porcelain surfaces for optimal bonding. *Compend. Contin. Educ. Dent.,* 2008, 29, 324-35.

[92] Lu, R; Harcourt, JK; Tyas, MJ; Alexander, B. An investigation of the composite resin/porcelain interface. *Aust. Dent. J.*, 1992, 37, 12-9.

[93] Versa-Link Porcelain Bonding: Material safety data sheet [online]. 2008 [cited 2010 May 30]. Available from: http://structuredweb.com/sw/swchannel/CustomerCenter/ documents/6470/2/Versalink_Porcelain_Bonding_and_Repair_System_70520_-_Rev._05-08.pdf.

[94] Ormco Porcelain Primer: Material safety data sheet [online]. 2005 [cited 2010 May 30]. Available from: http://www.ormco.com/msds/ormco/us/english/PorcelainRepairPrimer.pdf.

[95] Clearfil Ceramic Primer: Material safety data sheet [online]. 2008 [cited 2010 May 30]. Available from: www.kuraraydental.com/products/28/ceramic_primer.pdf.

[96] Shofu Porcelain Primer: Material safety data sheet [online]. 2008 [cited 2010 May 30]. Available from: www.shofu.com/Procelain_Primer_MSDS.pdf.

[97] AZ primer: Material safety data sheet [online]. 2007 [cited 2010 May 30]. Available from: www.shofu.com/AZ_Primer_MSDS.pdf.

[98] Soares, CJ; Soares, PV; Pereira, JC; Fonseca, RB. Surface treatment protocols in the cementation process of ceramic and laboratory-processed composite restorations: a literature review. *J. Esthet. Restor. Dent.*, 2005, 17, 224-35.

[99] Kim, BK; Bae, HE; Shim, JS; Lee, KW. The influence of ceramic surface treatments on the tensile bond strength of composite resin to all-ceramic coping materials. *J. Prosthet. Dent.,* 2005, 94, 357-62.

[100] Nagayassu, MP; Shintome, LK; Uemura, ES; Araújo, JE. Effect of surface treatment on the shear bond strength of a resin-based cement to porcelain. *Braz. Dent. J.*, 2006, 17, 290-5.

[101] Goia, TS; Leite, FP; Valandro, LF; Ozcan, M; Bottino, MA. Repair bond strength of a resin composite to alumina-reinforced feldspathic ceramic. *Int. J. Prosthodont.*, 2006, 19, 400-2.

[102] IPS e.max Ceram: Instruction for use. Schaan: *Ivoclar Vivadent AG*; 2009.

[103] IPS d.SIGN: Instruction for use. Schaan: *Ivoclar Vivadent AG*; 2008.

[104] Tsuo, Y; Yoshida, K; Atsuta, M. Effects of alumina-blasting and adhesive primers on bonding between resin luting agent and zirconia ceramics. *Dent. Mater J.*, 2006, 25, 669-74.

[105] Yang, B; Barloi, A; Kern, M. Influence of air-abrasion on zirconia ceramic bonding using an adhesive composite resin. *Dent. Mater*, 2010, 26, 44-50.

[106] Robbins, JW. Intraoral repair of the fractured porcelain restoration. *Oper. Dent.*, 1998, 23, 203-7.

[107] Rada, RE. Intraoral repair of metal ceramic restorations. *J. Prosthet. Dent.*, 1991, 65, 348-50.

[108] Barreto, MT; Bottaro, BF. A practical approach to porcelain repair. *J. Prosthet. Dent.*, 1982, 48, 349-51.

[109] Mukai, M; Fukui, H; Hasegawa, J. Relationship between sandblasting and composite resin-alloy bond strength by a silica coating. *J. Prosthet. Dent.*, 1995, 74, 151-5.

[110] Kern, M; Thompson, VP. Sandblasting and silica-coating of dental alloys: volume loss, morphology and changes in the surface composition. *Dent. Mater*, 1993, 9, 155-61.

[111] McCaughey, AD. Sandblasting and tin-plating surface treatments to improve bonding with resin cements. *Dent. Update*, 1992, 19, 153-7.

[112] Livaditis, GJ; Thompson, VP. Etched casting: an improved retentive mechanism for resin-bonded retainers. *J. Prosthet. Dent.*, 1982, 47, 52-8.

[113] Hero, H; Ruyter, IE; Waarli, ML; Hultquist, G. Adhesion of resins to Ag-Pd alloys by means of the silicoating technique. *J. Dent. Res.*, 1987, 66, 1380-5.

[114] Simonsen, R; Thompson, VP; Barrach, G. Etched cast restorations: clinical and laboratory techniques. Chicago: *Quintessence*; 1983.

[115] Thompson, VP; Del Castillo, E; Livaditis, GJ. Resin-bonded retainers. Part I: Resin bond to electrolytically etched nonprecious alloys. *J. Prosthet. Dent.*, 1983, 50, 771-9.

[116] Jackson, TR; Healey, KW. Chairside electrolytic etching of cast alloys for resin bonding. *J. Prosthet. Dent.*, 1985, 54, 764-9.

[117] Love, LD; Breitman, JB. Resin retention by immersion-etched alloy. *J. Prosthet. Dent.*, 1985, 53, 623-4.

[118] Livaditis, GJ. A chemical etching system for creating micromechanical retention in resin-bonded retainers. *J. Prosthet. Dent.*, 1986, 56, 181-8.

[119] Doukoudakis, A; Cohen, B; Tsoutsos, A. A new chemical method for etching metal frameworks of the acid-etched prosthesis. *J. Prosthet. Dent.*, 1987, 58, 421-3.

[120] McLean, JW; Sced, IR. The bonded alumina crown. 1. The bonding of platimun to aluminous dental porcelain using tin oxide coating. *Aust. Dent. J.*, 1976, 21, 119-28.

[121] Imbery, TA; Davis, RD. Evaluation of tin plating system for a high-noble alloy. *Int. J. Prosthodont.*, 1993, 6, 55-9.

[122] Kiatsirirote, K; Northeast, SE; van Noort, R. Bonding procedure for intraoral repair of exposed metal with resin composite. *J. Adhesive Dent.*, 1999, 1, 315-21.
[123] Bertolotti, RL; DeLuca, SS; DeLuca, S. Intraoral metal adhesion utilized for occlusal rehabilitation. *Quintessence Int.*, 1994, 25, 525-9.
[124] van der Veen, JH; Jongebloed, WL; Dijk, F; Purdell-Lewis, DJ; van de Poel, AC. SEM study of six retention systems for resin-to-metal bonding. *Dent. Mater*, 1988, 4, 266-71.
[125] van de Veen, H; Krajanbrink, T; Bronsdijk, B; van de Poel, F. Resin bonding of electroplated precious metal fixed partial dentures: one year clinical results. *Quintessence Int.*, 1986, 17, 299-301.
[126] Zidan, O. Etched base-metal alloys: comparison of relief patterns, bond strengths and fracture modes. *Dent. Mater*, 1985, 1, 209-13.
[127] Gates, WD; Diaz-Arnold, AM; Aquilino, SA; Ryther, JS. Comparison of the adhesive strength of a BIS-GMA cement to tin-plated and non-tin-plated alloys. *J. Prosthet. Dent.*, 1993, 69, 12-6.
[128] Inoue, K; Murakami, T; Terada, Y. The bond strength of porcelain to Ni-Cr alloy - the influence of tin or chromium plating. *Int. J. Prosthodont.*, 1992, 5, 262-8.
[129] Hansson, O. The silicoating technique for resin-bonded prostheses: clinical and laboratory procedures. *Quintessence Int.*, 1989, 20, 85-99.
[130] Peutzfeldt, A; Asmussen, E. Silicoating: evaluation of a new method of bonding composite resin to metal. *Scand. J. Dent. Res.*, 1988, 96, 171-6.
[131] Hansson, O. Strength of bond with comspan to three silicoated alloys and titanium. *Scand. J. Dent. Res.*, 1990, 98, 248-56.
[132] Ishijima, T; Caputo, AA; Mito, R. Adhesion of resin to casting alloys. *J. Prosthet. Dent.*, 1992, 67, 445-9.
[133] Tanaka, T; Nagata, K; Takeyama, M; Atsuta, M; Nakabayashi, N; Masuhara, E. 4-META opaque resin-a new resin strongly adhesion to nickel-chromium alloy. *J. Dent Res.*, 1981, 60, 1697-706.
[134] Matsumura, H; Nakabayashi, N. Adhesive 4-META/MMA-TBB opaque resin with poly(methyl mathacrylate)-coated titanium dioxide. *J. Dent. Res.*, 1988, 67, 29-32.
[135] Aboush, YEY; Jenkins, CBG. Tensile strength of enamel-resin-metal joints. *J. Prosthet. Dent.*, 1989, 61, 688-94.
[136] Diaz-Arnold, AM; Williams, VD; Aquilino, SA. Tensile strengths of three luting agents for adhesion fixed partial dentures. *Int. J. Prosthodont.*, 1989, 2, 115-22.
[137] Matsumura, H; Tanaka, T; Atsuta, M. Effect of acidic primers on bonding between stainless steel and auto-polymerizing methacrylic resins. *J. Dent.*, 1997, 25, 285-90.
[138] Swift, EJ Jr. New adhesive resins. A status report for the American Journal of Dentistry. *Am. J. Dent.*, 1989, 2, 358-60.
[139] Taira, Y; Kamada, K; Atsuta, M. Effects of primers containing thiouracil and phosphate monomers on bonding of resin to Ag-Pd-Au alloy. *Dent. Mater J.*, 2008, 27, 69-74.
[140] Atsuta, M; Matsumura, H; Tanaka, T. Bonding fixed prosthodontic composite resin and precious metal alloys with the use of a vinyl-thiol primer and an adhesive opaque resin. *J. Prosthet. Dent.*, 1992, 67, 296-300.
[141] Matsumura, H; Leinfelder, KF. Effect of an adhesive primer on the integrity of occlusal veneer-metal interface and wear of composite resin veneered restorations. *J. Prosthet. Dent.*, 1993, 70, 296-9.

[142] Yoshida, K; Taira, Y; Matsumura, H; Atsuta, M. Effect of adhesive metal primers on bonding a prosthetic composite resin to metals. *J. Prosthet. Dent.*, 1993, 69, 357-62.

[143] Watanabe, I; Matsumura, H; Atsuta, M. Effect of two metal primers on adhesive bonding type IV gold alloys. *J. Prosthet. Dent.*, 1995, 73, 299-303.

[144] Taira, Y; Imai, Y. Primer for bonding resin to metal. *Dent. Mater*, 1995, 11, 2-6.

[145] Petrie, CS; Eick, JD; Williams, K; Spencer, P. A comparison of 3 alloy surface treatments for resin-bonded prostheses. *J. Prosthodont.*, 2001, 10, 217-23.

[146] Matsumura, H; Shimoe, S; Nagano, K; Atsuta, M. Effect of noble metal conditioners on bonding between prosthetic composite material and silver-palladium-copper-gold alloy. *J. Prosthet. Dent.*, 1999, 81, 710-4.

[147] Scott, JA; Strang, R; McCrosson, J. Silane effects on luting resin bond to a Ni-Cr alloy. *J. Dent.*, 1991, 19, 373-6.

[148] Anagnostopoulos, T; Eliades, G; Palaghias, G. Composition, reactivity and surface interactions of three dental silane primers. *Dent. Mater*, 1993, 9, 182-901.

[149] Cochran, MA; Carlson, TJ; Moore, BK; Richmond, NL; Brackett, WW. Tensile bond strengths of five porcelain repair systems. *Oper. Dent.*, 1988, 13, 162-7.

[150] Chung, KH; Hwang, YC. Bonding strengths of porcelain repair systems with various surface treatments. *J. Prosthet. Dent.*, 1997, 78, 267-74.

[151] Özcan, M; van der Sleen, JM; Kurunmäki, H; Vallittu, PK. Comparison of repair methods for ceramic-fused-to-metal crowns. *J. Prosthodont.*, 2006, 15, 283-8.

[152] Kupiec, KA; Wuertz, KM; Barkmeier, WW; Wilwerding, TM. Evaluation of porcelain surface treatments and agents for composite-to-porcelain repair. *J. Prosthet. Dent.*, 1996, 76, 119-24.

[153] Panah, FG; Rezai, SM; Ahmadian, L. The influence of ceramic surface treatments on the micro-shear bond strength of composite resin to IPS Empress 2. *J. Prosthodont.*, 2008, 17, 409-14.

[154] Oh, WS; Shen, C. Effect of surface topography on the bond strength of a composite to three different types of ceramic. *J. Prosthet. Dent.*, 2003, 90, 241-6.

[155] Cavalcanti, AN; Foxton, RM; Watson, TF; Oliveira, MT; Giannini, M; Marchi, GM. Bond strength of resin cements to a zirconia ceramic with different surface treatments. *Oper. Dent.*, 2009, 34, 280-7.

[156] Kern, M; Barloi, A; Yang, B. Surface conditioning influences zirconia ceramic bonding. *J. Dent. Res.*, 2009, 88, 817-22.

[157] Lehmann, F; Kern, M. Durability of resin bonding to zirconia ceramic using different primers. *J. Adhes. Dent.*, 2009, 11, 479-83.

[158] Kitayama S, Nikaido T, Takahashi R, Zhu L, Ikeda M, Foxton RM, Sadr A, Tagami J. Effect of primer treatment on bonding of resin cements to zirconia ceramic. *Dent. Mater.*, 2010, 26, 426-32.

[159] Yun, JY; Ha, SR; Lee, JB; Kim, SH. Effect of sandblasting and various metal primers on the shear bond strength of resin cement to Y-TZP ceramic. *Dent. Mater.*, 2010, 26, 650-8.

[160] Vallittu, PK. Bonding of hybrid composite resin to the surface of gold-alloy used in porcelain-fused-to-metal restorations. *J. Oral Rehabil.*, 1997, 24, 560-7.

[161] Lynde, TA; Whitehill, JM; Coffey, JP; Meiers, JC. The bond strength of an adhesive resin luting cement to a variety of surface treatments of a high-palladium copper alloy. *J. Prosthodont.*, 1996, 5, 295-300.

[162] van der Veen, JH; Bronsdijk, AE; Siagter, AP; van de Poel, AC; Arends, J. Tensile bond strength of Comspan resin to six differently treated metal surfaces. *Dent. Mater,* 1988, 4, 272-7.
[163] Breeding, LC; Dixon, DL. The effect of metal surface treatment on the shear bond strengths of base and noble metals bonded to enamel. *J. Prosthet. Dent.*, 1996, 76, 390-3.
[164] Eder, A; Wickens, J. Surface treatment of gold alloys for resin adhesion. *Quintessence Int.,* 1996, 27, 35-40.
[165] Watanabe, F; Powers, JM; Lorey, RE. In vitro bonding of prosthodontic adhesives to dental alloys. *J. Dent. Res.,* 1988, 67, 479-83.
[166] Creugers, NH; Welle, PR; Vrijhoef, MM. Four bonding systems for resin-retained cast metal prostheses. *Dent. Mater*, 1988, 4, 85-8.
[167] Naegeli, DG; Duke, ES; Schwartz, R; Norling, BK. Adhesive bonding of composites to a casting alloy. *J. Prosthet. Dent.*, 1988, 60, 279-83.
[168] Moulin, P; Degrange, M; Picard, B. Influence of surface treatment on adherence energy of alloys used in bonded prosthetics. *J. Oral Rehabil.*, 1999, 26, 413-21.
[169] Abreu, A; Loza, MA; Elias, A; Mukhopadhyay, S; Looney, S; Rueggeberg, FA. Tensile bond strength of an adhesive resin cement to different alloys having various surface treatments. *J. Prosthet. Dent.*, 2009, 101, 107-18.
[170] Swartz, JM; Davis, RD; Overton, JD. Tensile bond strength of resin-modified glass-ionomer cement to microabraded and silica-coated or tin-plated high noble ceramic alloy. *J. Prosthodont.,* 2000, 9, 195-200.
[171] Barkmeier, WW; Latta, MA. Laboratory evaluation of a metal-priming agent for adhesive bonding. *Quintessence Int.,* 2000, 31, 749-52.
[172] Stoknorm, R; Isidor, F; Ravnholt, G. Tensile bond strength of resin luting cement to a porcelain-fusing noble alloy. *Int. J. Prosthodont.*, 1996, 9, 323-30.

In: Advances in Materials Science Research, Volume 7
Editor: Maryann C. Wythers

ISBN 978-1-61209-821-0
© 2012 Nova Science Publishers, Inc.

Chapter 4

NITROCELLULOSE IN PROPELLANTS: CHARACTERISTICS AND THERMAL PROPERTIES

Mª Ángeles Fernández de la Ossa[1,2], *Mercedes Torre*[1,2] *and Carmen García-Ruiz*[1,2]

[1]Department of Analytical Chemistry, Faculty of Chemistry, University of Alcalá, Ctra. Madrid-Barcelona km. 33.600, 28871 Alcalá de Henares (Madrid), Spain
[2]University Institute of Research in Police Sciences (IUICP), Pilot Plant of Fine Chemistry, University of Alcalá, Ctra. Madrid-Barcelona km. 33.600, 28871 Alcalá de Henares (Madrid), Spain

ABSTRACT

Nitrocellulose was discovered, by the German-Swiss chemist C.F. Schönbein, in the first half of the nineteenth century but remains to have a great interest today. Nitrocellulose has a similar aspect to cotton (white and fibrous texture). It is a nitrate cellulose ester polymer with β (1→4) bonds between monomers, produced from the nitration of cellulose. Its chemical formula is $[C_6H_7O_2(OH)_{3-x}(ONO_2)_x]_n$, where x indicates the hydroxyl groups exchanged by nitro groups. This macromolecule has different applications depending on their degree of nitration. Nitrocellulose with a low degree of nitration is applied in paints, lacquers, varnishes, inks, etc., while nitrocellulose with a high degree of nitration (>12.5%) is used in explosives. Within the nitrocellulose-containing explosives are included dynamites and propellants. Propellants containing nitrocellulose are smokeless gunpowders, which are widely used by the international military community for propelling projectiles. Depending on gunpowder's composition (active components), they can be classified as: i) single-base gunpowders, which contain mainly nitrocellulose, ii) double-base gunpowders, which contain nitrocellulose and other explosive substance (nitroglycerin, dinitroethylenglycol or dinitrotoluene), and iii) triple-base gunpowders, which are composed by nitrocellulose and two other explosive substances (nitroglycerin or dinitroethylenglycol and nitroguanidine).

Nitrocellulose characteristics, i.e., its high molar mass, complex structure, and unusual chemical behavior, make difficult to perform ordinary studies of this polymer. For these reasons, most works on the characterization of nitrocellulose in propellants have been based on measurements of physico-chemical properties such as molar mass distributions, viscosity, and specific refractive index of nitrocellulose by Size Exclusion Chromatography (SEC) with different types of detectors.

The thermal properties of nitrocellulose are important since it must be burned for its use as propellant. These properties have mainly been studied by determining, using spectrometric techniques (mainly infrared and mass spectrometry), the by-products released during the thermal degradation of nitrocellulose by means of pyrolysis or incineration processes. The most important kinetics parameters of thermal decomposition of nitrocellulose (activation energy, enthalpy, critical explosion and decomposition temperature, etc.) have been studied by typical thermal analytical techniques such as Differential Scanning Calorimetry (DSC) and ThermoGravimetry (TG) or Differential Thermal Analysis (DTA).

The aim of this book chapter is to provide an updated overview (from 1999 until nowadays) of the characteristics and thermal properties of nitrocellulose of high nitrogen content used in propellants.

1. INTRODUCTION

Nitrocellulose has a similar aspect to cotton, is white and has a fibrous texture. It is produced from cellulose. Cellulose is a polysaccharide formed of hundred to over ten thousand D-glucopyranose units linked by β (1→4) bonds [1]. This molecule is a natural polymer that reacts with nitric acid to give the nitrated cellulose ester polymer called nitrocellulose or cellulose nitrated. The general reaction is an esterification R-OH + HONO$_2$ → R-ONO$_2$ + H$_2$O, which is reversible and highly exothermic [2].

The precursor (cellulose) and the final product (nitrocellulose) present a similar structure but some hydroxyl groups have been changed by nitro groups. This replacement occurs in carbons C2, C3 and C6, which are the unique available places for nitro groups, being the rate for nitration C6>>C2≈C3 [2, 3]. Thus, in these positions nitro groups can be joined, giving a compound with the chemical formula of $[C_6H_7O_2(OH)_{3-x}(ONO_2)_x]_n$, where x indicates the hydroxyl groups exchanged by nitro groups. The degree of substitution (DS) designates the quantity of hydroxyl groups exchanged and can be calculated by the equation [4]:

$$Degree\ of\ substitution\ (DS) = \frac{3.6 \text{ x nitrogen content [\%]}}{31.13 - \text{nitrogen content [\%]}}$$

DS is one of the most important properties of nitrocellulose since it affects other properties such as solubility and viscosity and determines the applications of nitrocellulose. The lowest DS value is one, which means a nitrogen content of 6.76% in the nitrocellulose monomer. The highest DS value is 3; it means that all the hydroxyl groups are replaced with nitro groups producing a theoretical nitrogen content of 14.14% (Figure 1).

However, the maximum DS synthesized is usually lower than the maximum theoretically achievable, being reported a maximum DS of 2.9 (≈13.9% of nitrogen content) [3, 4, 5], because to achieve higher DS values is expensive and led to an unstable product.

Nitrocellulose solubility is inversely proportional to DS and to the degree of polymerization (DP, defined as the numbers of repeated units along the chain). In fact, while nitrocellulose is not dissolved by aliphatic and aromatic hydrocarbons, it can be dissolved by alcohols, esters or ketones, depending on its nitrogen content; for example, nitrocellulose with a nitrogen content from 10 to 12.6% is soluble in ether-alcohol mixtures. In nitrocellulose solutions, the polymer is swelled and viscosity increases but it does not form a saturated solution like ionic or molecular compounds, because it is a macromolecular substance that forms colloidal solutions. On the other hand, there is a directly proportional relationship between viscosity of nitrocellulose solutions and the nitrogen content of this macromolecule, which is also influenced by DP [2 -4].

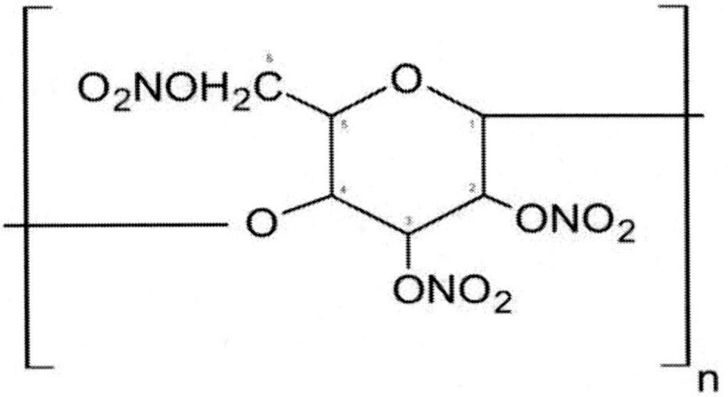

Figure 1. Theoretical chemical structure of a nitrocellulose polymer completely nitrated (degree of substitution, DS = 3, 14.14% of nitrogen content).

The first nitrocellulose was synthesized by the French chemist and pharmacist H. Braconnot, in 1832. He prepared an inflammable and unstable solid by the treatment of cotton or wood pulp (crude cellulose) with concentrated nitric acid (85%, v/v). This method produced a heterogeneous and unstable substance, called xyloidine, which was characterized by being a low nitrated cellulose (nitrogen content of 4-5% as maximum). Braconnot discovered the forerunner of nitrocellulose but it was C.F. Schönbein that created, 14 years later, nitrocellulose as a stable product. This German-Swiss chemist achieved it by the reaction between cotton and a mixture of nitric acid and sulfuric acid and patented it. At present, this method, with few variations, is used to generate commercial nitrocellulose [2, 5].

Nitrocellulose formulation did not suffer marked changes for over forty years until, 1884, when P. Vieille developed his famous "B gunpowder". It was the first smokeless gunpowder. He discovered how to suspend nitrocellulose as a colloid into alcohol-ether solutions allowing to the nitrocellulose to be the main component in gunpowders [6].

With respect to the way by which nitrocellulose is obtained, as it has been stated above, when using concentrated nitric alone, a heterogeneous and unstable nitrocellulose with a low nitrogen content was achieved [3]. The synthesis process was improved adding nitric acid vapors as pretreatment, previously to the addition of 98% nitric acid. In this case, the vapor phase helped to the generation of stable and homogeneous nitrocellulose with a higher nitration degree (nitrogen content ≈13.6%) [3].The reaction of cellulose with a mixture of nitric and sulfuric acid allowed to achieve a stable product with a high nitration degree (nitrogen content of 13.9% as maximum) [3]. The best proportion of these two acids was demonstrated to be 1:1 to 1:3 (nitric acid:sulfuric acid). However, it is important to take into account that sulfuric acid generates unstable sulfuric esters of cellulose.

A mixture of nitric and phosphoric acid in a ratio of 1:1 to 1:3 (nitric acid:phosphoric acid) produced nitrocellulose with a nitrogen content up to 13.7% [3]. By this way, the final product was a highly stable nitrocellulose, but phosphoric acid corrodes iron and steel, being a problem for nitrocellulose manufacture [2, 5].

Nitrocellulose with nitrogen content as high as 14% may be produced by mixing nitric and acetic acids [2]. The main disadvantage of this reaction was the appearance of acetyl nitrate in the reaction medium, which is a very unstable substance at high temperatures.

Another way to manufacture nitrocellulose consisted in the use of nitric acid and organic solvents, such as carbon tetrachloride, methyl nitrate or chloroform, which produced nitrocellulose with a high nitration degree (nitrogen content up to 13.4%) and large yields [3].

Nitrocellulose obtained by the above-mentioned methods is an acid and unstable substance that requires to be neutralized in order to stabilize it. The neutralization process is based on numerous washes with water at room temperature and at 100 °C, followed with washes at basic pH. Then, the trapped sulfuric acid on nitrocellulose fibers is released by broken of the fibers in a mill up to lengths from 0.2 to 0.5 mm. Finally, washes with water (1:10, nitrocellulose: water) at 140 °C under 3-4 atm of pressure are made. In conclusion, to obtain a stable product, long, expensive and complex stabilization procedures are needed [2].

Nowadays, nitrocellulose is usually prepared by the batch-type mechanical dipper process or by a continuous nitration processing method. Both procedures are based on the etherification reaction between the hydroxyl groups of cellulose and the nitric acid, in a mixture of nitric: sulfuric acids. Nitration and separation processes have changed little since the firsts nitrocellulose manufactures, about 100 years ago. However, the manufacture process has evolved in terms of the materials of manufacture equipments (changing lead and iron by stainless steel, to avoid corrosion and break down), transports of raw materials (from manual to automatic), experimental variables (flow, temperature, revolutions, etc.) and the quality control of the whole production process (from raw materials to final product). Dry nitrocellulose is very unstable, being necessary suspending it in water or other appropriate solvents for transport or storage [2]. Moreover, nitrocellulose decomposes for three mechanisms in ambient conditions: hydrolytic, thermal, and photochemical [7].

The nitrogen content of nitrocellulose defines the uses or applications of this polymer. Thus, films, inks, lacquers, and paints are manufactured using nitrocellulose with low nitrogen content (<12%). However, explosives are made with high nitrogen content nitrocellulose (>12%). Highly nitrated nitrocellulose, characterized by average molar masses from 20 to 250 kDa [2, 4, 7], takes part of the composition of some explosives such as dynamites and propellants. Dynamites, which are strong explosives mainly used for civil purposes, are composed by explosive components (ammonium nitrate and nitroglycol), combustible components (butyl phthalate, flour or sawdust, and nitrocellulose) and inert components (such as calcium carbonate). Propellants or gunpowders, which are used to move rapidly projectiles, are classified, according to their composition, as black gunpowders, homogeneous gunpowders, composed gunpowders, and high-explosive gunpowders [2, 5]. Black gunpowders contain mixtures of inorganic compounds such as potassium nitrate, sulfur, and coal [5]. Homogeneous gunpowders, also known as colloidal or nitrocellulose-based gunpowders, have nitrocellulose as common active component. Depending on the number of active components three types of homogeneous gunpowders can be distinguished: i) single-base gunpowders, which contain mainly nitrocellulose, ii) double-base gunpowders, which contain nitrocellulose and one more explosive substance (nitroglycerin, dinitroethylenglycol or dinitrotoluene), and iii) triple-base gunpowders, which are composed by nitrocellulose and two other explosive substances (nitroglycerin or dinitroethylenglycol and nitroguanidine) [5, 8]. Composed gunpowders are plastic matrices with inorganic oxidants that do not contain nitrocellulose [5]. High-explosive gunpowders contain a high explosive, energetic and non-energetic binding materials, a plasticizer agent, and a stabilizer. As example, low-vulnerability-ammunitions (LOVA) are high-explosive gunpowders mainly composed by hexogen and a cellulose derivative [2, 9].

Nowadays, the research on explosives is of great interest in forensic science. This situation is, in part, due to the terrorist attacks committed in the last years towards civil and military people. Then, studies on the characterization and the sensitive determination of explosive substances to identify the explosive used and to obtain information that could lead to the criminals are needed nowadays. In this sense, Forensic Analytical Chemistry, defined as a discipline to analyze proofs of all of kind crime-scenes, choosing the better chemical analysis in each case, depending on the sample and the future use of the analytical information [10], may be very helpful.

In the explosive field, the characterization and determination of nitrocellulose is, still nowadays, a challenge goal for forensic analytical chemistry due to the difficulties in its analysis, inherent to the polymeric nature of this energetic material. As a consequence, the aim of this book chapter is to provide an updated overview from 1999 until nowadays of the characteristics and thermal properties of nitrocellulose of high nitrogen content used in propellants.

2. CHARACTERIZATION OF NITROCELLULOSE

Nitrocellulose characteristics, such as its high molar mass, complex structure, and unusual chemical behavior, make difficult to perform ordinary studies of this polymer. In addition, this macromolecule is generally inside a complex matrix, mixed with other components that should be removed before its analysis. For these reasons, most studies on nitrocellulose are aimed to the characterization of nitrocellulose as polymer.

In this section, recent studies developed to evaluate the physico-chemical, thermal, mechanical and morphological properties of nitrocellulose of high nitrogen content are presented.

2.1. Physico-Chemical Properties of Nitrocellulose

Molar mass, viscosity and specific refractive index of nitrocellulose are determined by Size Exclusion Chromatography (SEC), with different types of detectors.

G. Heinemann [11] reported the advantages of using a triple detection system (combining refractive index, viscosity, and light-scattering detectors) in SEC. This set up allowed to obtain information about molar mass (as conventional SEC with refractive index or ultraviolet detector) and polymeric structures. SEC with triple detection was appropriate to obtain accurate results on the mass average molar mass (Mw, average value of the distribution of the different molar mass taking part of the polymer) and the number average molar mass (Mn, total mass of all molecules in a polymer sample divided by the total number of molecules present) without the need of external standards required in conventional SEC. However, a comparison between polystyrene standards and nitrocellulose was required to study the nitrocellulose structural properties by a Mark-Houwink-Plot (MHP), where the logarithm of the intrinsic viscosity is plotted versus the logarithm of the Mw. Due to a higher slope was obtained in the MHP plot for nitrocellulose than for polystyrene standards, a more open and stretched structure was attributed to nitrocellulose. In addition, for nitrocellulose samples of 240 and 440 kDa, values of Mw/Mn of 238.5/112 kDa and 443.8/102.2 kDa were obtained, respectively.

SEC with triple detection and simple detection using a calibration with standard polymers was also employed by A.F. Macdonald [12] to characterize nitrocellulose. Conventional standards, like polystyrene, polymethylmethacrylate, and polytetrahydrofuran were used to obtain the Mw of nitrocellulose by calibration. Well-characterized monodispersed standards of the same polymer of interest are necessary to obtain true values of Mw. However, this is not possible in the case of nitrocellulose and different values for Mw and Mn were obtained by calibrating with the above-mentioned standard polymers or by using SEC with triple detection. In addition, different Mw values were obtained depending on the standards employed, which confirmed the dependence of this determination from the standard used. In fact, polytetrahydrofuran standards presented the most similar values to those obtained by SEC with triple detection. In this work, a MHP comparison between nitrocellulose and polystyrene showed a similar trend to the work previously cited [11]. That is, a higher stiffness and a less coiled structure in nitrocellulose than in polystyrene standard were confirmed. This was attributed to the high intrinsic viscosity present in the different nitrocellulose samples studied (from cotton and wood).

In 2008, P. Deacon et al. [13] studied the reproducibility of SEC data obtained for nitrocellulose samples with different nitrogen content. It was demonstrated that different analysts presented low reproducible results (Mw and Mn data) while the same analyst gave an excellent reproducibility when the analysis was repeated over three months. It was also shown that Mw measurements were influenced by the moisture content of nitrocellulose, because lower Mw values were obtained for the driest samples. A more complete reproducibility study was performed later [14]. The assessment of the reproducibility obtained among nine laboratories from eight different countries using the same method (STANAG 4178 Ed.2) by SEC with different detectors (viscosity, refraction index, UV/Vis and/or light-scattering) was carried out for nitrocellulose samples having nitrogen contents ranging from 11.6 to 13.5%. Results obtained for these samples presented low reproducibility in Mw and Mn values. This was mainly attributed to the different drying methods used because dried samples leaded to low Mw and *vice versa*. Another factor leading to a lack of reproducibility was attributed to the non-definition of similar and good baseline in the chromatograms obtained, since the area measured affects the Mw determination. In conclusion, this work highlights how complicate is to generate data of quality for nitrocellulose by SEC, mainly among different analysts.

P.R. Deacon et al. [15] tackled the problem of the nitrocellulose behavior in solution. In this work, SEC with triple detection was used to study two types of samples: one with high nitrogen content nitrocellulose (propellant-grade) and, the other, with low nitrogen content nitrocellulose (lacquer-grade). Changes in Mw, viscosity, and refraction index of nitrocellulose samples were observed during the dissolution process until the sample dissolved completely, at which time, these parameters remained constant. Differences in Mw were also observed due to changes in the nitrocellulose concentration in solution. It was demonstrated that, when the concentration of nitrocellulose in solution was increased, the Mw was reduced, which allowed the authors to conclude that true Mw values only could be obtained at low polymer concentrations, except for nitrocellulose with low Mw values. This behavior was explained considering that, when the concentration of nitrocellulose in solution was increased, a reduction in the hydrodynamic volumes was produced, driving to a reduction in the Mw. Besides, these authors determined the dissolution time of nitrocellulose, by means of the refractive index detector, whose measurement is directly proportional to Mw. Interestingly, the dissolution time of nitrocellulose was also influenced by the nitrogen content of this polymer. Thus, high nitrogen content nitrocellulose required less dissolution time to achieve its complete dissolution. For example, the highly nitrated nitrocellulose (propellant-grade, nitrogen content of 12.6%) needed 72 hours for its dissolution in contrast with the lacquer-grade nitrocellulose sample (nitrogen content of 11.8%) that needed 168 hours.

The behavior of nitrocellulose in solution was also studied by J.M. Bellerby et al. [16]. These authors proved that the dissolution of nitrocellulose was not a reversible process since this polymer suffered a permanent change in its structure. Moreover, under stirring, the dissolution time decreased markedly and lower Mw were measured in comparison with no-stirred solutions ($\approx$10% lower values). This was explained by a change in the nitrocellulose structure, by breaking of polymer chains, with stirring. These authors also emphasized, in accordance with the studies of Deacon et al. [15], the influence of percentage nitrogen of nitrocellulose on its solubility, owing to possible internal associations, as hydrogen bonds, that create larger molecules.

High-quality analyses on the degradation process of nitrocellulose are very interesting in the explosive field to understand the behavior of this polymer and to obtain valuable information for the safe production, manipulation, transport, and storage of nitrocellulose-based explosives. In 2001, P.R. Deacon et al. [17] carried out an accelerating ageing experiment by using a thermal treatment for nitrocellulose (with a nitrogen content ranging between 11.7% and 12.2%) extracted from a Polymer Bonded Explosive (PBX). SEC with a triple detection system was used to study changes in molar mass, intrinsic viscosity, and refractive index, in order to investigate the kinetics of the degradation process. Nitrocellulose under ageing process reduced extremely its intrinsic viscosity (from 4.575 to 1.147 dL g^{-1}), as well as its Mw (from 875.1 to 122.4 kDa). Moreover, an activation energy value of 101 ± 11 kJ mol^{-1} for the degradation process of nitrocellulose was reported. This value was consistent with those obtained for similar degradation processes such as acid catalyzed hydrolysis of cellulose and nitrocellulose and thermolysis of nitrocellulose.

A.F. Macdonald et al. [12] also investigated the ageing process of nitrocellulose extracted from PBX made from two different sources (wood pulp and cotton) through the variation of the molar mass of this polymer during this process. The investigation proved that the more extreme were the conditions of ageing (higher temperatures and more time under thermal treatment), the greater was the change in Mw values. Thus, Mw values in the cotton derived nitrocellulose sample were reduced from 378.166 to 87.433 kDa after 783 days. Intrinsic viscosity also was reduced, changing the values for the same sample from 3.555 to 0.360 dL g^{-1}. These results were consistent with a random chain scission process that is one of the accepted models for the nitrocellulose degradation.

2.2. Thermal, Morphological, and Mechanical Properties of Nitrocellulose

G. Herder et al. [18] utilized Differential Scanning Calorimetry (DSC), Thermal Mechanical Analysis (TMA), and Dynamic Mechanical Analysis (DMA) to study the relaxation transitions values of nitrocellulose in a generic propellant (mainly consisted of pure nitrocellulose). Three small relaxation steps, designated as α, β and γ, at 40, -35, and -80 °C, respectively, were investigated. However, due to the dispersion of the data obtained by the three techniques, this study did not reveal the existence of relaxation transitions at these particular temperatures investigated.

A complete surface study of nitrocellulose contained in three nitrocellulose-based propellants was reported by using Scanning Electron Microscopy (SEM), Fourier Transform InfraRed Photoacoustic Spectroscopy (FTIR-PAS) and surface abrasion and posterior FTIR MicroReflectance Spectroscopy (FTIR-MR) analysis [19]. SEM images of burned samples showed that the region affected by the combustion process in these propellants was only a thin external layer of 5-10 μm (Figure 2), remaining the major part of the internal surface without any modification. FTIR-MR spectroscopy was applied to burned samples at different depths made by surface abrasion procedures that were carried out with an abrasive blaster. The depth of the material removed was calculated from the changes in the weight of the material, and knowing the density and diameter of the unburned material. By this technique it was also confirmed that at a depth greater than 10 μm, the material was not affect by burning.

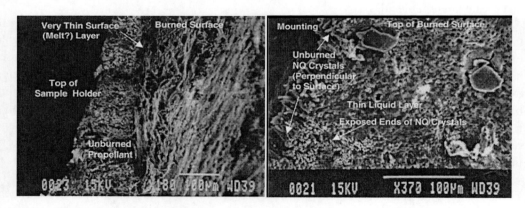

Figure 2. SEM photographs of burned surfaces (burned with a massive copper stub and quenched later) of nitrocellulose-based propellants. (Left) SEM photograph (x180) of cross-section of a propellant (59.5% of nitrocellulose with 13.04% of nitrogen content) burned in air at 0.5 MPa. (Right) SEM photograph (x370) of cross-section of a propellant (28% of nitrocellulose with 12.68% of nitrogen content) burned in air at 2.0 MPa. SEM conditions: JEOL Model JSM-820 instrument. Reprinted from [19]. Copyright (2001), with permission of Elsevier.

From FTIR-PAS and FTIR-MR spectra it was difficult to investigate the modifications in the structure of propellant samples because they also contained many other compounds that generated a great quantity of signals. The main information obtained for nitrocellulose analysis was related with the material affected by the burning process where a weak signal at 1730 cm^{-1} appeared, which was assigned to the carbonyl groups formed in burned nitrocellulose.

The surface characterization of a propellant sample containing 59.5% of nitrocellulose with a nitrogen content of 13.04% was performed by J. Newberry et al. [20] using Macro-ATR, Micro-ATR, and FTIR Specular Reflectance Spectroscopy. Ignition of the sample was carried out by a plasma source. In spectra obtained by Macro-ATR and Specular Reflectance FTIR (not shown) one band about 1700 cm^{-1} was observed, which was characteristic of carbonyl groups produced by the oxidation of the nitrate esters of nitrocellulose. However a higher signal-to-noise was observed by Macro-ATR FTIR, which allowed a better study of the nitrocellulose degradation in the propellant ignited by plasma.

Porosity (pore size and pore size distribution) of nitrocellulose was studied by thermoporometry using Differential Scanning Calorimetry (DSC) [21]. Nitrocellulose presented the typical difficulties of hydrophobic substances to achieve a complete penetration of water in its porous, for this reason nitrocellulose in water was stirred or boiled to help the water penetration. Nitrocellulose containing different nitrogen content (12.4 and 13.2%) was studied, being the sample of the highest nitrogen content in granular form. Pore size of nitrocellulose containing 12.4% of nitrogen was of 6.8 ± 0.9 nm while granular nitrocellulose with a nitrogen content of 13.2% was of 15.4 ± 2.0 nm. Since cellulose presented a pore size of 6 nm, the pore size value for the lowest nitrogen content nitrocellulose suggested that the structure after cellulose nitration was not modified. The highest pore size for granular nitrocellulose was justified for the increase in the nitrogen content; nevertheless the different forms of nitrocellulose also may contribute to such changes. Moreover, values of 29.0 ± 0.6 nm (multiple value of pore size) for both nitrocellulose samples were obtained and were attributed to a layer structure of this polymer. A similar behavior was also obtained with benzene as solvent.

Finally, the characterization of nitrocellulose contained in triple base propellants used in tank gun ammunition attending its mechanical properties was made by R.R. Sanghavi et al. [22]. In propellants dissolved in a mixture of acetone and alcohol, the mechanical parameters of compression strength and percentage compression were measured. These properties decreased with the nitrogen content of nitrocellulose contained in propellants. For example, the compression strength for a sample containing nitrocellulose with a nitrogen content of 12.2% was 395 kg cm^{-2}, while for a sample containing nitrocellulose with a nitrogen content of 12.6% it was 316 kg cm^{-2}. The same trend was observed with the percentage compression, which decreased from 35 to 31%.

3. DEGRADATION OF NITROCELLULOSE CONTAINED IN EXPLOSIVES

There are also studies on different properties of nitrocellulose which are important for its use as explosive: the stability and the decomposition processes of nitrocellulose. Processes of decomposition of nitrocellulose contained in explosive formulations are of particular interest from the point of view of safety in the manufacture, storage, manipulation, and use of explosives. In this context, studies on kinetics and thermochemical processes involved in the degradation of explosives are quite important. In fact, several investigations on thermal, biological, and mechanical degradation have been developed for nitrocellulose-based explosives.

3.1. Thermal Decomposition of Nitrocellulose

Thermal analyses of nitrocellulose are the most widely performed since it must be burnt for its use. Some of these analyses deal with the study of by-products released in the thermal degradation of nitrocellulose by pyrolysis or incineration processes. The effect of nitrogen content on thermal stability and decomposition of nitrocellulose in explosives was investigated by DSC and ThermoGravimetry-Differential Thermal Analysis (TG-DTA) in four nitrocellulose samples (nitrogen content ranging from 12.5 to 13.9%) in solid state [23]. Critical explosion temperature and half-life were the thermal parameters obtained in this work. Also kinetics parameters like activation energy and frequency factor of thermal decomposition were acquired. The study of DSC curves proved that thermal stability of nitrocellulose decreased at higher nitrogen content and lower heating rate. Kinetics parameters were calculated in this work assuming that the nitrocellulose decomposition process followed a first-order kinetics. Activation energy about 155 kJ mol^{-1}, critical explosion temperature of 196 °C, and half-life about five years at 50 °C were determined for the nitrocellulose sample containing the highest nitrogen content (13.9%), which was the most unstable one. For this sample, a temperature of 192 °C was defined, from TG-DTA curve data, as the starting temperature for its thermal decomposition.

The thermal decomposition of nitrocellulose fibers with a nitrogen content of 13.9% and possessing micro- and nano-particle sizes was studied by M.R. Sovizi et al. [24]. First, two different particle sizes images for the nitrocellulose fibers were confirmed by SEM (see Figure 3).

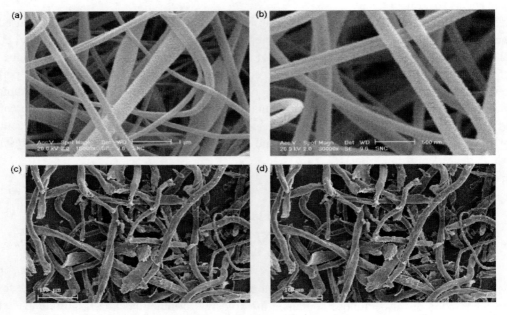

Figure 3. SEM images of nitrocellulose fibers. (a) nano-fibers sample (x 1.5x10^4), (b) nano-fibers sample (x 3.0x10^4) and (c and d) microfibers sample. SEM conditions: Philips XL30 series instrument using a gold film for loading the dried particles. Gold films were prepared by a Sputter Coater model SCD005 made by BAL-TEC. Reprinted from [24]. Copyright (2009), with permission of Elsevier.

Dependence between the thermal decomposition, studied by DSC and TG-DTA, and the nitrocellulose particle size was found. In addition, nitrocellulose micro-fibers presented higher thermal stability than nitrocellulose nano-fibers, due to their higher activation energy, enthalpy, decomposition temperature, and critical explosion temperature values, determined for the micro-fibers in comparison with the nano-fibers. First-order kinetics was also assumed in this work for the determination of the above-mentioned kinetics parameters. Thus, nitrocellulose nano-fibers were proposed for more exigent devices due to their higher heat sensitivity.

The thermal decomposition process of nitrocellulose was also studied by Accelerating Rate Calorimetry (ARC) and simultaneous TG-DTA, coupled to FTIR and Mass Spectrometry (TG-DTA-FTIR-MS) to study the thermal degradation of nitrocellulose samples (nitrogen content of 13.15%) in different environments [25]. TG-DTA curves showed usual mass loss while in FTIR spectra signals due to gasses release during the thermal decomposition of nitrocellulose were observed. For air and helium environments, similar activation energy values were obtained by DTA (170 ± 4 kJ mol^{-1} in air and 169 ± 6 kJ mol^{-1} in He) and TG (166 ± 2 kJ mol^{-1} in air and 174 ± 2 kJ mol^{-1} in He), which indicated an independence between the activation energy and the gas atmosphere. However, the activation energy values obtained by isothermal ARC measurements at different temperatures were lower than those previously reported (86 ± 4 kJ mol^{-1} for argon and 122 ± 8 kJ mol^{-1} in air). In addition, a significant dependence on gas environment was observed for the parameter measured. The authors justified these results by a previous oxidation process of nitrocellulose in air that does not exist in an argon atmosphere. In air, an increase of pressure from 0.1 to 1.8 MPa generated a decrease in the onset temperature (decrease in the thermal stability of nitrocellulose), whereas a later increase in pressure did not changed significantly the onset temperature and thermal stability of nitrocellulose. Nevertheless, in argon atmosphere the opposite behavior was observed: a raise in pressure from 0.27 to 1.91 MPa improved the thermal stability of nitrocellulose but also a later increment in pressure did not modify strongly this behavior. Also kinetics parameters of thermal decomposition of nitrocellulose such as the activation energy (Ea) and the pre-exponential factor (A) was obtained in this study and were compared with those given in previous works. Three different kinetic processes were suggested: first-order, autocatalytic, and second-order. While first-order and autocatalytic processes were proposed in previous works, most measurements obtained in this study were included in a second-order line. It is important to remark that kinetic data compared were obtained by different reaction models, being not comparable and making difficult the establishment of the real process occurring in the thermal degradation of nitrocellulose.

Thermal decomposition of nitrocellulose was proposed by R.K. Campbell et al. [26] as a strategy to decrease the nitrogen content of nitrocellulose with the aim to transform it into a non hazardous material. Waste fines of nitrocellulose generated during the manufacturing process of this compound have explosive properties, being difficult their treatment or their application in alternative uses. In this work, two types of nitrocellulose (pulp and lint, with a nitrogen content of 13.5% and 13.1%, respectively) were heated at 130, 140, and 150 °C. The faster decrease in nitrogen was observed at the highest temperature, although at this temperature increased the explosion risks from nitrocellulose samples. Contrary to previous works, the application of an air flow (which implies the presence of an O_2 atmosphere) and the water content of samples did not modify significantly the nitrocellulose decomposition rate. However, the plot of mass loss against the nitrocellulose nitrogen content was of special interest because the value measured for this magnitude exceeded the limits corresponding to removal of all nitro groups of nitrocellulose, which leaded to the idea that the degradation of the polymer structure by other mechanisms was also produced. This fact was confirmed by Gas Chromatography (GC), where CO_2 was detected and there were not sources of carbon other than nitrocellulose in the analyzed environment. Thermal stability tests were carried out to confirm the non explosive characteristics of samples after their thermal degradation.

The nitrocellulose melting process was studied by Modulated Differential Scanning Calorimetry (MDSC) because from sixties/seventies years there was not agreement about the real process. N. Binke et al. [27] obtained various photographs of nitrocellulose during heating processes and proved that the melting and decomposition processes coexisted. For high nitrogen content nitrocellulose molecules (from 11.9 to 14.1%) the decomposition temperature was lower than the melting temperature, which indicated that the melting process was produced on a mixture of nitrocellulose and its decomposition products. This fact was confirmed by FTIR spectra, since nitrocellulose suffered a decrease in $O-NO_2$ peak by heating at 20 K min^{-1}, which proved that nitrocellulose was decomposed partly before melting (see Figure 4).

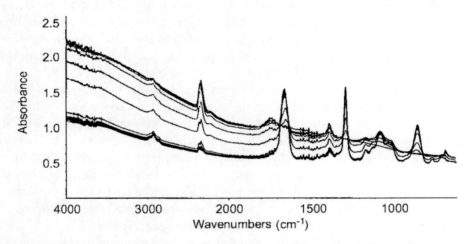

Figure 4. FTIR spectra obtained from the thermal decomposition process of nitrocellulose (12.97% of nitrogen content) at a heating rate of 20 K min^{-1}. Collecting time spectra from bottom to top is 5.36, 5.93, 6.50, 6.78, 7.06, 7.35, 7.63, 7.94, 8.19, 8.48, 8.76, 9.33, 12.67 minutes, respectively. FTIR conditions: Nicolet Spectrometer (60SXR) with a deuterated triglycine sulfate (DTGS) detector using 4 cm^{-1} of resolution (8 scans/file). Reprinted from [27]. Copyright (1999), with permission of Springer.

Another two interesting papers studying the kinetic of the melting process of highly nitrated nitrocellulose were published by the same research group [28, 29]. The TG curve obtained for the initial 50% of mass-loss of highly nitrated nitrocellulose was described by a first order autocatalytic equation, whereas for the latter 50% mass-loss two other mechanisms were suggested [28]. In addition, the critical temperature, measured by non-isothermal DSC and two different methods, was about 182 °C [29]. This value is consistent with the value of 192 °C previously reported for a nitrocellulose containing a nitrogen content of 13.9% and with the affirmation that, at higher nitrogen content lower thermal stability [23].

The ignition mechanism of nitrocellulose during isothermal storage at 393 K was studied by K. Katoh et al. [30]. In this work, nitrocellulose with a nitrogen content of 12% was stored at 393 K up to 35 hours in 4.7% (v/v) NO_2/air, or an O_2/N_2 mixture (dry air), at different O_2 partial pressures, to study its thermal behavior. After storage, gases released during this treatment were analyzed by GC. During nitrocellulose storage under dry air, a decrease in the O_2 pressure and a heat generation was observed, suggesting an autoxidation mechanism by a first-order reaction with respect to the O_2 decrease. During nitrocellulose storage in a NO_2/air atmosphere, similar values of heat reaction were obtained (450 J g^{-1}) in comparison with those obtained under dry air atmosphere (460 J g^{-1}). Moreover, the heat conversion presented a linear relationship with the storage time in both atmospheres, which leaded to the fitting of these systems to a first-order reaction when O_2 decreased. Rate constant and induction period calculated by this way showed significant differences: in a NO_2/air ambient the calculated rate constant was 1.5×10^{-4} s^{-1} while in dry air this constant was 7.7×10^{-5} s^{-1}, observing a lower induction period in NO_2 atmosphere (from 9.2 to 3.4 hours). According to these data, the authors proposed a spontaneous ignition mechanism where NO_2 had a strong influence in the initiation process but a low effect in the autoxidation reaction. This proposal was reinforced by the fact that an increase in the amount of nitrocellulose, which supposed an increment in the NO_2 content, generated a decrease in the induction time without modifying the heat released.

R.I. Hiyoshi et al. [31] studied the thermal degradation of nitrocellulose films (13% of nitrogen content) by pyrolysis (300 °C under 0.1 MPa) in air and Ar atmospheres using T-Jump/FTIR spectroscopy. The air atmosphere was created by combining 80% N_2 and 20% O_2, to avoid interferences provided by the strong IR absorption of CO_2 present in natural air. The samples were analyzed by FTIR spectroscopy under both atmospheres. Products of decomposition in air and Ar were identified as CO, CO_2, NO, NO_2, H_2O, HCN, and CH_2O, being CO the most widely produced. According to these results, it was obvious that the atmosphere did not interfere in the measurements. The authors attributed these results to the low vapor pressure of nitrocellulose (not measurable), which implied that nitrocellulose decomposed primarily in the condensed phase on the T-Jump filament and the products thus formed were relatively independent of the surrounding atmosphere. Moreover, higher vapor pressure explosives as pentaerythritol tetranitrate (1.1×10^{-7} MPa at 100 °C) and nitroglycerin (3.1×10^{-5} MPa at 90 °C) were studied in this work, showing a dependence with atmosphere. In this case, a vaporization-decomposition process occurred due to the higher vapor pressure of these compounds. Hence greater mixing of the gases during decomposition process was exhibit, modifying the gases in each environment.

The analysis of pyrolysates of nitrocellulose by T-Jump/Time-Of-Flight Mass Spectrometry (T-Jump/TOF-MS) with an electron ionization (EI) source [32] was carried out by mixing nitrocellulose with diethyl ether or acetone and coating on a T-Jump filament. Later it was heated at $\approx 1.3 \times 10^5$ K s^{-1}, during about 9 ms, and time-resolved spectra were recorded at different times and temperatures. In these spectra, above m/z 100 no signals appeared while the main signals were observed between 15 and 60 m/z values. Peaks were assigned to OH, H_2O, N_2, NO, NO_2, CO, HCN, HNO, HCO_2, and hydrocarbons (characteristic of nitrocellulose pyrolysis) were also detected (Figure 5). In addition, it was observed that the decomposition process started at 575 K (301.85 °C), because signals different to blank runs appeared.

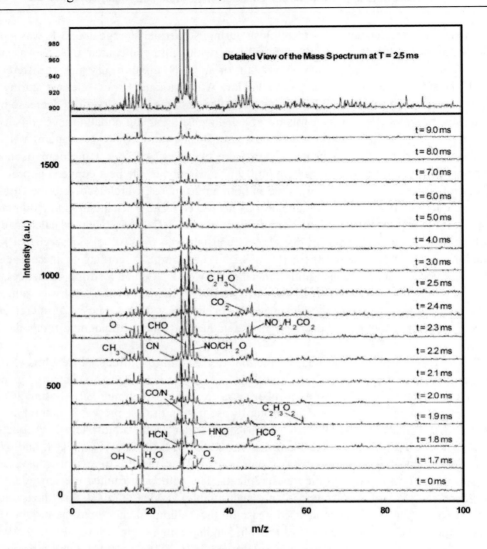

Figure 5. Time-resolved mass spectra of nitrocellulose. Heating rate 1.3×10^5 K s^{-1}. EI/TOFMS conditions: Electron beam was nominally operating at 70 eV and 1 mA, with the background pressure in the TOF chamber at $\approx 10^{-7}$ Torr. 95 spectra were sampled with a temporal resolution of 100 μs per spectra (1.0×10^4 Hz), only 17 were plotted in this figure. A more detailed view of spectra at 2.5 ms (top of figure) is also showed. Reprinted from [32]. Copyright (2008), with permission of Wiley.

D.M. Cropek et al. [33] analyzed the pyrolysates of a double-base propellant composed by nitrocellulose (with a nitrogen content of 13.4%) and nitroglycerin using GC coupled with mass spectrometry (GC-MS). In addition to nitrocellulose, the other components of the propellant were investigated independently, to evaluate their contribution to the propellant behavior. Owing to the diverse mass of by-products obtained by pyrolysis, two different GC configurations were used, one for light molar mass (LMW) gases and another for heavy molar mass (HMW) gases. Main signals of the pyrogram obtained for the propellant in the LMW configuration corresponded to CO, NO, CO_2, and H_2O ($\approx$70% of the total peak area), while in the HMW configuration main signals were observed for not identified light gases ($\approx$90% of the total peak area) (see Figure 6).

Nitrocellulose in Propellants: Characteristics and Thermal Properties

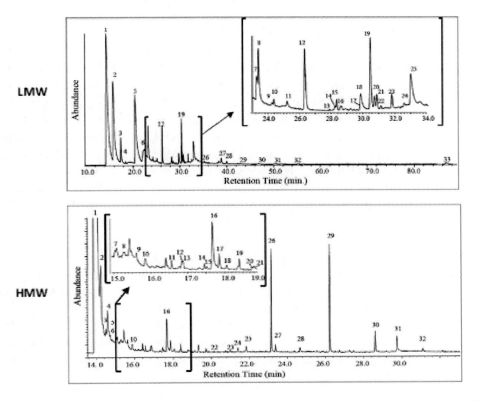

Figure 6. Pyrograms for (top) light molar mass (LMW) gases and (bottom) heavy molar mass (HMW) gases from a double-base propellant (mainly composed by nitrocellulose and nitroglycerin). GC and MS conditions: GC column of 50 m x 0.32 mm i.d. x 10 μm film thickness. GC oven began at 40 °C for 13 minutes, ramped to 200 °C at 10 °C min^{-1}, and stayed at 200 °C for 60 minutes. Injector port was held at 200 °C and used in the splitless mode. Detector port was held at 280 °C. Mass detector scanned from 10 to 400 amu. Identification of the pyrolysate peaks was accomplished by comparing mass spectral data to library standards. Peaks identification: (top) Peak 1: CO, NO (34.3% total peak area); peak 2: CO_2 (18.7% total peak area) and peak 5: H_2O (16.1% total peak area). (bottom) Peak 1: non identified light gases (89.9% total peak area) Reprinted from [33]. Copyright (2001), with permission of US Army Corp. of Engineers.

Comparing the double-base propellant and the nitrocellulose pyrograms (not shown), it was observed that nitrocellulose was the main by-products source generated by the propellant pyrolysis. Approximately, the 50% of products proceeded from nitrocellulose, since 25 products were common in propellant and nitrocellulose. It is remarkable that the only two compounds appearing in the nitrocellulose pyrogram ($C_2H_4O_2$ and $C_5H_4O_2$, for LMW and HMW analysis, respectively), were presented in propellant pyrograms, which indicated that they were produced only during the nitrocellulose pyrolysis. In addition, these comparative studies were useful to investigate which products were obtained by the decomposition of the propellant and which were obtained by later reactions produced by the generated gases.

The same authors [34] collected and characterized the incinerator emissions released during the incineration treatment of the above-mentioned double-base propellant and nitrocellulose wastes. Owing to the different emissions produced, they were sampled in four fractions: volatile organic compounds (VOC), semivolatile organic compounds (SVOC), hydrogen cyanide (HCN), and continuous emission monitors (CEM). VOC and SVOS fractions were analyzed by GC-MS under different conditions. CEM fraction, which was constituted by total hydrocarbons (THC), NO_x, O_2, CO_2, and CO gases, was characterized by various EPA methods (based on nondispersive infrared spectroscopy, chemiluminescence, paramagnetic properties, and heated flame ionization) and measured continuously over an entire incineration run. Main compounds analyzed in VOC fraction of both samples were toluene and benzene. Most abundant components in the SVOC fraction were unknown hydrocarbons, 2-fluorobiphenyl (in nitrocellulose samples) and di-n-propyladipate and 2-nitrodiphenylamine, which are two minor components acting as additives in the propellant. No significant amounts of HCN were detected in neither of samples. About the CEM fraction, levels of NO_x, CO, and CO_2 increased during the incineration of the double-base propellant, and then these gases were assigned as the incineration by-products. As it was expected, O_2 was consumed during the process. It was also observed that nitrocellulose generated different compounds than the propellant analyzed, which was attributed to the important role of the additives in the incineration process of gunpowders.

D.M. Cropeck et al. [35], compared pyrolysis and incineration results obtained in previous works [33, 34] and concluded that pyrolysis analysis gave information about the remaining components in the incineration process, the products coming from the incomplete combustion of each component separately or in combination with other species, as well as those components that contribute to the most troublesome emissions. For example, in the case of SVOC emissions, 2-nitrodiphenylamine (unwanted by-product) was one of the principal released compounds despite being a minority component. Hence additives dominated the pyrolysis behavior in this case.

3.2. Biological and Mechanical Degradation of Nitrocellulose

Biological decomposition processes are focused to remove nitro groups of nitrocellulose and to transform waste nitrocellulose into a non hazardous material. Biotransformation under denitrifying and sulfidogenic conditions using an activated sludge inoculum was performed by D.L. Freedman et al. [36] to turn into a non hazardous material the waste nitrocellulose. The variation of the nitrogen content in nitrocellulose was followed by a digestion/titration method. Nitrocellulose reduced in such conditions presented, approximately, 1% of nitrogen less than virgin nitrocellulose (the exact decrease in nitrogen content was from 13.1-13.2% to 12.2-12.4%). According to these results, it is necessary the presence of electron donors to achieve the reduction of nitrocellulose (methanol in denitrifying treatment and lactate in sulfidogenic conditions) because, without them, there were not changes in the nitrogen content of samples. In addition, a study of possible changes in the structure of nitrocellulose subjected to these treatments was made by FTIR spectroscopy. After the denitrifying treatment (Figure 7) monomeric alcohol signals about 3400-3500 cm^{-1} were detected in the FTIR spectra, may be due to the presence of hydroxyl groups, which were generated by the loss of nitro groups in the nitrocellulose.

Nevertheless, under the sulfidogenic conditions the FTIR spectra of samples were not different to those of virgin samples. About explosive properties of nitrocellulose treated, neither of the two conditions applied were able to transform the explosive samples into non-hazardous materials, in spite of a decrease of flammable and explosive properties were observed.

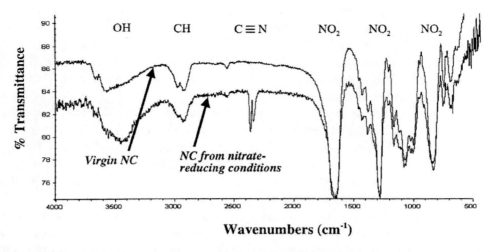

Figure 7. FTIR spectra of virgin nitrocellulose and nitrocellulose from nitrate-reducing conditions. FTIR conditions: Nicolet Spectrometer (Impact 410 IR). Reprinted from [36]. Copyright (2002), with permission of Elsevier.

Mechanochemical processes are made to assess the effect of mild stress on the nitrocellulose degradation, in order to evaluate its mechanochemical stability. S. Vyazovkin et al. [37] used a pioneer set-up based on Dynamic Mechanical Analysis (DMA) combined with Mass Spectrometry (MS) to study the degradation suffered by a film made from a nitrocellulose fiber sample with a nitrogen content of 12.4%. Experiments were carried out at 150 and 160 °C and frequencies from 100 to 600 Hz, increased manually in 100 units during 30 minutes. Temperatures above 150 °C were necessary to begin the degradation process since at lower temperature values more time was necessary to observe some degradations products by MS spectra. Nevertheless, the study of blank runs proved that there were not signals corresponding to degradation products when temperature increased, being possible to consider that only a mechanochemical degradation was present. From the different temperatures tested, nitrocellulose degradation needed more time at 150 °C. The decomposition of nitrocellulose was followed through MS peaks appearing at m/z 30 (NO$^+$ and CH$_2$O$^+$ ions), because they were the most intense MS peaks of the spectra and corresponded to well known products formed during the thermal degradation of nitrocellulose. Both ions were detected in the all the experiments performed. Finally, reflected light micrographs were obtained by a polarized light microscope to take pictures of fractured films. Curiously, the examination showed that the fracture front looked similar to that of a broken window glass with a few cracks running under an angle to the front.

4. CONCLUSION

Although the discovery of nitrocellulose was carried out two hundred years ago, today there are still no analytical methodologies to enable its determination with good accuracy and precision in explosive samples. In fact, main difficulties in the analysis of this polymeric compound are its high molar mass, structural complexity and unusual behavior in solution. Of special interest is the analysis of nitrocellulose of high nitrogen content (≥12%), used in the manufacture of explosives, particularly in nitrocellulose-based gunpowders.

Most studies on high nitrogen content nitrocellulose deal with their characterization as a polymer. Regarding this subject, Size Exclusion Chromatography (SEC) is the main separation technique used for the physicochemical characterization of this polymer without previous treatment. This technique provides mainly information on molar mass, viscosity and specific refractive index of nitrocellulose.

In addition, due to the explosive nature of nitrocellulose, several studies found in the literature are focus on the determination of thermal properties of this macromolecule. Techniques such as ThermoGravimetry (TG), Differential Scanning Calorimetry (DSC) and Differential Thermal Analysis (DTA) are the most commonly used with this purpose, while Scanning Electron Microscopy (SEM) and Fourier-transform InfraRed Spectroscopy (FTIR) are employed for the morphological study of nitrocellulose.

Other important aspect of nitrocellulose contained in gunpowders is its stability and the possibility of decomposition by means of thermal, mechanical or biological processes. Most studies dealt with thermal decomposition of nitrocellulose, which is carried out mainly by means of thermal analytical techniques: TG, DSC, ThermoGravimetry-Differential Thermal Analysis (TG-DTA), Accelerating Rate Calorimetry (ARC), Modulated Differential Scanning Calorimetry (MDSC), and Pyrolysis. This degradation process was followed through the formation of by-products of reaction and the release of gases, utilizing with this aim, the analysis by FTIR, Gas Chromatography (GC), Mass Spectrometry (MS) and coupled techniques (FTIR-MS and GC-MS).

ACKNOWLEDGMENTS

Authors thank to the Ministry of Science and Innovation the project CTQ2008-00633-E. Mª Ángeles Fernández de la Ossa thanks to University of Alcalá her pre-doctoral grant.

REFERENCES

[1] Nelson, D.L.; Cox, M.M. *Lehninger Principios de bioquímica*; ISBN: 9788428214865; Omega: Barcelona, Spain, 2009; pp 235-270.
[2] Monforte, M. *Las Pólvoras y sus aplicaciones.Vol. 1*; UEE Explosivos: Madrid, Spain, 1992; pp 304-360.
[3] Saunders, C.W.; Taylor L.T. *J. Energ. Mater*. 1990, *8*, 149-203.
[4] Dow Wolff Cellulosics. (2008). Nitrocellulose. http://msdssearch.dow.com/Published LiteratureDOWCOM/dh_014c/0901b8038014c3b9.pdf?filepath=dowwolff/pdfs/noreg/ 822-00007.pdf&fromPage=GetDoc.
[5] Rasines, R.; López, M.; Torre, M.; García, C. *An. Quím*. 2009, *4*, 265-270.
[6] Helmenstine, T. Paul Vieille Biography. http://chemistry.about.com/od/famouschemists /p/paul-vieille-bio.htm.
[7] Selwitz, C. *Cellulose nitrate in conservation*; ISBN: 0-89236-098-4; The Getty Conservation Institute: USA, 1998; pp 1-67.

[8] López-López, M.; Fernández de la Ossa, M.A.; Sáiz Galindo, J.; Ferrando, J.L.; Vega, A.; Torre, M.; García-Ruiz, C. *Talanta*. 2010, *81*, 1742-1749.
[9] Ouellet, N.; Brochu, S.; Lussier, L. *Appl. Spectrosc.* 2002, *56*, 125-133.
[10] Cruces-Blanco, C.; Gamiz-Gracia, L.; Garcia-Campana, A.M. *Trends Anal. Chem.* 2007, *26*, 215-226.
[11] Heinzmann, G. *Int. Annu. Conf. ICT. 33rd*. 2002, *58*, 1-5.
[12] Macdonald, A.F. *Int. Annu. Conf. ICT. 34th*. 2003, *126*, 1-10.
[13] Deacon, P.; Macdonald, A.; Gill, P.; Mai, N.; Bohn, M.; Pontius, H. *Int. Annu. Conf. ICT. 39th*. 2008, *68*, 1-12.
[14] Deacon, P.; Macdonald, A.; Gill, P.; Mai, N.; Bohn, M.A.; Pontius, H.; van Hulst, M.; de Klerk, W.; Baker, C. *Int. Annu. Conf. ICT. 40th*. 2009, *81*, 1-14.
[15] Deacon, P.R.; Garman, R.N.; Macdonald, A.F.; Baker, C.A. *Int. Annu. Conf. ICT. 37th*. 2006, *151*, 1-7.
[16] Bellerby, J.M.; Deacon, P.R.; Gill, P.P. *Int. Annu. Conf. ICT. 37th*. 2006, *71*, 1-11.
[17] Deacon, P.R.; Kennedy, G.R.A.; Lewis, A.L.; Macdonald, A.F. *Symp. Chem. Probl. Connected Stab. Explos. 12th*. 2004, 195-204.
[18] Herder, G.; de Klerk, W.P.C. *J. Therm. Anal. Calorim.* 2006, *85*, 169-172.
[19] Schroeder, M.A.; Fifer, R.A.; Miller, M.S.; Pesce-Rodriguez, R.A.; McNesby, C.J.S.; Singh, G. *Combust. Flame*. 2001, *126*, 1569-1576.
[20] Newberry, J.; Kaste, P.J. *IEEE Trans. Magn*. 2003, *39*, 253-256.
[21] Ksiazczak, A.; Radomski, A.; Zielenkiewicz, T. *J. Therm. Anal. Calorim.* 2003, *74*, 559-568.
[22] Sanghavi, R.R.; Pillai, A.G.S.; Velapure, S.P.; Singh, A. *J. Energ. Mater*. 2003, *21,* 87-95.
[23] Pourmortazavi, S.M.; Hosseini, S.G.; Rahimi-Nasrabadi, M.; Hajimirsadeghi, S.S.; Momenian, H. *J. Hazard. Mater*. 2009, *162,* 1141-1144.
[24] Sovizi, M.R.; Hajimirsadeghi, S.S.; Naderizadeh, B. *J. Hazard. Mater*. 2009, *168,* 1134-1139.
[25] Turcotte, R.; Acheson, B.; Armstrong, K.; Kwok, Q.S.M.; Jones, D.E.G.; Paquet, M. *Proc. Int. Pyrotech. Semin. 33rd*. 2006, 351-361.
[26] Campbell, R.K.; Freedman, D.L.; Kim, B.J. *Environ. Eng. 1999 Proc. ASCE-CSCE Natl. Conf.* 1999, 246-253.
[27] Binke, N.; Rong, L.; Xianqi, C.; Yuan, W.; Rongzu, H.; Qingsen, Y. *J. Therm. Anal. Calorim.* 1999, *58,* 249-256.
[28] Binke, N.; Rong, L.; Zhengquan, Y.; Yuan, W.; Pu, Y.; Rongzu, H.; Qingsen, Y. *J. Therm. Anal. Calorim.* 1999, *58,* 403-411.
[29] Rong, L.; Binke, N.; Yuan, W.; Zhengquan, Y.; Rongzu, H. *J. Therm. Anal. Calorim.* 1999, *58,* 369-373.
[30] Katoh, K.; Le, L.; Kumasaki, M.; Wada, Y.; Arai, M.; Tamura, M. *Thermochim. Acta*. 2005, *431,* 161-167.
[31] Hiyoshi, R.I.; Brill, T.B. *Propellants, Explos., Pyrotech*. 2002, *27,* 23-30.
[32] Zhou, L.; Piekiel, N.; Chowdhury, S.; Zachariah, M.R. *Rapid Commun. Mass Spectrom*. 2009, *23*, 194-202.
[33] Cropek, D.M.; Kemme, P.A.; Day, J.M. (2001). Pyrolytic decomposition studies of AA2, a double-base propellant. http://owww.cecer.army.mil/techreports/Cropek Pyrolosis_Decomposition/Cropek_Pyrolysis_decomposition.pdf.

[34] Cropek, D.M.; Day, J.M.; Kemme, P.A. (2001). Incineration by-products of AA2. NC fines, and NG slums. http://owww.cecer.army.mil/techreports/cropek_incineration/cropek_incineration.pdf.
[35] Cropek, D.M.; Kemme, P.A.; Day, J.M.; Cochran, J. *Environ. Sci. Technol.* 2002, *36*, 4346-4351.
[36] Freedman, D.L.; Cashwell, J.M.; Kim, B.J. *Waste Manage.* 2002, *22*, 283-292.
[37] Vyazovkin, S.; Dranca, I.; Lang, A.J. *Thermochim. Acta.* 2005, *437*, 75-81.

Chapter 5

STUDIES ON ENERGY PROPERTIES, THERMAL BEHAVIORS, COMBUSTION PROPERTIES AND BURNING RATE PREDICTION OF COMPOSITE MODIFIED DOUBLE BASE PROPELLANT CONTAINING HNIW

Si-Yu Xu, Feng-Qi Zhao, Jian-Hua Yi, Hong-Xu Gao, and Rong-Zu Hu

Xi'an Modern Chemistry Research Institute, Xi'an, Shaanxi, China

ABSTRACT

A series of composite modified double base propellants containing HNIW (HNIW-CMDB propellant) are prepared. Their properties including energy properties, thermal behaviors, non-isothermal decomposition reaction kinetics, combustion properties and burning rate prediction are investigated. The results show that the energy property parameters of HNIW-CMDB propellant, such as theory specific impulse, characteristic velocity, oxygen coefficient, combustion temperature, heat of explosion at constant volume, heat of combustion and the average relative molecular mass of combustion gas increase linearly with increasing HNIW content. The burning rate of HNIW-CMDB propellants with HNIW content less than 50% increases linearly with increasing pressure between 2MPa and 22MPa, and the effect of HNIW content on burning rate is negligible under the same pressure. The apparent activation energy (E_a), pre-exponential constant (A) and mechanism function of thermal decomposition reaction under both 4 MPa and 7MPa are obtained by non-isothermal method. There is obvious dark zone above the burning surface of HNIW-CMDB propellant when HNIW content is less than 28%. There are some luminous flame-lets from burning surface in the dark zone. There is no obvious dark zone above the burning surface of HNIW-CMDB propellant when HNIW content is more than 28%. The temperature distribution of combustion wave of HNIW-CMDB propellant is obtained under 6MPa. The model for predicting the burning rate of HNIW-CMDB propellant is established based on one-dimension gas phase reaction theory. The burning rates of HNIW-CMDB propellant are calculated and the results are in agreement with the tested data.

Keywords: HNIW, HNIW-CMDB propellant, combustion property, burning rate prediction, thermal behavior

1. INTRODUCTION

Hexanitrohexaazaisowurtzitane (HNIW) is an ecologically safe, high-energy density material with a cage structure. Compared with other nitramines such as hexogen (RDX) and octogen (HMX), HNIW has six $N-NO_2$ groups in its polycyclic structure (RDX and HMX have three and four $N-NO_2$ groups, respectively), resulting in an increase in both density and heat of formation. So HNIW can increase the specific impulse and density of solid propellant greatly. Today it is considered as the most powerful explosive and has great application value in solid propellant field. In view of its superior performance, HNIW can be regarded as a deputy of the next generation propellant raw material replacing various energetic compounds such as RDX and HMX [1]. Now, there are some reports on the synthesis, polymorphism, spectroscopy, combustion characteristics and thermal behavior of HNIW [2-14]. A large number of papers have been concerned with the combustion properties of pure HNIW and HNIW based glycidyl azide polymer (GAP) propellant [10-13]. However, there are very few publications on composite modified double base propellant containing HNIW (HNIW-CMDB propellant). The presently reported research is undertaken to elucidate the effect of HNIW content on properties of HNIW-CMDB propellant, such as energy properties, thermal behavior, non-isothermal decomposition reaction kinetics, combustion properties etc.

2. EXPERIMENTAL

2.1. Materials

HNIW is prepared by Beijing Institute of Technology according to the reported procedure [2]. The formulations of propellant samples used in experiments are listed in Table 1, named as sample **1-5**. These propellants are designed according to certain double base propellant in which the double base (DB) binder is replaced gradually by HNIW. The samples are prepared by solventless extrusion technology including absorbing, dehydrating, extrusion and cutting.

Table 1. Formulations of propellant samples

Sample	Composition	Mass ratio
1	NC+NG/HNIW/ additives	84.7:8.0:7.3
2	NC+NG/HNIW/ additives	74.7:18.0:7.3
3	NC+NG/HNIW/ additives	64.7:28.0:7.3
4	NC+NG/HNIW/ additives	54.7:38.0:7.3
5	NC+NG/HNIW/ additives	44.7:48.0:7.3
6	NC+NG/HNIW/ additives	34.7:58.0:7.3
7	NC+NG/HNIW/ additives	24.7:68.0:7.3
8	NC+NG/HNIW/ additives	14.7:78.0:7.3
9	NC+NG/HNIW/ additives	4.7:88.0:7.3

2.2 Apparatus and Experimental Conditions

The DSC experiments are carried out on a differential scanning calorimeter (Model DSC910S, TA Co., USA). The accuracies of temperature and heat flow are 0.1 K and 1 kW·kg^{-1}, respectively. Samples with mass less than 2.00 mg are heated at the heating rates (β) of 5, 10, 20, 25 K·min^{-1}. Pressure is obtained by filling nitrogen gas (purity, 99.999%) in the heating furnace. Thermogravimetric (TG) analysis of HNIW-CMBD propellant samples are carried out on a thermogravimetric analyzer (Model TGA2950, TA Co., USA). The accuracies of temperature and mass are 0.1 K and 1%, respectively. Samples are under atmospheric pressure with a heating rate of 10 K·min^{-1} and a nitrogen gas flow of 40 mL·min^{-1}.

The burning rate is determined by using a Crawford bomb pressurized with nitrogen gas. The sample is cut into strands φ5 mm×150 mm. The profile of the tested propellant strand is coated repeatedly six times by polyvinyl alcohol and aired. The surrounding temperature is 298.2 K while testing.

A single color frame amplification photography technology is used to record the flame structure of propellant burning steadily at different pressure. The sample is cut into rectangular strand as 1.5 mm×4 mm×25 mm. The tested propellant strand is fixed in burner with observation windows. The burner is pressurized to the desired pressure with nitrogen gas. In order to make photographs be recorded vividly, nitrogen gas is purged from the bottom and exhausted from the ceiling to circulate the air while testing.

Temperature distribution of combustion wave of HNIW-CMDB propellant is studied under different pressure by the Π mode miniature thermocouples. The miniature thermocouple is embedded in tested propellant strand. The profile of the propellant strand is coated twice repeatedly by polyvinyl alcohol and aired. The prepared propellant strand is then settled in burner. The burner is pressurized to the desired pressure with nitrogen gas. The signal emitted by thermocouple is recorded by data collecting system of HP Company. The temperature versus distance curve and its differential curve can be obtained by collected data. Because the temperature along combustion wave of propellant is different, the changing rate of temperature versus distance on the bounder of two combustion zones is larger, and the bounder of every combustion zone is corresponded to the peak of differential curve of temperature versus distance theoretically. So the width and temperature of every combustion zone can be obtained by this method.

3. RESULTS AND DISCUSSION

3.1. Energy Characteristics of HNIW-CMDB Propellant

Influence of Both Pressure and HNIW Content on Energy Characteristics

In order to investigate the influence of HNIW content on energy characteristics of propellant, the energy characteristics of propellant containing different HNIW mass ratio (W_{HNIW}) are calculated under 2, 7, 10, 20 and 30 MPa, respectively. Energy characteristics calculated contain oxide and fuel ration (Φ), combustion temperature (T_c), the average relative molecular mass of combustion products (M_c), characteristic velocity (C^*), theory specific impulse (I_{sp}), heat of explosion (Q_v) and heat of combustion (Q_c). The calculation program is Energy Calculation Star (ECS) made by Xi'an Modern Chemistry Research Institute. The calculated results are showed in Figure1. The calculated propellants contain nitrocellulose (NC), nitroglycerin (NG), di(2-nitrooxyethyl)nitramine (DINA), centralite II (C_2) and vaseline (VSL). These propellants are designed according to certain double base propellant in which the NC (12.0%N) and NG are replaced gradually by HNIW, and NC and NG decrease in certain proportion. According to the manufacturing process of solventless extrusion technique, the HNIW content in propellants are less than 50% in mass.

The calculated results show that the energy characteristics increase linearly with increasing HNIW content. When NC and NG in double base propellant are replaced by HNIW whose mass ratio is 48% in propellant, the I_{sp}, C^*, T_c, M_c, Q_c and Q_v increase 140.7 N·s·kg^{-1}, 83.5 m·s^{-1}, 351.9 K, 0.92, 0.009, 412.2 kJ·kg^{-1} and 300.8 kJ·kg^{-1}, respectively. The energy characteristics, such as I_{sp}, C^* and T_c, increase with increasing chamber pressure. However, the increase of M_c is insignificant.

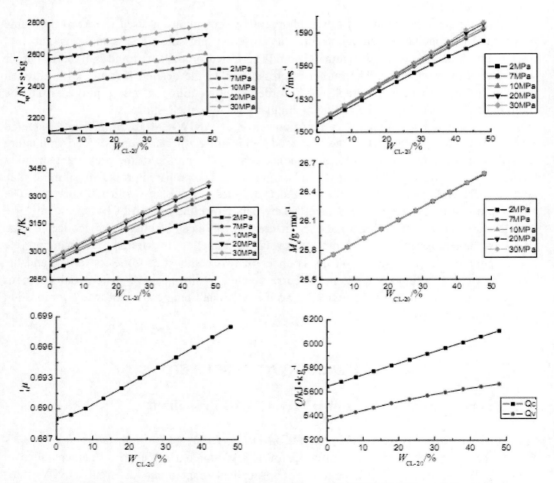

Figure 1. The plots of energy characteristics vs HNIW content in HNIW-CMDB propellant.

Comparison In Energy Characteristics of Propellants Containing Different Energetic Materials

The physic-chemical properties of several energetic materials are listed in Table 2 [5,15-17], such as symbol, formula, relative average molecular mass (m), density (ρ), standard molar enthalpy of formulation (ΔH_f^θ). The materials contain ammonium dinitramide (ADN), 3,4-Dinitrofurazanfuroxan (DNTF), N-guanylurea dianitramide (FOX-12), 1,1-Diamino-2,2-dinitroethylene (FOX-7), Hexogen (RDX) and 3,3-Trinitroazetidine (TNAZ). The energy characteristics of these materials under 7 MPa are calculated by ECS. The calculated results are listed in Table 3.

Table 2. Physic-chemical properties of several materials

Materials	Formula	m	ρ/g·cm^{-3}	ΔH_f^θ/kJ·mol^{-1}
ADN	H$_4$N$_4$O$_4$	124.06	1.80	-133.01
HNIW	C$_6$H$_6$N$_{12}$O$_{12}$	438.19	2.04	+429.42
DNTF	C$_6$N$_8$O$_8$	312.12	1.90	+644.30
FOX-12	C$_2$H$_7$N$_7$O$_5$	209.12	1.80	-355.30
FOX-7	C$_2$N$_4$H$_4$O$_4$	148.08	1.89	-133.80
RDX	C$_3$H$_6$N$_6$O$_6$	222.12	1.81	+61.55
TNAZ	C$_3$H$_4$N$_4$O$_6$	192.09	1.84	+59.75

The physic-chemical data show that relative molecular mass (*m*) and density of HNIW are the largest among the several new energetic materials. The standard molar enthalpy of formulation of HNIW is lower than that of DNTF and higher than those of others. The calculation results show, for mono-propellant, the values of I_{sp}, C^*, and Q_c of HNIW are higher than those of others except those of DNTF and TNAZ. The values of T_c and M_c of HNIW are higher than those of others except that of DNTF. The value of its Q_v is only lower than those of RDX and TNAZ. The value of its Φ is only lower than that of ADN. From the comparison it can be concluded that the integrated energy characteristics of HNIW-based mono-propellant are excellent.

Table 3. Energy characteristics of different energetic materials under 7 MPa

Name	Φ	I_{sp}/N·s·kg^{-1}	C^*/m·s^{-1}	T_c/K	M_c	Q_v/kJ·kg^{-1}	Q_c/kJ·kg^{-1}
ADN	2.000	2275.23	1440.40	2587.81	24.81	3768.99	2826.42
HNIW	0.800	2678.89	1651.26	3599.42	29.19	5685.18	6676.60
DNTF	0.667	2705.65	1672.38	4072.34	31.21	5498.75	7107.86
FOX-12	0.667	2101.53	1334.74	2135.87	23.24	3984.48	3752.73
FOX-7	0.667	2351.19	1489.62	2794.77	24.68	5024.85	5017.42
RDX	0.667	2609.04	1644.50	3281.48	24.68	5941.63	6198.09
TNAZ	0.750	2705.78	1668.24	3550.09	27.43	6098.52	6925.27

Note: Data of Qv and Qc are independent of pressure.

In order to investigate the influences of energetic material category and content on the energy characteristics of propellant, the energy characteristics of CMDB propellants containing different energetic materials in Table 3 are calculated by ECS under 7 MPa. These propellants are designed based on certain double base propellant in which the NC (12.0%N) and NG are replaced gradually by energetic materials, and NC and NG decrease in certain proportion. The calculated results are shown in Figure 2.

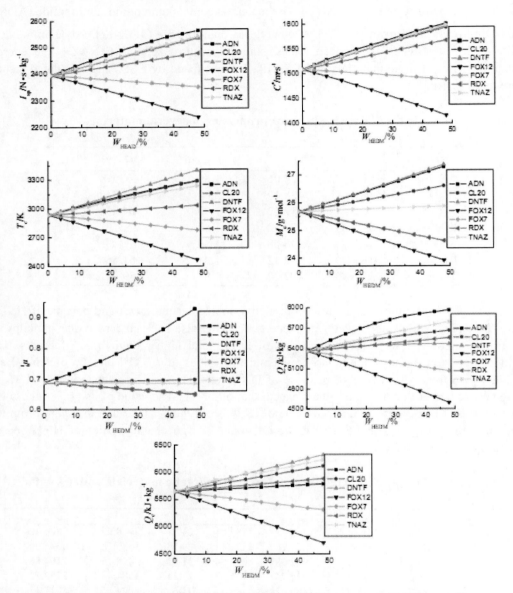

Figure 2. The plots of energy characteristics vs energetic materials content in CMDB propellant.

The calculated results show that the values of energy characteristics of every energetic material are different. These values, except Φ and Q_v of ADN, have linear changes with increasing the energetic materials mass ratio (W_{HEAD}). The values of I_{sp} and C^* increase when different energetic materials are added into propellants except FOX-7 and FOX-12. The increasing order is ADN>HNIW>DNTF>TNAZ>RDX. The energy materials such as ADN,

HNIW, DNTF and TNAZ will enlarge T_c and M_c. The enlargement order is ADN>DNTF>HNIW>TNAZ. RDX can increase T_c a little and decrease M_c significantly. FOX-7 and FOX-12 will decrease T_c and M_c. HNIW and ADN can increase Φ and Q_v, the latter is better in performance. Except for FOX-7 and FOX-12, these energetic materials can increase Q_c. The enhancing performance order is DNTF>TNAZ>HNIW>RDX.

3.2. Thermal Behaviors of HNIW

DSC and TG curves of the thermal decomposition of HNIW are shown in Figures. 3 and 4, respectively. There is only one exothermic peak in the DSC curve at about 524 K. The peak temperature increases from atmospheric pressure to 2 MPa and decreases from 2 MPa to 6 MPa. There is no obvious endothermic peak of crystal transformation. Figure 4 shows that there are two mass loss stages during decomposition between 506.4 K and 513.4 K with a mass loss of 7.5%. The second mass loss was between 513.4 K and 533.6 K with a mass loss of 83.1%.

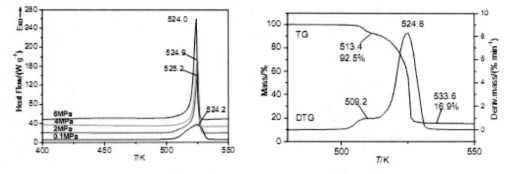

Figure 3. DSC curves of HNIW under different pressures Figure4 TG-DTG curve of HNIW under atmospheric pressure.

3.3. Thermal Behavior of HNIW-CMDB Propellant

The DSC and TG curves of samples 1, 3 and 5 are shown in Figure 5 to 10. By comparing Figure 5 with Figure 7 and Figure 9, several observations can be made. With the increase of HNIW content in propellants, the number of the peak temperature of the exothermic decomposition increases from one to three at atmosphere. Under 2~6 MPa pressure, with the increase of pressure, the peak temperature of exothermic decomposition shifts towards lower temperature for sample 1, and a small exothermic peak caused by decomposition of HNIW appears after the main exothermic peak in the DSC curves of Figure 7 and Figure 9. Comparing DSC curves of sample 3 in Figure 7 with those of sample 5 in Figure 10, the small exothermic peak enlarges as the content of HNIW in samples 3 and 5 increasing.

For samples 1, 3 and 5, the peak shape becomes narrower and the height of the exothermic peak increases with the increase of pressure, indicating that high pressure and

HNIW content enable the thermal decomposition reaction of the control propellant to take place easily and to accelerate quickly.

Figures. 6, 8 and 10 show that there are three mass loss stages (stages I, II and III) in TG curves, corresponding to three peaks in DTG curves. Stage I begins at about 350 K, with summit peak at about 450 K, accompanied with 21.5%-37.0% mass loss. It is mainly attributed to the volatilization and decomposition of nitroglycerine (NG). Stage II begins at about 477 K, with summit peak at about 483 K, accompanied with 27.9%-41.2% mass loss. It is mainly attributes to the decomposition of nitrocellulose (NC), corresponding to the obvious peak in DSC curve in the temperature range. Stage II is caused by the main exothermic decomposition reaction. Stage III begins at about 504 K, with summit peak at about 517 K, accompanied with 4.1%-31.2% mass loss. It is mainly attributed to the decomposition of the partly HNIW, corresponding to the obvious peak in DSC curve in the temperature range. There are a few residues at the end of decomposition.

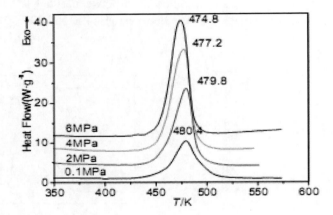

Figure 5. DSC curves of sample 1.

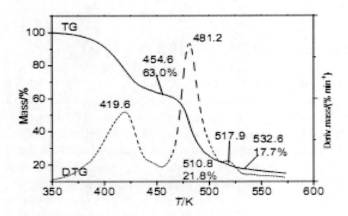

Figure 6. TG-DTG curve of sample 1 under 0.1 MPa.

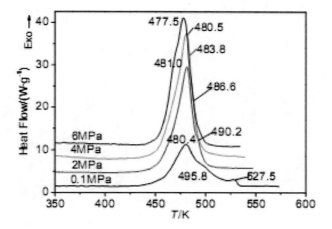

Figure 7. DSC curves of sample 3.

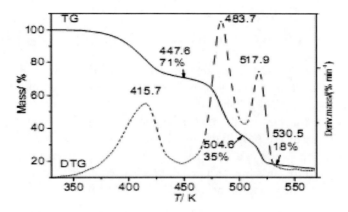

Figure 8. TG-DTG curve of sample 3 under 0.1 MPa.

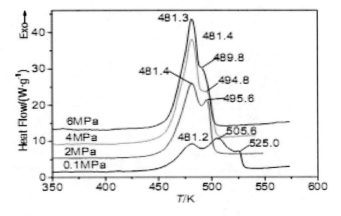

Figure 9. DSC curves of sample 5.

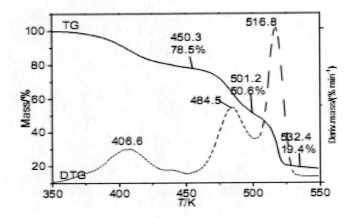

Figure 10. TG-DTG curve of sample 5 under 0.1 MPa.

3.4. Non-Isothermal Reaction Kinetics of HNIW-CMDB Propellant

The DSC curves at different heating rates under both 4 MPa and 7 MPa for HNIW-CMDB propellant sample are shown in Figure11. There is only one exothermic peak on each DSC curve under 4 MPa and 7 MPa, which appears very different with that under 0.1 MPa. It indicates that pressure influences thermal decomposition process strongly, and concentrates the exothermic district of the propellant sample. With the increase of heating rate, the onset temperature, peak temperature and final temperature of the exothermic decomposition of the propellant sample shift towards higher temperature. The exothermic peak temperature at same heating rate decreases with pressure increasing.

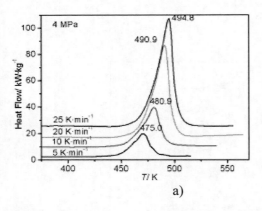
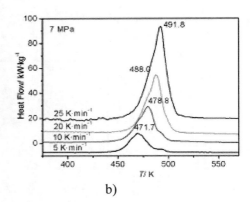

Figure 11. DSC curves for sample at different heating rates under different pressure: a) at 4 MPa; b) at 7 MPa.

In order to study the thermal decomposition mechanism of the main exothermic reaction stage (main exothermic peak) under propellant working pressure and obtain the corresponding kinetic parameters [apparent activation energy (E_a), pre-exponential constant (A)] and the most probable kinetic model function, the DSC curves at heating rates of 5, 10,

20 and 25 K·min^{-1} are dealt by mathematic means, and five integral methods [Equations (1)-(5)] and one differential methods [Equation (6)] listed in Table 4 are employed [18-21].

Table 4. Kinetic analysis methods

Method	Equation	
Ordinary-integral	$\ln[G(\alpha)/T^2] = \ln[(AR/\beta E)(1-2RT/E)] - E/RT$	(1)
Mac Callum-Tanner	$\lg[G(\alpha)] = \lg(AE/\beta R) - 0.4828E^{0.4357} - (0.449+0.217E)/(0.001T)$ (E/kcal·mol^{-1})	(2)
Šatava-Šesták	$\lg[G(\alpha)] = \lg(A_S E_S/\beta R) - 2.315 - 0.4567 E_S/RT$	(3)
Agrawal	$\ln[G(\alpha)/T^2] = \ln\{(AR/\beta E)[1-2(RT/E)]/[1-5(RT/E)^2]\} - E/RT$	(4)
Flynn-Wall-Ozawa	$\lg\beta = \lg\{AE/[RG(\alpha)]\} - 2.315 - 0.4567 E/RT$	(5)
Kissinger	$\ln(\beta_i/T_{pi}^2) = \ln(A_k R/E_k) - E_k/RT_{pi}$, $i = 1,2,\cdots 4$	(6)

Where α is the conversion degree of sample reacted; T the temperature (K) at time of t; T_0 the temperature of the initial point at which DSC curve deviates from the baseline; R the gas constant; $f(\alpha)$ and $G(\alpha)$ are the differential model function and the integral model function, respectively, and the means of E_a, A, β and T_p are mentioned before. The data needed for the equations of the integral and differential methods, i, α_i, β, T_i, T_p, $(d\alpha/dT)_i$, $i = 1, 2, 3 \bullet \bullet \bullet$, are obtained from the DSC curves and summarized in Table 5.

The values of E_a are obtained by Ozawa's method [Equation (5)] with α changing from 0.01 to 1.00 as shown in Table 5. From Figure12 we can see that activation energies change little with the increase of conversion degree. In the range of 0.01~0.85 (α) under 4 MPa and 0.10~0.80 (α) under 7 MPa, activation energies change even faintly, which means that the decomposition mechanism of the processes does not transferred in essence or the transference could be ignored. So, it is feasible to study the reaction mechanism and kinetics in these ranges.

Forty-one types of kinetic model functions in Ref.[18] and the original data tabulated in Table 5 are put into Equations (1)~(6) for calculations, respectively. The values of E_a, $\lg(A/s^{-1})$, linear correlation coefficient (r) and standard mean square deviation (Q) under 4 MPa and 7 MPa can be calculated on the computer with the linear least squares method at various heating rates of 5, 10, 20 and 25 K·min^{-1} The most probable mechanism function is selected by the better values of r, and Q based on the following four conditions: (1) the values of E_a/(kJ•mol^{-1}) and $\lg(A/s^{-1})$ selected are in the ordinary range of the thermal decomposition kinetic parameters for solid materials [E_a/(kJ•mol^{-1})=80~250 and $\lg(A/s^{-1})$=7~30]; (2) linear correlation coefficient (r) is greater than 0.98; (3) the values of E_a/(kJ•mol^{-1}) and $\lg(A/s^{-1})$ obtained with the differential and integral methods are approximately the same; (4) the mechanism function selected must be in agreement with the tested sample state. The results satisfying the conditions mentioned above are listed in Table 6.

Table 5. Data for decomposition processes of sample 3 at different heating rates from DSC curves

α	T / K under 4 MPa β / K·min⁻¹				T / K under 7 MPa β / K·min⁻¹			
	5	10	20	25	5	10	20	25
0.01	440.0	446.2	456.3	459.8	440.9	449.1	452.8	455.8
0.02	444.5	451.0	460.9	464.2	445.0	453.4	457.3	461.0
0.03	447.4	453.9	463.8	467.0	447.5	456.0	460.1	464.2
0.04	449.3	456.0	466.0	469.0	449.2	457.8	462.1	466.6
0.05	450.8	457.6	467.6	470.7	450.6	459.2	463.6	468.4
0.06	451.9	458.9	469.0	472.0	451.7	460.3	464.9	469.9
0.07	452.9	460.0	470.2	473.2	452.6	461.4	466.0	471.2
0.08	453.8	461.0	471.3	474.3	453.4	462.3	466.9	472.3
0.09	454.6	461.9	472.2	475.2	454.2	463.1	467.8	473.2
0.10	455.3	462.7	473.0	476.1	454.8	463.8	468.6	474.1
0.11	456.0	463.4	473.8	476.9	455.4	464.5	469.3	474.9
0.12	456.6	464.1	474.6	477.7	456.0	465.1	469.9	475.6
0.13	457.2	464.8	475.2	478.4	456.5	465.7	470.6	476.3
0.14	457.8	465.4	475.9	479.1	457.0	466.2	471.1	476.9
0.15	458.3	466.0	476.5	479.7	457.5	466.7	471.7	477.5
0.16	458.8	466.6	477.1	480.4	457.9	467.2	472.2	478.1
0.17	459.2	467.2	477.7	480.9	458.3	467.7	472.7	478.6
0.18	459.7	467.7	478.2	481.5	458.7	468.1	473.1	479.1
0.19	460.1	468.2	478.7	482.1	459.0	468.5	473.6	479.6
0.20	460.5	468.7	479.2	482.6	459.3	468.9	474.0	480.1
0.21	460.9	469.2	479.7	483.1	459.7	469.3	474.4	480.6
0.22	461.2	469.6	480.1	483.6	460.0	469.7	474.8	481.0
0.23	461.6	470.1	480.6	484.1	460.3	470.1	475.2	481.4
0.24	462.0	470.5	481.0	484.5	460.5	470.4	475.5	481.8
0.25	462.3	470.9	481.5	485.0	460.8	470.8	475.9	482.2
0.26	462.7	471.3	481.9	485.4	461.1	471.1	476.3	482.6
0.27	463.0	471.7	482.3	485.9	461.4	471.5	476.6	483.0
0.28	463.3	472.1	482.6	486.3	461.7	471.8	476.9	483.4
0.29	463.6	472.5	483.0	486.7	461.9	472.1	477.3	483.7
0.30	464.0	472.9	483.4	487.1	462.2	472.5	477.6	484.1
0.31	464.3	473.2	483.8	487.5	462.5	472.8	477.9	484.4
0.32	464.6	473.6	484.1	487.8	462.7	473.1	478.2	484.7
0.33	464.9	474.0	484.5	488.2	463.0	473.4	478.5	485.1
0.34	465.2	474.3	484.8	488.6	463.3	473.7	478.8	485.4
0.35	465.4	474.6	485.2	489.0	463.5	474.0	479.1	485.7
0.36	465.7	475	485.5	489.3	463.8	474.3	479.4	486.0
0.37	466.0	475.3	485.8	489.7	464.1	474.5	479.7	486.3

Table 5. (Continued)

α	T / K under 4 MPa β / K·min⁻¹				T / K under 7 MPa β / K·min⁻¹			
	5	10	20	25	5	10	20	25
0.38	466.3	475.6	486.1	490.0	464.3	474.8	480.0	486.6
0.39	466.5	475.9	486.5	490.4	464.6	475.1	480.3	487.0
0.40	466.8	476.3	486.8	490.7	464.8	475.4	480.6	487.2
0.41	467.1	476.6	487.1	491.0	465.1	475.7	480.8	487.5
0.42	467.3	476.9	487.4	491.3	465.4	475.9	481.1	487.8
0.43	467.6	477.2	487.7	491.7	465.6	476.2	481.4	488.1
0.44	467.8	477.5	488.0	492.0	465.9	476.5	481.6	488.4
0.45	468.1	477.7	488.3	492.3	466.1	476.8	481.9	488.6
0.46	468.3	478	488.6	492.6	466.4	477.0	482.2	488.9
0.47	468.5	478.3	488.8	492.9	466.6	477.3	482.4	489.2
0.48	468.7	478.5	489.1	493.1	466.9	477.5	482.6	489.4
0.49	469.0	478.8	489.4	493.4	467.1	477.8	482.9	489.7
0.50	469.2	479.1	489.6	493.7	467.4	478.0	483.1	490.0
0.51	469.4	479.3	489.8	493.9	467.6	478.3	483.3	490.2
0.52	469.6	479.5	490.1	494.2	467.9	478.5	483.5	490.4
0.53	469.8	479.8	490.3	494.4	468.1	478.7	483.7	490.7
0.54	470.0	480	490.5	494.6	468.3	479.0	484.0	490.9
0.55	470.2	480.2	490.7	494.8	468.6	479.2	484.2	491.1
0.56	470.4	480.4	490.9	495.0	468.8	479.4	484.4	491.3
0.57	470.6	480.6	491.1	495.2	469.1	479.6	484.6	491.6
0.58	470.8	480.9	491.3	495.4	469.3	479.9	484.8	491.8
0.59	471.0	481.1	491.5	495.6	469.5	480.1	485.0	492.0
0.60	471.2	481.3	491.6	495.7	469.8	480.3	485.2	492.2
0.61	471.4	481.5	491.8	495.9	470.0	480.5	485.4	492.4
0.62	471.6	481.7	491.9	496.0	470.3	480.7	485.6	492.6
0.63	471.8	481.9	492.1	496.2	470.5	480.9	485.8	492.8
0.64	472.0	482.1	492.2	496.3	470.8	481.1	486.0	493.0
0.65	472.2	482.3	492.4	496.4	471.1	481.3	486.2	493.2
0.66	472.4	482.5	492.5	496.6	471.3	481.5	486.4	493.4
0.67	472.6	482.7	492.7	496.7	471.6	481.8	486.6	493.6
0.68	472.8	482.8	492.8	496.8	471.8	482.0	486.8	493.8
0.69	473.0	483.0	492.9	497.0	472.1	482.2	487.0	494.0
0.70	473.2	483.2	493.1	497.1	472.4	482.4	487.2	494.2
0.71	473.4	483.4	493.2	497.2	472.6	482.7	487.4	494.4
0.72	473.6	483.6	493.3	497.3	472.9	482.9	487.6	494.6
0.73	473.8	483.8	493.4	497.4	473.2	483.1	487.8	494.8
0.74	474.0	484.0	493.6	497.6	473.5	483.4	488.0	495.0
0.75	474.2	484.2	493.7	497.7	473.9	483.6	488.2	495.2

α	T / K under 4 MPa β / K·min⁻¹				T / K under 7 MPa β / K·min⁻¹			
	5	10	20	25	5	10	20	25
0.76	474.4	484.4	493.8	497.8	474.2	483.8	488.4	495.4
0.77	474.7	484.6	494.0	497.9	474.5	484.1	488.6	495.6
0.78	474.9	484.8	494.1	498.1	474.9	484.4	488.9	495.8
0.79	475.1	485.1	494.3	498.2	475.3	484.6	489.1	496.1
0.80	475.4	485.3	494.4	498.3	475.8	484.9	489.4	496.3
0.81	475.6	485.6	494.6	498.5	476.2	485.2	489.6	496.6
0.82	475.9	485.9	494.7	498.6	476.7	485.5	489.9	496.8
0.83	476.1	486.2	494.9	498.8	477.2	485.9	490.1	497.1
0.84	476.4	486.5	495.0	499.0	477.7	486.3	490.4	497.4
0.85	476.8	486.8	495.2	499.2	478.3	486.6	490.7	497.7
0.86	477.1	487.2	495.4	499.4	478.9	487.0	491.1	498.0
0.87	477.5	487.6	495.6	499.6	479.5	487.5	491.4	498.3
0.88	477.9	488.0	495.9	499.9	480.0	488.0	491.8	498.6
0.89	478.4	488.4	496.2	500.2	481.0	488.5	492.2	499.0
0.90	478.9	488.8	496.5	500.6	481.9	489.2	492.7	499.4
0.91	479.4	489.3	496.8	500.9	482.9	489.8	493.2	499.8
0.92	479.9	489.9	497.2	501.4	484.0	490.5	493.8	500.3
0.93	480.5	490.6	497.7	501.9	485.3	491.3	494.4	500.8
0.94	481.1	491.5	498.3	502.5	486.5	492.1	495.1	501.4
0.95	482.0	492.7	498.9	503.2	487.9	492.9	495.9	502.1
0.96	483.0	494.1	499.7	504.1	489.2	493.8	496.7	502.9
0.97	484.4	496.1	500.7	505.3	490.5	494.9	497.7	503.9
0.98	486.5	499.1	502.1	506.9	492.0	496.3	499.1	505.2
0.99	489.4	504.9	504.3	509.9	494.0	498.4	501.2	507.1
1.00	495.4	521.9	515.8	526.2	506.0	506.0	509.7	517.7
	T_p=475.0 K	T_p=480.9K	T_p=490.9K	T_p=494.8K	T_p=471.7K	T_p=478.8K	T_p=488.0K	T_p=491.8K

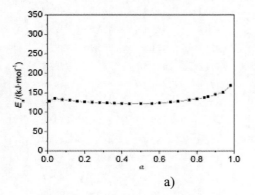
a)

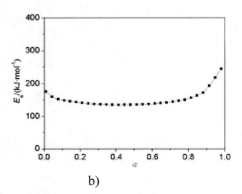
b)

Figure 12. $E_\alpha \sim \alpha$ curves under different pressure obtained by Ozawa's method: a) at 4 MPa; b) at 7 MPa.

Table 6. Kinetic parameters for the decomposition process of sample under 4 and 7 MPa

Method	β /(K·min^{-1})	4 MPa E_a /(kJ·mol^{-1})	lg(A/s^{-1})	r	Q	7 MPa E_a /(kJ·mol^{-1})	lg(A/s^{-1})	r	Q
Ordinary-integral	5	156.6	15.2	0.9998	0.0225	148.2	14.3	0.9949	0.1543
	10	142.0	13.5	0.9992	0.0789	149.9	14.4	0.9997	0.0124
	20	151.2	14.5	0.9971	0.2701	155.9	15.2	0.9998	0.0091
	25	147.8	14.1	0.9963	0.3715	148.8	14.3	0.9993	0.0225
Mac Callum-Tanner	5	156.8	15.2	0.9999	0.0040	148.4	14.2	0.9962	0.0289
	10	142.2	13.5	0.9992	0.0143	150.2	14.4	0.9996	0.0022
	20	151.7	14.5	0.9972	0.0509	156.4	15.2	0.9997	0.0017
	25	148.3	14.1	0.9971	0.0700	149.4	14.3	0.9994	0.0042
Šatava-Šesták	5	156.3	15.2	0.9998	0.0040	148.3	14.3	0.9955	0.0286
	10	142.5	13.5	0.9991	0.0146	150.0	14.4	0.9996	0.0021
	20	151.4	14.5	0.9973	0.0507	155.9	15.2	0.9997	0.0016
	25	148.1	14.1	0.9960	0.0701	149.3	14.3	0.9994	0.0043
Agrawal	5	156.6	15.2	0.9997	0.0225	148.2	14.3	0.9950	0.1542
	10	142.0	13.5	0.9990	0.0785	149.9	14.4	0.9996	0.0122
	20	151.2	14.5	0.9969	0.2701	155.9	15.2	0.9997	0.0091
	25	147.8	14.1	0.9957	0.3715	148.9	14.3	0.9994	0.0227
Mean		149.5	14.3			150.8	14.5		
Flynn-Wall-Ozawa		147.8		0.9918		146.7		0.9972	
Kissinger		147.4	15.8	0.9908		146.3	15.8	0.9969	

The values of E_a/(kJ·mol^{-1}) and lg(A/s^{-1}) obtained from a single non-isothermal DSC curve are in good agreement with the values calculated by Kissinger's method and Ozawa's method. Therefore, we conclude that the reaction mechanism of the main exothermal decomposition process of the sample under 4 MPa and 7 MPa are the same one. It is classified as random nucleation and then growth, and the mechanism function is the Avramic Erofeev Equation with $n=2/3$, and $G(\alpha) = [-\ln(1-\alpha)]^{2/3}$, $f(\alpha) = (3/2)(1-\alpha)[-\ln(1-\alpha)]^{1/3}$. Substituting $f(\alpha)$ with $(3/2)(1-\alpha)[-\ln(1-\alpha)]^{1/3}$, E_a/kJ·mol^{-1} with 149.5 and lg(A/s^{-1}) with 14.3 into Equation (7)

$$d\alpha/dt = Af(\alpha)e^{-E/RT} \qquad (7)$$

the kinetic equation of exothermal decomposition reaction under 4 MPa may be described as:

$$d\alpha/dt = 10^{14.5}(1-\alpha)[-\ln(1-\alpha)]^{1/3}e^{-17981.7/T}$$

Substituting $f(\alpha)$ with $(3/2)(1-\alpha)[-\ln(1-\alpha)]^{1/3}$, E_a/(kJ·mol^{-1}) with 150.8 and lg(A/s^{-1}) with 14.5 into Equation (7), the kinetic equation of exothermal decomposition reaction under 7 MPa may be described as:

$$d\alpha/dt = 10^{14.7}(1-\alpha)[-\ln(1-\alpha)]^{1/3}e^{-18138.1/T}$$

3.5. The Burning Rate of HNIW-CMDB Propellant

The burning rate curves of HNIW-CMDB propellants determined by a Crawford bomb are shown in Figure 13. It can be seen from Figure 13 that the five curves are almost overlapped. They almost increase linearly from 4.6 mm·s^{-1} under 2 MPa to 26.5 mm·s^{-1} under 22 MPa. Although some papers [22-24] reported that the burning rate of pure HNIW was high, the burning rate of HNIW-CMDB propellant changes insignificantly between 2 MPa and 22 MPa with the increase of HNIW content under 50%.

Table 7. Burning rate pressure exponent of HNIW-CMDB propellant

sample No.	burning rate pressure exponent 8~18 MPa
CL01	0.76
CL02	0.80
CL03	0.80
CL04	0.82
CL05	0.85

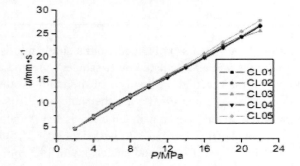

Figure 13. Burning rate curves for HNIW-CMDB propellants.

The burning rate pressure exponents of propellants have been calculated between 8 MPa and 18 MPa and listed in Table 7. Table 7 shows that the burning rate pressure exponents of HNIW-CMDB propellant are high, approximately 0.8, and increase as a whole with HNIW content increasing in propellant.

3.6. Flame Structure of HNIW-CMDB Propellant

The flame structure photos of HNIW-CMDB propellant under different pressures recorded by single color frame amplification photography technology are shown in Figures. 14-17. Figures. 14 and 15 show that there is distinct dark zone above the burning surface in flame structure of HNIW-CMDB propellant in which HNIW content is less than 28%. The width of the dark zone decreases with pressure increasing for the same propellant, and luminous flame zone becomes brighter. The width of dark zone decreases with HNIW content decreasing. Although there is dark zone in flame structure of HNIW-CMDB propellant (HNIW content is less than 28%), it is obviously different to that of double base (DB) and

RDX-CMDB propellants (The flame structure of RDX-CMDB propellant is showed in Figure 18). The dark zone of HNIW-CMDB propellant is special. There are a lot of bright luminous flame jets ejecting from the burning surface in dark zone. The special dark zone of HNIW-CMDB propellant is termed as mixed dark zone. The number of bright luminous flame jet increases with HNIW content increasing. Figure 14 shows that there are lots of light lines along flame of sample 1 under three different pressures. It can be seen from dark zone under 2 MPa of sample 1 that the light lines are airflows ejecting from burning surface, and the airflows eject from the shinning balls on the burning surface. There is no obvious airflow in the mixed dark zone and the airflows are replaced by bright luminous flame jets between mixed dark zone and flame zone for sample 1 under 4 MPa and 6 MPa. There are airflows in flame structures of sample 3 and RDX-CMDB propellants, while the airflows of the former are more obvious than those of the latter. Airflow isn't observed in flame structure of DB propellant. The reason of the above phenomena is that the thermal decomposition products of nitramine propellants cannot be ignited by the feedback heat from luminous flame zone when this type of propellant combusts under low pressure. The shining balls in the burning surface are the heated carbonaceous material produced by the nitramine and DB binder during combustion. The shining balls in burning surface of HNIW-CMDB propellant are obvious. It is known by papers [25-27] that there are quantities of residue during thermal decomposition of HNIW. As pressure increasing, the distance between flame zone and burning surface decreases and the transported heat between the two zones increases, and some airflows combust during the thermal decomposition. So a lot of bright luminous flame jets replace airflows in the mixed dark zone.

From Figures. 16 and 17, it can be seen that there is no obvious mixed dark zone in the flame structure of HNIW-CMDB propellant under different pressure when HNIW content is between 28% and 48%. The flame structure is similar to that of composite modified double base containing ammonium perchlorate (AP-CMDB) propellant. The luminous flame zone of sample 5 is nearer to burning surface than that of sample 4 under same pressure. There are a lot of bright luminous flame jets ejecting from burning surface. It can also be observed in flame structures of DB and RDX-CMDB propellants.

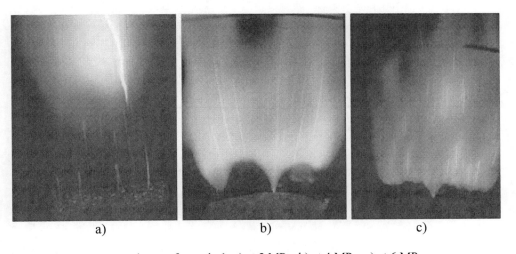

Figure 14. Flame structure photos of sample 1: a) at 2 MPa; b) at 4 MPa; c) at 6 MPa.

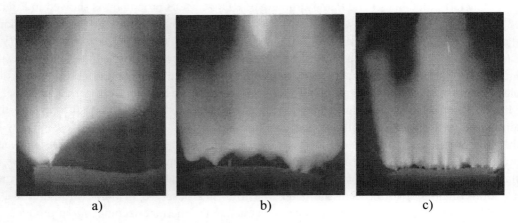

Figure 15. Flame structure photos of sample 3: a) at 2 MPa; b) at 4 MPa; c) at 6 MPa.

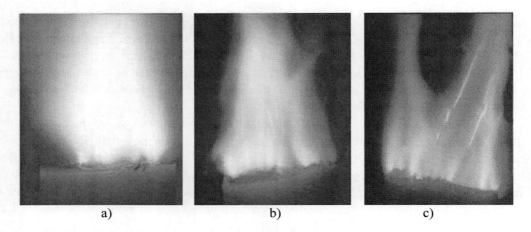

Figure 16. Flame structure photos of sample 4: a) at 2 MPa; b) at 4 MPa; c) at 6 MPa.

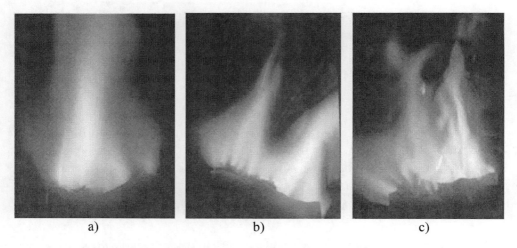

Figure 17. Flame structure photos of sample 5: a) at 2 MPa; b) at 4 MPa; c) at 6 MPa.

3.7. Combustion Wave Distribution of HNIW-CMDB Propellant

The combustion wave distribution curves and their analyzed results of samples 1-5 are shown in Figure 19 and Table 3. Table 8 shows that the temperature of burning surface (T_s) and combustion flame maximum temperature (T_{max}) of HNIW-CMDB propellants increases with pressure increasing. The temperature of fizz zone (T_f) and mixed dark zone (T_d) increases as a whole with pressure increasing though they are fluctuant. The width of fizz zone (L_f) and mixed dark zone (L_d) decrease as a whole with pressure increasing though they are fluctuant, too. The reason of the data fluctuating on T_f, T_d, L_f and L_d relates to the experiment error.

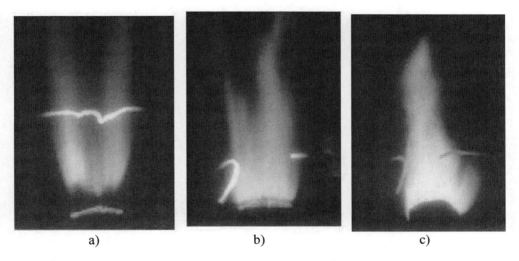

a)   b)   c)

Figure 18. Flame structure photos of RDX-CMDB propellant (DB/RDX/others=64/30/6): a) at 1 MPa; b) at 3 MPa; c) at 5 MPa.

It can be seen from Figure 19 that there is no obvious fizz zone in the combustion wave distribution curves of sample 4 and sample 5. It can't be concluded whether there is fizz zone in the flame structure of the two propellant samples from the phenomena or not. Because the two propellant samples contain quantities of DB binder. The DB binder would appear fizz zone in its flame structure. The reason there isn't fizz zone observed in the flame structure of the two propellant samples is that the number of bright luminous flame jets on the burning surface of HNIW-CMDB propellant increases with HNIW content increasing. The exothermal heat of bright luminous flame jets covers that of DB binder above the burning surface. So there isn't obvious peak in the differential temperature versus distance curves by temperature detection.

It is known there is obvious temperature changing near mixed dark zone in combustion wave distribution curves of sample 4 and sample 5. But mixed dark zone isn't observed in flame structure photos of the two propellant samples. The two phenomena seem to be inconsistent. Through analysis we found that they are contradictory to a certain extent. The flame structures are obtained by vision. The bright luminous flame jets above the burning surface in the flame structure would conceal the dark zone produced by DB binder when HNIW-CMDB propellant combusts, especially when HNIW content is higher such as in sample 4 and sample 5. While the thermocouple has the capacity of detecting real temperature, it can record the temperature changing in mixed dark zone.

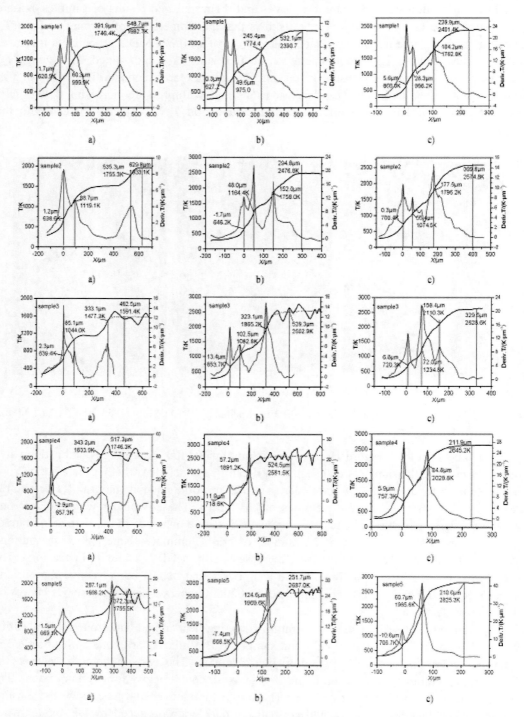

Figure 19. Combustion wave distribution curves of HNIW-CMDB propellants: CL01-CL05; a) at 2 MPa; b) at 4 MPa; c) at 6 MPa.

Table 8. Combustion wave distribution data of HNIW-CMDB propellants

No.	P/MPa	T_s/°C	L_f/μm (fizz zone)	T_f/°C (fizz zone)	L_d/μm (mixed dark zone)	T_d/°C (mixed dark zone)	T_{max}/°C
CL01	2	620.9	58.6	999.9	331.6	1746.4	1982.7
	4	627.2	49.3	975	195.8	1774.4	2390.7
	6	666	22.7	996.2	75.9	1762.8	2401.4
CL02	2	638.6	87.5	1119.1	446.6	1755.3	1933.1
	4	646.2	49.7	1164.4	104	1758	2476.8
	6	700.4	55.1	1074.5	122.1	1796.2	2574.8
CL03	2	639.4	82.8	1044	248	1477.3	1591.4
	4	653.7	89.1	1082.8	220.6	1865.2	2502.9
	6	720.3	65.2	1234.8	86.4	2110.3	2628.6
CL04	2	657.3				1633.9	1746.3
	4	718.6				1891.2	2581.5
	6	757.3				2029.8	2645.2
CL05	2	669.1				1666.2	1755.5
	4	666.5				1969.6	2687
	6	706.7				1965.6	2825.3

3.8. Numerical Simulation for Burning Rate of HNIW-CMDB Propellant

One-Dimension Gas Phase Reaction Theory

In the initial stage of propellant combustion, the main gaseous combustion products are divided into five classes, which are represented respectively by [NO$_2$], [CH$_2$O], [CHO], [CH] and [CO]. According to their chemical properties and the function they act in the dominant combustion reaction, the gases are also classified respectively as oxidizers, fuel, unstable groups, less reactive hydrocarbons groups, and less reactive CO groups in the burning surface of propellants.

It is almost impossible to measure accurately the quantities of the five gases present under a given combustion condition, but the composition of the propellant and the chemical structure of each component are definite. Therefore, the relationship between the gaseous composition and the chemical structure of the components in propellant can be developed. The five gases can also represent the chemical groups of the propellant and its components. The quantities of the five classes of gases of a propellant can be calculated by adding the quantity of the five classes of gases of each component together.

The unstable group [CHO] represents the class of gases that can split. The form of [CHO] is different at different pressures. When the pressure is very low, [CHO] is present in the vicinity of the burning surface in the form of a polymer, such as (CHO)$_n$. When the pressure is very high, [CHO] splits into two groups, [CO] and [H] (categorized to [CH]). The natural split process of [CHO] can be described as the following reaction:

$$[CHO] \longrightarrow [CO]+[H]$$

The relationship between the natural split process of the unstable group [CHO] and pressure can be described by the following equation:

$$\eta(p) = 2 - e^{0.6931(1-p/p^*)} \qquad (8)$$

At the characteristic pressure (p^*) the quantities of the five classes of gases are symbolized as B_5, B_4, B_3, B_2 and B_1. They are also called the chemical structure parameters. Then, the molecular formula of 1 kg propellant can be written as $[NO_2]_{B5}$, $[CH_2O]_{B4}$, $[CHO]_{B3}$, $[CH]_{B2}$ and $[CO]_{B1}$.

We symbolized the mole fraction of gaseous oxidizer $[NO_2]$ near the burning surface as $\theta_0(p)$. If $b_4=B_4/B_5$, $b_3=B_3/B_5$, $b_2=B_2/B_5$, $b_1=B_1/B_5$, then, at any pressure,

$$\theta_0(p) = \frac{1}{b_1 + b_2 + b_3 \cdot \eta(p) + b_4 + 1} \qquad (9)$$

The binder of HNIW-CMDB propellant is NC/NG. It is a typical double base series propellant. So the combustion properties of HNIW-propellant are similar to those of DB propellant and RDX-CMDB propellant, but not be identified by the investigation of flame structures and combustion wave distributions. The flame structures of HNIW-CMDB propellant show there is obvious mono-propellant flame [28]. It is known that the combustion process of HNIW-mono-propellant is different with single ring nitramine such as RDX and HMX. Its burning rate is higher than that of single ring nitramine because of different combustion properties except for combustion products. This difference of combustion properties are described by characteristic pressure (p_c^*), and the p_c^* is 15 MPa. The HNIW-mono-propellant flame burns singly before the products diffuse to the double base binder flame. So its typical combustion property is described by modifying the parameter $\theta_0(p)$ of the one-dimension gas phase reaction theory and the modified equation is Equation (7):

$$\theta_c(p) = \frac{1}{b_1 + b_2 + b_{31} \cdot \eta_1(p) + b_{3c} \cdot \eta_c(p) + b_4 + 1} \qquad (10)$$

where b_{31} is the quantity of [CHO] group produced by double base binder and other auxiliaries, $\eta_1(p)$ is the reaction function of b_{31} during propellant combustion and it is similar to that in Equation (8), b_{3c} is the quantity of [CHO] group produced by HNIW, $\eta_c(p)$ is the reaction function of b_{3c} and its equation is as follow:

$$\eta_c(p) = 2 - \exp[0.6931(1 - p/p_c^*)] \qquad (11)$$

There are two kinds of premixed flame above the burning surface of HNIW-CMDB propellant. They are called DB flame and HNIW flame and the ratio of them changes with the content of propellant. The two flames compete during combustion and the compete function $[S(x)]$ is described in the following equation:

$$S(x) = D_1 x^2 - D_2 x + D_3 \qquad (12)$$

where x is the HNIW content in HNIW-CMDB propellant, D_1, D_2 and D_3 are coefficients simulated by testing data.

Studies on Energy Properties, Thermal Behaviors, Combustion Properties ... 243

The burning rate of HNIW-CMDB propellant is a function of pressure and propellant compositions. At 293.1 K the expression for the burning rate is as follow.

$$u(p) = Kp\vartheta_0^2(p,X)S(x)/\rho_P \tag{13}$$

where K is an experience constant simulated by experiment data and changes with the type and using range of energetic materials. ρ_p is the density of gun propellant.

The Thermal Decomposition of HNIW at Initial Combustion

The initial thermal decomposition bonds of ε-HNIW shown in Figure 20 are calculated by quantum chemical non-restriction Hartree-Fock self-consistency field PM3 (UHF-SCF-PM3) molecule orbit method (MO) and Mulliken overlap population method. According to calculation the Mulliken bond series of C1-C3, C7-N2, N2-N13 (pentabasic ring) and N12-N28 (hexahydric ring) is minimum. So these bonds are the broken bonds as heated. The broken activity energies of N-NO₂ bonds at branch chains, C-C bonds and C-N bonds at cage skeleton are 96.2 kJ·mol⁻¹and 100.2 kJ·mol⁻¹, 132.5 kJ·mol⁻¹, 189.3 kJ·mol⁻¹, respectively. So the broken activity energy of N-NO₂ bonds is less than those of C-C bonds and C-N bonds. These results are consistent with those reported [29-32]. These data indict that the initial broken bonds of ε-HNIW as heated are N-NO₂ bonds at branch chains. There are two kinds of N-NO₂ bonds at branch chains of ε-HNIW. Their broken activity energies are similar. But the calculation results of static state and dynamic state show that the –NO₂ bonds at hexahydric ring is more stable than that at pentabasic ring. So the initial homolysis bond of ε-HNIW is the N-NO₂ of pentabasic ring. By analyzing the testing results shown the initial decomposition step of HNIW is the homolysis of N-NO₂ at N9 point (shown in Figure 20). The homolysis of N-NO₂ bond will weaken the C10-N11 bond at hexahydric ring of HNIW, and many reactions of the backbone cage can occur. Two that are consistent are shown in reaction Equations (14) and (15).

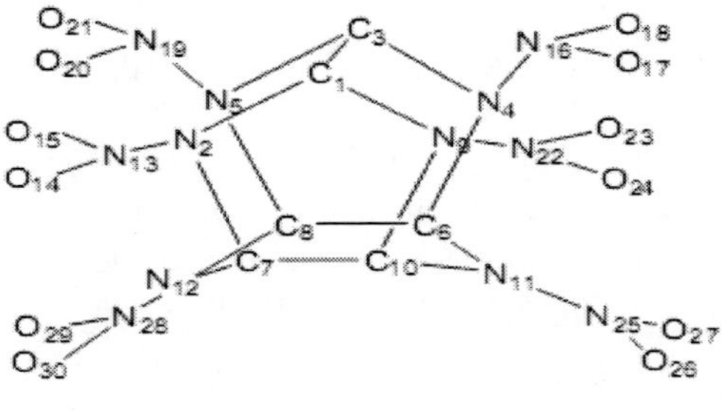

Figure 20. The formula structure and atom number of ε-HNIW.

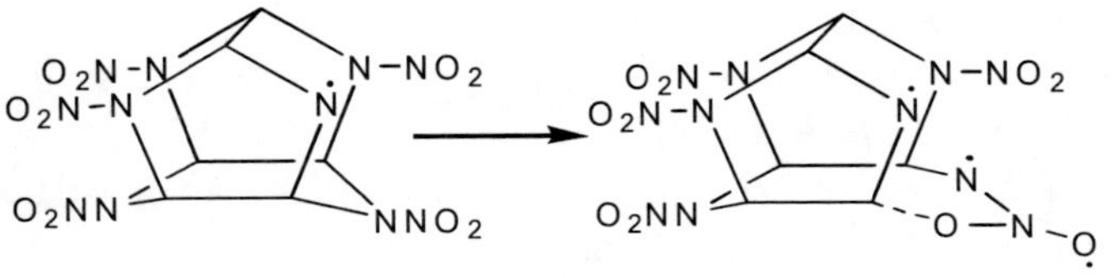

$$\tag{14}$$

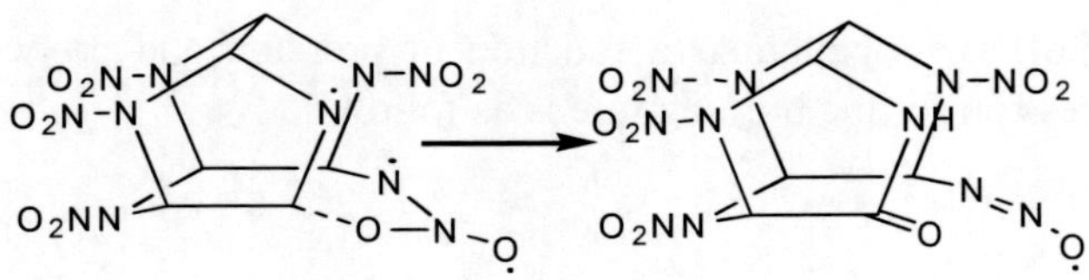

(15)

With temperature increasing, the NO₂ produced by HNIW decomposition will react with other residues, and form NO, CO, CO₂ and residues containing more C and H and less N and O, such as melon etc.

According to the above analysis, the thermal decomposition mechanism of HNIW at initial combustion stage can be expressed as follow:

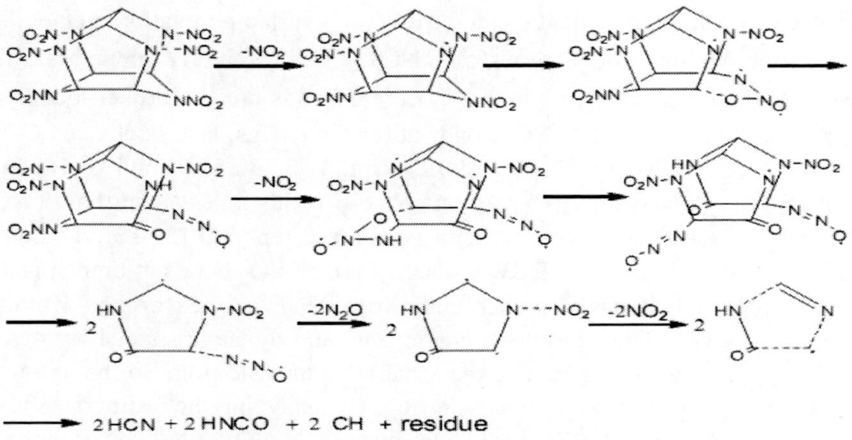

Figure 21. The thermal decomposition of HNIW.

According to the above thermal decomposition process hypothesis, the equation of thermal decomposition for HNIW at initial combustion stage can be expressed as follows:

$$C_6H_6N_{12}O_{12}(s) \rightarrow 4NO_2(g)+2N_2O(g)+2HCN(g)+2HNCO(g)+2CH(g)+residue \quad (16)$$

Based on one dimension gas phase reaction theory, the thermal decomposition products of HNIW such as NO_2 and N_2O are ascribed to oxidizer group ([NO_2]), HCN is ascribed to unstable group ([CHO]), HNCO and CH are ascribed to less reactive hydrocarbons group ([CH]). The thermal decomposition results of HNIW show that there are 9%-15% solids residues in mass, so the quantity of five kinds of groups such as oxidizer group, hydrocarbons group, unstable group, and hydrocarbons group are 11.77 mol·kg⁻¹, 3.96 mol·kg⁻¹, 7.85 mol·kg⁻¹ and 0 mol·kg⁻¹, respectively. These groups are added into parameter data base of burning rate program. The burning rates of HNIW-CMDB propellant at different pressure can be calculated.

Calculation of HNIW-CMDB Propellant Burning Rate

The burning rates of HNIW mono-propellant and samples 1-9 are calculated by Equations.(8)-(13). The calculation and experimental data of burning rate for HNIW-CMDB propellant formulation are shown in Figures.22 and 23.

Studies on Energy Properties, Thermal Behaviors, Combustion Properties ... 245

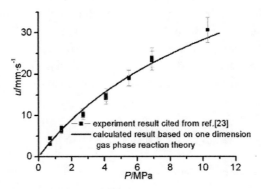

Figure 22. The u vs p curve of HNIW monopropellant.

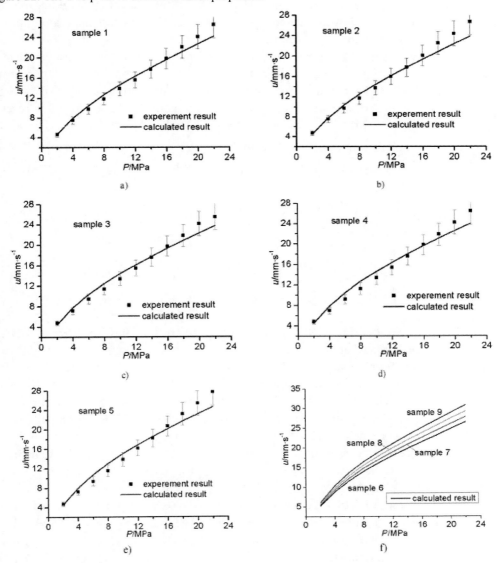

Figure 23. Calculated results of burning rate of the non-catalyst HNIW-CMDB propellants: a) sample 1; b) sample 2;c) sample 3; d) sample 4;e) sample 5;f) samples 6-9.

From Figures.22 and 23, it can be seen that the calculated burning rates show good agreement with the experimental values over the pressure range from 2 to 22 MPa. Most of the calculated results locate in the ±10% error bars of experimental values when HNIW content is less then 50%. The burning rate of HNIW-CMDB propellants increases with increasing the HNIW content. But the increasing extent is small.

4. CONCLUSIONS

(1) Theory specific impulse, characteristic velocity, oxygen coefficient, combustion temperature, heat of explosion at constant volume, heat of combustion and the average relative molecular mass of combustion gas of HNIW-CMDB propellant increase linearly with HNIW content increasing, showing the energy properties of CMDB propellant are improved synthetically with HNIW content increasing.

(2) The decomposition reaction of HNIW-CMDB propellant under atmosphere can be divided into three stages: the volatilization and decomposition of NG, exothermic decomposition of NC, and exothermic decomposition of HNIW. There is only one exothermic decomposition peak under pressure of 4 and 7 MPa. The exothermic decomposition reaction mechanisms at 4 and 7 MPa obey random nucleation and then growth rule. The kinetic parameters of the reaction are: E_a = 149.5 kJ·mol^{-1}, A=10$^{14.3}$ s^{-1} under 4 MPa and E_a = 150.8 kJ·mol^{-1}, A=10$^{14.5}$ s^{-1} under 7 MPa. The kinetic equation of exothermic decomposition reaction can be expressed as:

for 4 MPa, $d\alpha/dt = 10^{14.5}(1-\alpha)[-\ln(1-\alpha)]^{1/3} e^{-17981.7/T}$

and

for 7 MPa $d\alpha/dt = 10^{14.7}(1-\alpha)[-\ln(1-\alpha)]^{1/3} e^{-18138.1/T}$.

(3) The flame structure of HNIW-CMDB propellant is similar with those of DB and RDX-CMDB propellants when the added amount of HNIW is less than 28%, having obvious dark zone above the burning surface of HNIW-CMDB propellant, and some bright luminous flame jets ejecting from burning surface in dark zone. The flame structure of HNIW-CMDB propellant is similar with that of AP-CMDB propellant when HNIW content is greater than 28%, no obvious dark zone above the burning surface of HNIW-CMDB propellant. The approximate temperature and width of several combustion zones with pressure range from 2MPa to 6MPa are attained by the differential curve method of treating the temperature distribution of combustion wave of HNIW-CMDB propellant obtained by the Π mode miniature thermocouple.

(4) The model for predicting the burning rate of HNIW-CMDB propellant based on one-dimension gas phase reaction theory is established. The burning rate of HNIW-CMDB propellant is predicted. The burning rate obtained is in agreement with the tested result within ±10% error. With HNIW content increasing, the burning rate of the non-catalyzed HNIW-CMDB propellants increases, but the increasing extent isn't as large as we thought before.

ACKNOWLEDGMENTS

This work received the equipment assistance of National Key Laboratory of Science and Technology on Combustion and Explosion.

REFERENCES

[1] M. Golfier, H. Graindrge, Y. Longevalle, et al. *Proceeding of the 29th International Annual Conference of ICT*, Kalsruhe, Germany, Jume 30-July 3,1998,3/1-3/18.
[2] J. H. Kim, Y. C. Park, Y. J. Yim, et al. *Journal of Chemical Engineering Japan*, 1998(2):237-241.
[3] B. R. Wardle, C. J. Hinshaw, P. Braithwaitem, et al. *Proceedings of the 27th ICT Conference ON Porpellants, Explosives and Pyrotechnics*, Karlsruhem 1996,C12-1-11.
[4] H. E. Johnston, R. B. Wardle. USP5,874,574, 1999.
[5] Y. X. Ou, H. P. Jia, Y. J. Xu, et al. *Science in China(Series B)*, 1999(2):217-224.
[6] Y. X. Ou, Y. J. Xu, B. Chen, et al. *Youji Huaxue*, 2000,20(4):556-559
[7] M. F. Foltz, C. L. Coon, et al. *Propellants, Explosives, Pyrotechnics*, 1994(19):19-25.
[8] T. P. Russell, P. J. Miller, et al. *Journal of Physics Chemical*, 1993(97):1993-1997.
[9] V. V. Nedelko, N. V. Chukanov, B. L. Korsounskii, et al. *Proceeding of the 31th International Annual Conference of ICT*, Karlsruhem Germany, June 27-June 30,2000,9-1-9-8.
[10] V. V. Nedelko, N. V. Chukanov, B. L. Korsounskii, et al. *Propellants, Explosives, Pyrotechnics*, 2000(25):255-259.
[11] A. I. Atwood, P. O. Curran, M. L. Chanm, R. Reed, et al. *Inernational Symposium on Energetic Materials Technology*, 1994:70-75.
[12] M. Golfler, H. Graindorge, Y. Longevialie, et al. *Proceedings of 29th International Conference of ICT*, 1998,3-1-3-16.
[13] H. M. An, Y. F. Liu, X. M. Li. *Chinese Journal of Explosives and Propellants*,2001,24(1):36-37.
[14] N. Kubota. Propellants and Explosives Thermochemical Aspects of Combustion[M].*WILEY-VCH GmbH*, Weinheim, Germany, 2002.
[15] H. Ostark, A. U. Bemm. *Journal Energetic Material*, 2000,(18):123～138.
[16] Y. Luo, H. X. Gao, F. Q. Zhao, et al. *Energetic materials*,2005,13(4):225-228.
[17] J. Z. Li, X. Z. Fan, B. Z. Wang. *Energetic Materials*, 2004,12(5): 304-308.
[18] R. Z. Hu, Q. Z. Shi. *Thermal analysis kinetics (Second version)*. Beijing:Science Press,2008.1.
[19] J. H. Yi, F. Q. Zhao, H. X. Gao, S. Y. Xu, M. C. Wang, R. Z. Hu. *Journal of Hazardous Materials*,2008,153:261-268.
[20] J. H. Yi, F. Q. Zhao, S. Y. Xu, H. X. Gao, R. Z. Hu. *Chemical Research in Chinese Universities*, 2008, 24(5), 608-614.
[21] J, H. Yi, F. Q. Zhao, S. Y. Xu, L. Y. Zhang, H. X. Gao, R. Z. Hu. *Journal of Hazardous Materials*,165(2009)853-859.
[22] A.I. Atwood, P.O. Curran M.L. Chanm, R. Reed, et al. *Inernational Symposium on Energetic Materials Technology*, 1994:70-75.

[23] M. Golfler, H. Graindorge, Y. Longevialie, et al. *Proceedings of 29th International Conference of ICT*, 1998,3-1-3-16.
[24] H. M. An, Y. F. Liu, X. M. Li. *Chinese Journal of Explosives and Propellants*,2001,24(1):36-37.
[25] D. G. Patil, T. B. Brill. *Combustion and Flame,*1991,(87):145-151.
[26] D. G. Patil, T. B. Brill. *Combustion and Flame*,1993,(92):456-458.
[27] S. Lobbëcke, M. A. Bohn, A. Pfeil, et al. *Proceeding of the 29th International Annual Conference of ICT*, Kalsruhe, Germany, Jume 30-July 3,1998,145-1-145-15.
[28] S. Y. Xu. *Investigations on combustion property and prediction of composite modified double base propellant containing CL-20*. Xi'an: Xi'an Modern Chemistry Research Institute,2006,3.
[29] J. Zhang, H. M. Xiao, X. D. Gong, et al. *Chinese Journal of Energetic Materials*, 2000,8(4):149-153.
[30] Z. Wen, S. H. Tian, P. J. Zhao, et al. *Chinese Journal of Energetic Materials*,1999,7(3):110-114.
[31] Y. X. Ou, H. P. Jia, B. R. Chen, et al. *Chinese Journal of Energetic Materials*,1999,7(2):49-52.
[32] H. M. Xiao. *The structure and property of energetic compounds*. Beijing: National Defence Industry Press, 2004.12.

In: Advances in Materials Science Research, Volume 7
Editor: Maryann C. Wythers
ISBN 978-1-61209-821-0
© 2012 Nova Science Publishers, Inc.

Chapter 6

CONJUGATED POLYMERS: PROPRIETY AND APPLICATION IN THE DEVELOPMENT OF ELECTROCHROMIC, ELECTROLUMINESCENT AND PHOTOVOLTAIC DEVICES

Marcos Roberto de Abreu Alves, Hállen Daniel Rezende Calado, Claudio Luis Donicci and Tulio Matencio

Pos-Graduacao em Quimica, Universidade Federal de Minas Gerais,
Belo Horizonte, MG, Brazil

INTRODUCTION

In this chapter, we initially introduce the general topic of principle electrochemistry techniques applied in the synthesis and characterization of electrochromic materials. We then go on to discuss conjugated polymers and the electronic structure and electrochemical proprieties of these materials. Conjugated polymers have the advantage of being synthesized with a modifiable structure to provide the electronic and mechanical properties that we desire. In this context, we will approach synthesis processes in a third topic, showing how we can control the electronic properties of a conducting polymer by modifying its main structure and side chains. As a fourth and final topic to be discussed in this chapter, we describe the application of conjugated polymers in developing electrochromic devices (as well as the desired properties of these materials and usual characterization techniques for electrochromic devices), electroluminescent devices and photovoltaics.

1. GENERAL ELECTROCHEMICAL TECHNIQUES - IMPORTANT CONCEPTS

1.1. Cyclic Voltammetry

Cyclic voltammetry (CV) is a technique that has great importance in the study of electrochemical reactions that occur on the surface of an electrode, where its primary application is as a qualitative analysis. The use of CV is mainly related to the study of electron transfer between the working electrode and the electroactive species, and the study and elucidation of mechanisms of heterogeneous electrochemical reactions (reactions that occur at the electrode/solution interface) [1].

This technique is based on the application of a linear variation of potential in the form of a triangular wave (Figure 1a), resulting in a current as a function of potential (Figure 1b). One must be careful to work in a range where the working electrode and the electrolyte are not electroactive, and where the solvent is stable. Furthermore, the initiation of the scan occurs where there is no electrochemical reaction of the electroactive species. Thus, the potential is varied in the anodic or cathodic direction until electroactivity of the species is observed. At this point, the direction of the potential scan is reversed back to the initial value [2]. Figure 2 illustrates an electrochemical cell with conventional three-electrode configuration used for electrochemical studies.

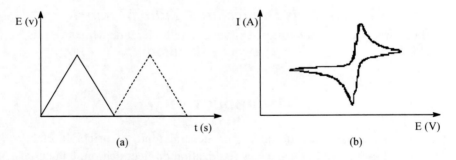

Figure 1. Schematic representation of the signals applied and measured during cyclic voltammetry: (a) Scanning potential versus time and (b) Profile of current as a function of potential.

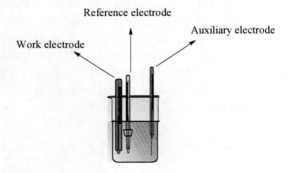

Figure 2. Electrochemical cell with three-electrode configuration.

The main parameters of the cyclic voltammetry curves are the: anodic peak potential, $E_{p,a}$, cathodic peak potential, $E_{p,c}$, anodic peak current, $I_{p,a}$, and cathodic peak current, $I_{p,c}$, see Figure 3 [3].

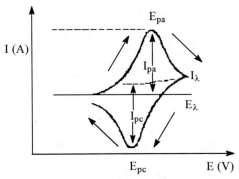

Figure 3. Cyclic voltammogram: E_λ is the potential of reversal and I_λ is the current of reversal.

The reversibility of electrochemical systems may exhibit three different behaviors: reversible, quasi-reversible, and irreversible. These different behaviors will be briefly discussed below.

1.1.1. Reversible Systems

The peak current, I_p, can be obtained by the Randles-Sevcik equation (Equation 1):

$$I_p = (2.69 \times 10^5) n^{\frac{3}{2}} A C D^{\frac{1}{2}} v^{\frac{1}{2}} \qquad (1)$$

where n is the number of electrons involved in the electrochemical reaction, A is the electrode area (cm^2), C is the concentration (mol.cm^{-3}), D is the diffusion coefficient (cm^2.s^{-1}), and v is the scan rate (V.s^{-1}) [1].

Identifying systems that may involve reversible reactions use the following parameters:

$$I_p \propto v^{\frac{1}{2}} \qquad (2)$$

E_p independent of scan rate

$$\left| E_p - E_{\frac{p}{2}} \right| = \frac{56.6}{n} \, mV \qquad (3)$$

$$\left| \frac{I_{p,a}}{I_{p,c}} \right| = 1 \qquad (4)$$

$$\Delta E_p = E_{p,a} - E_{p,c} = \frac{0.059}{n} \, V \qquad (5)$$

where E_p is the peak potential, $E_{p,a}$ is the anodic peak potential, $E_{p,c}$ is the cathodic peak potential, $E_{p/2}$ is the half peak potential, ΔE_p is the separation between the peak potential for a reversible system, n is the number of electrons involved in the reaction, $I_{p,a}$ is the anodic peak current, $I_{p,c}$ is the cathodic peak current, and v is the scan rate.

The formal potential, $E^{0'}$, for the reversible redox reactions can be determined with good approximation through the media of the potential of anodic and cathodic peak (Equation 6).

$$E^{0'} = \frac{1}{2}(E_{p,a} + E_{p,c}) \tag{6}$$

Other parameters can be used as criteria of reversibility, such as the Coulombic Efficiency, CE (Equation 7), and the peak width at half height, $\frac{\Delta E_p}{2}$ (Equation 8).

$$EC = \frac{Q_a}{Q_c} \times 100 \tag{7}$$

$$\Delta E_{\frac{p}{2}} = \frac{0,090}{n} V \tag{8}$$

where Q_a is the anodic charge and Q_c is the cathodic charge.

The anodic and cathodic charges can be obtained by calculating the area under the curves of the respective anodic and cathodic peaks, as shown in Equation 9.

$$Q_c = Q_a = \int \frac{IdE}{v} \tag{9}$$

In an ideal reversible process, it is expected that the charge involved in the oxidation process is equal to the charge used in the reduction process, so the coulombic efficiency should be equal to 100%.

1.1.2. Quasi-Reversible and Irreversible Systems

In quasi-reversible systems, the current is controlled by mass transport and charge transfer. The voltammograms for such systems have the potential peaks further away when compared to the voltammograms of reversible systems, as can be seen in Figures 3 and 4a.

In the case of an irreversible process, the current is controlled only by charge transfer. It is observed in the voltammograms a wide separation occurring between peak potentials, as well as a decrease in the height of individual peaks. These systems are also characterized by the change in peak potential with scan rate. Figure 4b illustrates a typical voltammogram for an irreversible system.

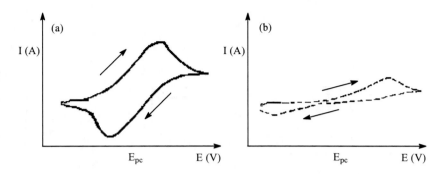

Figure 4. Cyclic voltammogram: (a) quasi-reversible system and (b) irreversible system.

Additional Information

The peaks observed in the voltammetric curves correspond to the current and potential maxima involved in the electrochemical reactions that occur inside the electrolytic cells. This phenomenon is interpreted as follows: at the beginning, for example, for the anodic potential, no exchanging of electrons occurs between the electrode and the electroactive species, so no current is observed; with increasing potential, current begins to be recorded, which indicates that the electroactive species is oxidized at the electrode surface, and the concentration of the electroactive species in the electrical double layer is kept constant by diffusion from the bulk solution. After a given amount of time, the exchange of electrons between the electrode and the electroactive species is so fast (due to the increased potential) that the diffusion of electroactive species present in the solution to the electrical double layer is no longer sufficient for the concentration of the electroactive species on the electrode surface be kept constant. This leads to a decrease in the concentration of the electroactive species in this region and consequently a decrease in current, resulting in the appearance of the anodic peak. The interpretation for the formation of the cathodic peak (in the reverse scan) is the same [3].

1.2. Chronoamperometry

Chronoamperometry is an electroanalytical technique that involves applying a constant external potential on the working electrode and recording the current, I, generated in the process of reduction or oxidation of the electroactive species present in the system versus time, t (chronoamperogram). The charge involved in this process can be determined by integration of the curve of current versus time (Equation 10).

$$Q = \int_0^t i\,dt \tag{10}$$

Figure 5a shows the applied signal, and Figure 5b shows the current response obtained as a function of time for a study using chronoamperometry. In this technique, the current curve as a function of time reflects the changes in the electrode surface in relation to concentration gradients.

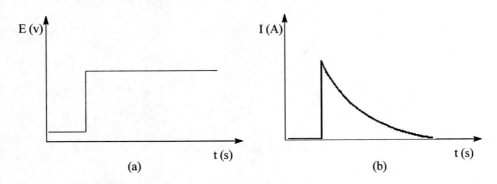

Figure 5. Chronoamperometry: (a) applied signal and (b) signal response (current versus time).

The chronoamperometry is often applied in the study of mechanisms of electrode processes and in studying the diffusion of electroactive species. This technique can also be employed to carry out the synthesis of electroactive polymers by applying a constant potential equal to the oxidation potential of monomer. Thus, it applies an anodic potential at the working electrode, and we can control the film thickness by controlling the deposition time of the polymer. Moreover, chronoamperometry is employed to cause double jumps in potential in thin films to study spectroelectrochemical properties, and assess the potential for application of materials in electrochromic devices. More information about these spectroelectrochemical studies and their importance will be presented in a later section.

2. Conjugated Polymers

Since the 1960s, is has been known that certain organic compounds with conjugated double bonds may exhibit semiconducting properties, but the interest in this type of material was not immediate due to the fact that the first conjugated polymers (CPs) presented themselves as insoluble, infusible, and consequently having low technological potential. The discovery that treatment of polyacetylene with acids or Lewis bases leads to an increase in conductivity of this material by up to eleven orders of magnitude [4], aroused the interest of the scientific community regarding the development of CPs, which combine into a single material the electrical properties of a metal or semiconductor with the advantages of a polymer, such as processability and flexibility.

CPs are also known as "synthetic metals" due to the fact of having magnetic, electronic, and optical properties similar to those of metals. These materials have conductivity associated with the presence of a π-conjugated system in its structure. Such materials become conductive through of the addition doping agents (chemical doping), and the term doping is used in analogy to inorganic semiconductor materials. The doping of CPs is the result of addition of counter ions in the structure of the material at different oxidation states. When a polymer is in the oxidized state, it means that the material presents positive charges resulting from the release of electrons and creation of defects. For electroneutrality in the polymer to be maintained, negatively charged species are attracted into the polymer structure. This process is known as p-type doping. The opposite phenomenon (for reduced polymers) causes the entry of cations into the material structure (n-type doping) [5].The conductivity of these polymers can be modulated by controlling the amount and type of dopant. Polythiophene and polypyrrole are examples of polymers that can undergo these types of doping. These polymers possess a π-conjugated backbone all along their length (Figure 6), and are usually insulators or semiconductors in their undoped forms. However, they can be transformed into a conductive material, comparable to a metal, when they undergo the doping process [6].

The properties of organic polymers depend on their structural characteristics, often having complicated morphological structures depending on the synthesis method and synthesis [6]. The conditions under which these materials are electrosynthesized influence the electrical and physical-chemical properties of the final material. Even before this stage, the monomer concentration, and temperature and nature of the electrolyte are some of the parameters that must be controlled to obtain polymers with satisfactory properties. Kabasakaloglu et al. (1999), in their studies of the electrochemical properties of polythiophenes, observed that electrochemical properties and conductance values differed depending on the type and concentration of the supporting electrolyte.

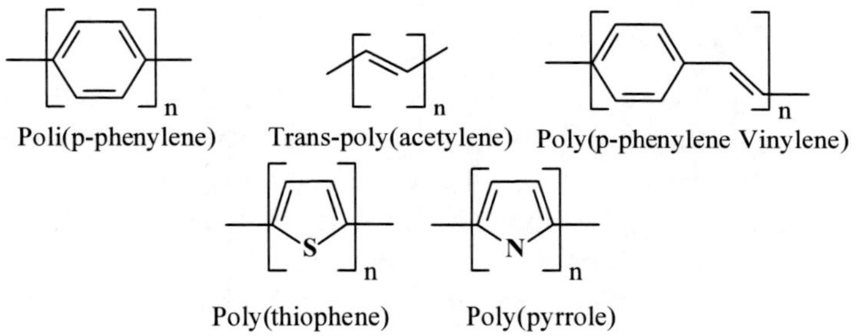

Poli(p-phenylene) Trans-poly(acetylene) Poly(p-phenylene Vinylene)

Poly(thiophene) Poly(pyrrole)

Figure 6. Structure of conjugated polymer examples.

Changes in the optical properties of these materials are the result of changes in the electronic states of conducting polymers. Thus, in modulating the electronic state of the polymer, one can obtain materials with specific applications such as electrochromic devices, light emitting diodes, and photovoltaic cells, among others. Based on this principle, optoelectronic devices are built according to their electronic and optical properties [7].

Another interesting property of CPs is the fact that they undergo reversible oxidation [8], which enables their application in batteries. Moreover, many of them have color changes associated with the applied potential or the level of oxidation, enabling applications in electrochromic devices [9].

2.1. Electronic Properties

The electronic behavior of materials can be outlined simply in three different ways, Figure 7:

a). Conductive - when the valence band is partially filled in, allowing the electrons to move freely between the energy states within the band.

b). Semiconductive - when the separation is such that electrons can be excited, thermally, from the valence band to the conduction band.

c). Insulating - when the separation corresponding to the energy gap (Eg) between the valence and conduction bands is very large and the electrons are confined in the valence band.

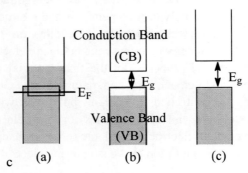

Figure 7. Behavior of materials in terms of energy bands.

where E_F = Fermi energy, E_g = energy gap (which corresponds to the forbidden region).

The CPs have aromatic or poly(enes) structure, where the simplest structure is the poly(acetylene). The true nature of the electrical conductivity in doped and undoped poly(acetylenes) is still subject to speculation. Various interpretations have been developed to explain the properties of these polymers. An ideal infinite chain of poly(acetylene) (CH)x, in which electrons fill half of the higher energy band, leading to a metallic one-dimensional structure is often used [10]. However, the finite length of the chains makes (CH)x an intrinsic semiconductor with an energy gap of 2 eV, due to changes in its structure of double and single bonds.

When neutral, the polymers show semiconductor or insulator behavior, presenting a significant conductivity only when oxidized (i.e., doped).

In contrast with the inorganic doping process, the charge carriers in conjugated polymers are not electrons or holes, but species interacting with the polymer network, known as solitons (in the case of (CH)x), or polarons and bipolarons (in the case of aromatic polymers); Figure 8 [11].

The formation of a polaron occurs when you remove (or add) an electron from a polymer chain, which creates a defect associated with a lattice distortion (caused by a relocation restriction), which corresponds to the localized energy states in the gap energy. The polaron corresponds to a radical ion with spin = 1/2. A bipolaron is formed when a second electron is removed, this situation is obtained with a diion with spin = 0. The formation of a polaron or a bipolaron requires approximately equal energies, which is supported by theoretical calculations. It was also noted that, despite the coulombic repulsion between two like charges (which is compensated for by the presence of a dopant with opposite charges), the formation of a bipolaron is thermodynamically more favorable than the formation of two polarons because this species has a greater gain in ionization energy [12,13].

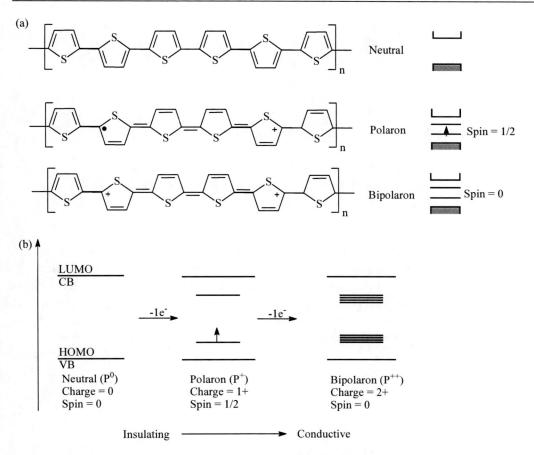

Figure 8. Structural representation (a) and energy (b) describing the formation of a polaron and the bipolaron by successive removal of two electrons in the polymer chain, exemplified here for a poly(heterocyclic). P indicates polymer.

In Figure 8b, the gap energy is the difference between the conduction band (CB) or LUMO and valence band (VB) or HOMO. In the energy gap is observed the formation of sub-bands to the extent that electrons are removed from the structure, leading to the formation of polaronic states.

2.2. Electrochemical Properties

The ability to reversibly oxidize the polymer between the neutral and oxidized states (p-type doping) and/or neutral and reduced (n-type doping) makes poly(thiophene) materials with high electroactivity [14]. Poly(thiophenes) are electro-oxidated readily and reversibly, while n-doping is harder to obtain and less stable [15].

The charge process, by applying an appropriate potential, involves the simultaneous transport of charge and ionic species in the polymer matrix. The final doped conducting state is characterized by the presence of charge compensation by ions along the polymer chain.

Redox reactions can be outlined as follows:

$$\text{doping } p \quad P^0 + A^- \underset{}{\overset{Ep^{0'}}{\rightleftharpoons}} P^+, A^- + e^- \tag{11}$$

$$\text{doping } n \quad P^0 + M^+ + e^- \underset{}{\overset{En^{0'}}{\rightleftharpoons}} P^-, M^+ \tag{12}$$

where: P^0 = neutral polymer; P^+ = oxidized polymer; P^- = reduced polymer

$Ep^{0'}$ = formal redox potential associated with the process of p-type doping
$En^{0'}$ = formal redox potential associated with the process of n-type doping
A^- = anion
M^+ = cation

The voltamperometric theoretical behavior for mixed doping can be seen in Figure 9, where we can identify two redox processes. The first, at positive potentials, is associated with the p-type doping, and the second, in the region of negative potentials, is associated with the process of n-type doping. These curves can be obtained by studying each region separately or by making a single scan starting from the initial potential (Ei-p), E = 0V, and going to the final potential of the system (Ef-p), then reversing the scanning direction until reaching the initial value of the potential system associated with n-type doping (Ei-n) and reversing the scanning direction once more to regain the potential Ei-p.

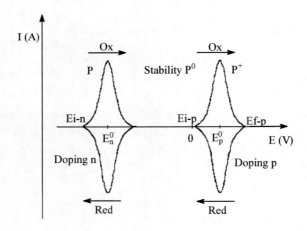

Figure 9. Theoretical voltammetric curves for mixed doping.

The energetic interpretation for the behavior shown in Figure 9 can be seen in Figure 10.

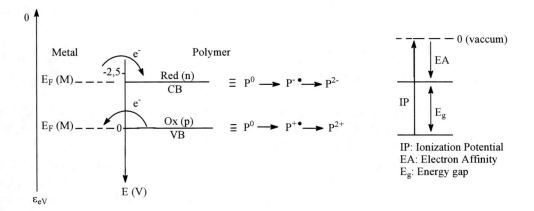

Figure 10. Interpretation of energy in Figure 9.

The species P$^{+\cdot}$ and P$^{-\cdot}$ correspond to polaron species formed during the process of p- and n-type doping, respectively. Species P^{2+} and P^{2-} correspond to bipolaron species formed during the process of p- and n-type doping, respectively. In fact, the electrochemical mechanism is more complex because it involves two steps; some authors [16] proposed a model reaction involving three species with two transfers monoelectronic (n = 1) and an interaction potential due to coulombic repulsions. Each species is chemically equivalent to six monomeric units: neutral (P^0), cation radical (P$^{+\cdot}$), and dication (P^{2+}), which leads to changes in the Nernst equation:

$$E = E_1^{0'} + \frac{RT}{F} \ln\left(\frac{C(P^{+\cdot})}{C(P^0)}\right) + \frac{\Delta\mu(P^{+\cdot})}{F} \tag{13}$$

$$E = E_2^{0'} + \frac{RT}{F} \ln\left(\frac{C(P^{2+})}{C(P^{+\cdot})}\right) + \frac{\Delta\mu(P^{2+}) - \Delta\mu(P^{+\cdot})}{F} \tag{14}$$

where, the potential of the interaction has a classical linear dependence on concentration:

$$\frac{\Delta\mu(P^{+\cdot})}{F} = B_{11} \frac{C(P^{+\cdot})}{C_T} + B_{12} \frac{C(P^{2+})}{C_T} \tag{15}$$

$$\frac{\Delta\mu(P^{2+})}{F} = B_{21} \frac{C(P^{+\cdot})}{C_T} + B_{22} \frac{C(P^{2+})}{C_T} \tag{16}$$

where:

$$C_T = C(P) + C(P^{+\cdot}) + C(P^{2+}) \tag{17}$$

BIJ: interaction coefficient (11: cation/cation; 12: cation/dication; 21: dication/cation; 22: dication/dication)
F: Faraday constant
C: concentration of species
μ: interaction potentials due to coulombic interactions

It is possible, in principle, to obtain important electronic information from these data. In fact, neglecting the solvation energy, we see that the ionization potential IP is roughly equivalent in energy to the thermodynamic redox potential for p-type doping (E_p^0). The electron affinity EA is approximately the energy equivalent of the thermodynamic redox potential of n-type doping (E_n^0). The energy of the energy gap Eg = IP – EA. However, the real voltammetric curves show some deviations from ideality that make it difficult for the extraction of some electrochemical parameters (h: yield coulomb, Q: charge, E0': formal redox potential, Dap: apparent diffusion coefficient) but do not invalidate the method for obtaining electronic data.

Among the various factors of deviation, one can cite the conformational changes during the redox processes, the molecular weight distribution, interactions between electroactive sites, as already described, the variation of resistance and capacitance of the films as a function of doping, and the complexity in the transport of matter within the polymer film.

Below, we show a simplified procedure for p-type doping, which could be the redox reaction scheme considering the different ion exchanges:

$$[P^{y^+}, M_{y^{\prime\prime}}^+, A_{y^\prime}^-] + x_a A_s^- \rightleftharpoons [P^{(y^\prime + y_a + y_b)^+}, A_{y^\prime} + x_a . M_{y^\prime - y_a}^-] + x_b M^+ + (y_a + y_b)e^-$$

where y = y' – y'' and s: solution

These multiple exchanges interfere with quantitative analysis of kinetic data (diffusion coefficient). Besides these factors, we know that the electrochemical data are sensitive to experimental conditions and the nature of the electrode, the electrolyte composition, and the film thickness [17,18]. Under certain circumstances, there may be problems with stability and reproducibility associated with polymer degradation (reduction of conjugation length and the number of active sites).

3. Main Methods for Obtaining Polythiophene and Derivatives

3.1. Chemical Synthesis

The chemical polymerization of thiophene and derivatives is often performed by the method of polyaddition. The coupling between the monomers occurs through the generation of radical cations by an initiator (oxidizing agent) and attack of these radical-cations by the other monomers present in the reaction system, passing through the stages of initiation, propagation, and termination (Figure 11) [19]. The oxidizing agents most commonly used are the metal halides (ferric chloride with a molar ratio 4:1 ($FeCl_3$: monomer) under an inert atmosphere) [20,21,22]. The regioregularity of couplings can be improved by monomers with substituents in position three of the thiophene ring. Obtaining copolymers can also be accomplished through the formation of radical ions employing the ferric chloride as initiator agent [23].

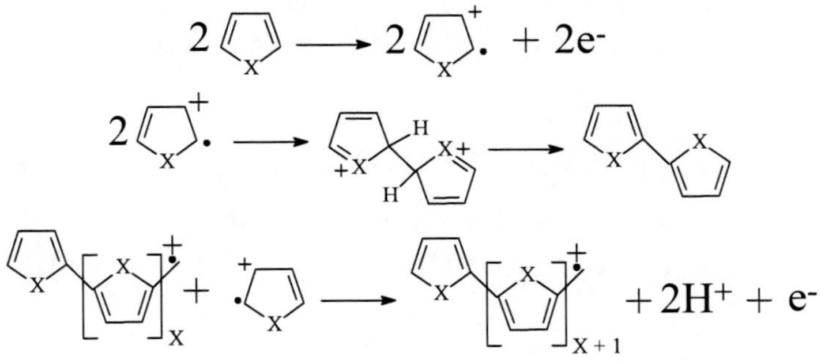

Figure 11. General mechanism for the polymerization of heteroaromatic monomers.

Polythiophenes and its derivatives may also be obtained via Grignard coupling made from monomers 2,5-dihalogenothiophene and its derivatives, Figure 12 [24].

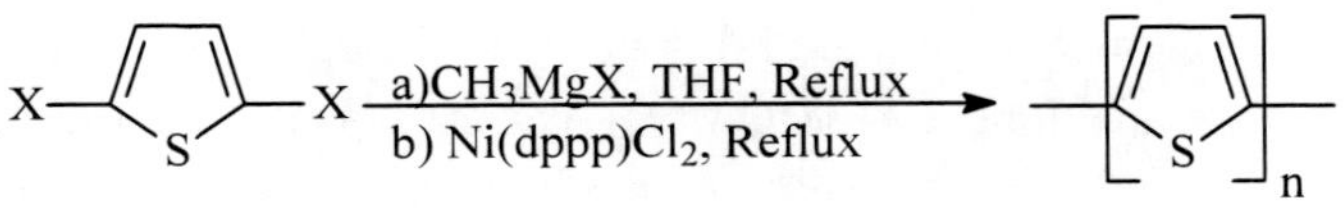

X = Br

Figure 12. Scheme of the polymerization of thiophene via Grignard coupling.

Polymerization using FeCl$_3$ as an oxidizing agent is a simple method, however, the polymer may have significant regiochemical defects. Nevertheless, it is possible to obtain regioregular materials through this methodology [21,25].

3.2. Electrochemical Synthesis

Electrochemical synthesis offers several advantages compared to chemical synthesis, such as higher speeds of synthesis and direct production of polymers on the electrode surface in both the doped and undoped states, among others [26,27]. Importantly, the conditions under which these materials are electrosynthesized directly influence the properties of final polymer material. Thus, the concentration of monomer, along with the temperature and nature of the electrolyte medium are some of the parameters that must be controlled to produce polymers with technologically attractive properties [28].

Electropolymerization is carried out in a conventional electrochemical cell with a three-electrode configuration. The monomers are dissolved in a suitable solvent and in low concentration, which may vary according to the ease of polymerization of the monomer. The solution must contain a supporting electrolyte at a concentration of 0.1 mol.L^{-1}. The electrochemical techniques used to perform the polymerizations are: Cyclic Voltammetry [29,30,31], Chronopotentiometry [32], and Chronoamperometry [33,34, 35].

Polymerization by cyclic voltammetry is the result of scanning, at a constant rate, a potential range in which the maximum oxidation potential of the monomer is included. Applying a constant potential equal to or above the oxidation potential of monomer for a given time is called chronoamperometry. Polymers obtained by applying a constant current are the result of the chronopotentiometry technique. As result of these three techniques, thin films of oxidized polymers form directly on the working electrode.

4. Conjugated Polymers: Potential for Application in the Development of Electrochromic, Electroluminescent, and Photovoltaic Devices

4.1. Electrochromism

Electrochromism is the result of electronic absorption bands generated in the visible region due to changes in redox states of the materials and means that the colors of the materials can vary between transparent and colored states, or between two colored states [36], i.e., the materials exhibit changes in their optical properties under extraction/insertion of charge [37]. Based on this principle, optoelectronic devices are constructed [38].

A conjugated polymer, in its neutral state, has absorption of electromagnetic radiation related to electronic transitions π-π* (HOMO $\rightarrow$ LUMO) as the electronic transition of higher intensity (λmax). When we vary the oxidation state in these materials, i.e., cause the injection/extraction of charge, the emergence of new levels within the Eg (polarons and bipolarons) occurs. The electronic transitions for these new energy levels (polarons and bipolarons) are more energetically favorable, resulting in a lower energy absorption region of the electromagnetic spectrum. This fact is responsible for the phenomenon of electrochromic conjugated polymers, because these electronic transitions of energy levels within the Eg lead to a change in color of the material. Importantly, the electrochromic phenomenon also occurs in inorganic materials, e.g., tungsten oxide (WO3), titanium oxide (TiO2), prussian blue, among others, [39]; however, we will focus on the electrochromic effect in conjugated polymers derived from beta-substituted thiophene.

According to Mortimer (1997) and Granqvist (2005), an electrochromic device can be constructed through the union of 5 layers between two substrates, such as glass or a flexible polyester (glass being the most common), because these layers must be transparent to visible light. A substrate and transparent conductive layer, such as ITO, can be commercially obtained (glass plates with conductive In2O3: Sn deposited on its surface). Thus, there are three layers between two plates of ITO, a film of stored ions, an ionic conductor (electrolyte), and an electrochromic film, as shown in Figure 13.

When we apply an appropriate potential between the two transparent conductive films (the ITO plates), ions can migrate into the storage layer of ions to the electrochromic film (cations - in the case of applying a potential to the cathode and anions - in the case of applying an anodic potential) through the electrolyte, causing a doping in electrochromic material, as a way to balance the load generated due to injection (negative charge) or withdrawn (positive charge) of electrons from the electrode material during application of potential. This process of doping is accompanied by the creation of energy levels within the energy gap of electrochromic films, which in turn results in a change of color of the electrochromic film due to changes in electronic properties (mainly Eg), accompanied by variation the optical properties. If at the end of a pulse of current, a permanence of color occurs due to the new oxidation state of the material, this phenomenon is known as "memory effect" [36].

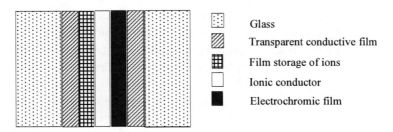

Figure 13. Schematic of an electrochromic device.

For a CP to be applicable in an electrochromic device, it must have certain characteristics such as good processability, high optical contrast, flexibility, and a fast response time [40]. Materials such as poly(3,4-ethylenedioxythiophene) (PEDOT) and its derivatives [29,30] and poly(3-chlorothiophene) [40] are examples of conjugated polymers with such properties.

Some of these properties can be measured using spectroelectrochemistry techniques (simultaneous measurements of optical and electrochemical properties). The spectroelectrochemical properties of materials can be obtained by measuring the transmittance (%T) at a fixed wavelength (λ), which can be chosen as the maximum wavelength (λ_{max}) of the absorbing material or a λ that corresponds to a certain color, or could refer to the maximum absorption in the region corresponding to transitions for the polaron and bipolaron states during a period where the polymer film suffers double jumps in potential (chronoamperometry) [41].

The response time (τ) is a measure of the rate of change of color of the material between the oxidized and reduced state and vice versa. It is desirable that this response time is short for potential applications. A good electrochromic material must provide a response time of less than 100 ms; however, depending on the final application, this value can vary and can come to a little over 1 s for applications such as smart windows. Thus, the possibility of applying these electrochromic materials directly depends on the rate with which the material responds to the electrical stimulus (a change in its oxidation state) and, consequently, a change in color.

Information on the optical contrast and coloration efficiency can be obtained from the very measure presented above. The optical contrast (Δ%T) is defined as the change in %T between the reduced and oxidized state [42]. The electrochromic efficiency (η) is defined as $\log(\%T_{re}/\%T_{ox})/Q$, where %$T_{re}$ is transmittance for the reduced state, %T_{ox} is transmittance for the oxidized state, and Q is the total charge involved in the oxidation/reduction of the polymer film. The obtained value of η provides a factor of proportionality between the charge density used in electrochemical processes (reduction/oxidation) and optical absorption at a particular wavelength, where the value is an inherent property of the electrochromic material studied.

Other features such as optical memory and electrochromic stability (or ciclability) are of great relevance for the assessment of potential applications of these materials. The optical memory of these materials comes from the fact they have the ability to maintain a state of color (absorption) for a certain period of time after removal of the applied electric field, i.e., color stability under open circuit. The time that this material remains without alteration in its color under open circuit is called optical memory or memory of an open circuit. The stability of electrochromic materials is directly linked to the electrochemical stability of the material because the irreversible processes and even the electrochemical degradation resulting from the application of high potentials is directly reflected in the decrease in optical contrast. For an electrochromic material to be regarded as promising for the development of devices, it is recommended that the material present stability (ciclability) of about 10^6 cycles, i.e., performing studies of double jumps in potential (chronoamperometry) of these materials must be able to withstand 10^6 cycles of charge and discharge. However, several other factors must also be taken into consideration, such as the ability to form thin, homogenous films, low resistance to charge transfer across interfaces, and good thermal stability.

4.2. Luminescence

The luminescence of polymeric materials occurs mainly through the formation of singlet excitons. When electrons are excited from the CP, either by an electrical stimulus or the incidence of light with an appropriate wavelength, they may be promoted from the ground state - π orbital, HOMO, the excited state - π* orbital, LUMO. Next, the excited electrons undergo relaxation and return to the ground state via three different processes: (a) loss of energy in the form of light (fluorescence), (b) nonradiative relaxation through the activation of phonons (vibrational relaxation) (internal conversion, IC) or (c) by an intersystem crossing (ISC) between singlet excited state and a triplet excited state then relaxing through the IC to the excited triplet state of lowest energy with subsequent radiative transition to the ground state (Phosphorescence) or IC [FALEIROS, 2007]. Figure 14 illustrates the phenomena of radiative and non-radiative transitions as described above.

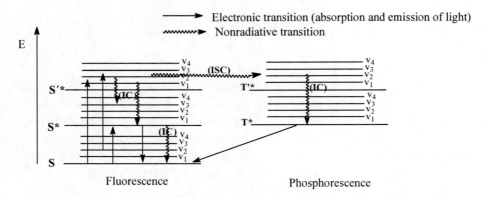

Figure 14. Transitions possible by electronic interaction between a CP and a photon (visible light or ultraviolet), where S represents the electronic levels and v the vibrational levels.

The process of optical excitation followed by luminescence is called photoluminescence (PL). The process of injecting electrons and holes generated by the polymer light emission due to recombination within the polymer matrix is known as electroluminescence. Even though electrons, after being excited by light with adequate power, or by injection of electrons and holes, are at energy levels higher than the LUMO (E_{00}), they suffer no radiative transitions until they reach the energy level E00 (Rule Kasha). Then, these electrons relax to the different vibrational energy sublevels of the ground state by emitting light. For CPs, the values of Eg varies between 1 and 3 eV, which classifies these materials a semiconductors.

An electroluminescent device can be constructed by placing one or more layers of a PC between two electrodes, a metal with a low work function (Al, Ca) that can inject electrons into the conduction band of the polymer (cathode) and an electrode that provides a higher work function (Au, ITO) so that it can inject holes in the valence band (anode) under the influence of an electric field (Figure 15). Electrons and holes injected into the polymer recombine radioactively, generating electroluminescence.

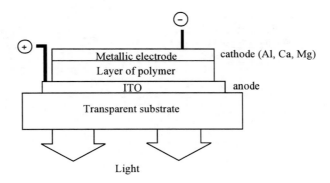

Figure 15. Schematic of a device based on light emitting electroluminescent CP. The ITO layer has a dual function as an electrode and as a transparent window for transmitting emitted light.

An interesting fact is that most polymers used in organic light emitting diodes (OLEDs), which exhibit electroluminescent properties, also have the ability to photoluminesce [5], which shows the importance of PL measurements on conjugated polymers.

4.3. Photovoltaic Cells

Organic electronics is one of the most promising new fields of technology. The possibility of obtaining flexible, low-cost optoelectronic materials opens up the range of applications for CPs in various areas such as solar cells. Grätzel solar cells [43] are constructed using titanium oxide (TiO2) as the active layer and organic dyes that act as sensitizers in the process of converting solar energy into electrical energy, placed between two electrodes. These devices still have a limited efficiency in converting solar energy into electrical energy, caused primarily by low absorption of visible light by the active layer and the low mobility of charge carriers. The limitation caused by low light absorption can be overcome with the addition of conjugated polymers that have high absorption of visible light and thus act as photosensitizers [44]. Moreover, these materials can be applied to the development of flexible, low-cost devices. A good conversion efficiency of solar energy into electrical energy employing conjugated polymers as sensitizers can be achieved through a rational planning of the structure of the polymer chain to obtain polymers with high absorption of light in the visible region (at a smaller band gap).

Solar cells can be constructed using conjugated polymers. The photoexcitation of conjugated polymers leads to the formation of bound electron-hole pairs called excitons. This electron-hole, in turn, must be decoupled so that the electron is available for generating electrical current. However, this separation occurs through the presence of strong electric fields (interface metal-polymer, for example). However, the dissociation of excitons and, consequently, the availability of electron to generate electricity occurs inefficiently, which makes solar cells with a monolayer of conjugated polymers (Figure 16a) show low efficiency of converting solar energy into electricity. Thus, researchers have concentrated on obtaining devices that favor dissociation of excitons use of "traps" of electrons (electron acceptor compounds).

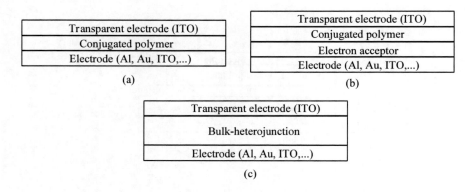

Figure 16. Three device architectures of conjugated polymer-based photovoltaic cells: (a) single-layer; (b) bilayer (heterojunction) and (c) bulk-heterojunction.

Some cells are constructed using two semiconductor layers (forming heterojunctions, Figure 16b), where one layer is p-type (electron donor - conjugated polymer) and another n-type (electron acceptor) [45]. One common example is a heterojunctions formed by CPs with derivatives of fullerene (C60, electron acceptors), which has shown efficiency of 4% using this configuration. However, the diffusion of excitons (about 20 nm) is the main limitation to separating the charges (only excitons generated near the n-p interface can decouple and contribute to the generation of electricity).

This limitation has been circumvented by obtaining disperse heterojunctions (bulk-heterojunctions) (Figure 16c), in which mixing between the donor and recipient layers results in the formation of interpenetrated networks among the materials. Carbon nanotubes have also been used to obtain "nanocomposites of charge transfer", where carbon nanotube network having a high charge mobility are generated in the polymer matrix, where they act as traps for electrons, in addition to contributing positively to the mechanical and thermal properties of the composite [46].

However, the search for new materials and architectures for photovoltaic devices based on conjugated polymers is still a field of research that has grown substantially in the last decade, whereas the search for high-efficiency (10%) conjugated polymer photovoltaic cells has been an object of study for several research groups.

REFERENCES

[1] Brett, C. M. A.; Brett, A. M. O. *Electrochemistry: principles, methods and applications*. New York: Oxford University Press, 1993.
[2] Wang, J. *Analytical Electrochemistry*. 2ª Ed. New York: Wiley-VCH, 2000.
[3] Kissinger, P. T.; Heineman, W. R.; *J. Chem. Educ.* 1983, 60, 702-706.
[4] Chiang, C. K.; Fincher, C. R.; Park, Y. W.; Heeger, A. J.; Shirakawa, H.; Louis, E. J.; Gau, S. C.; Macdiarmid, A. G.; *Phys. Rev. Lett.* 1977, 39, 1098-1101.
[5] Carpi, F.; Rossi, D. *Optics & Laser Technology* 2006, 38, 292-305.
[6] Kabasakaloglu, M.; Kiyak, T.; Toprak, H.; Aksu, M. L.; *Appl. Surf. Sci.* 1999, 152, 115–125.
[7] Lee, H. J.; Park, S. . *Phys. Chem.* B 2004, 108, 16365-16371.
[8] Ofer, D.; Crooks, R. M.; Wrighton, M. S. *J. Am. Chem. Soc.* 1990, 112, 7869-7879,.
[9] Mert, O.; Sahin, E.; Ertas, E.; Ozturk, T.; Aydin, E. A.; Toppare, L. *J. Electroanal. Chem.* 2006, 591, 53-58.
[10] Seymour, R. B.; Carracher, C. E.; Structure-Property Relationship in Polymer; ed. Plenum Press: New York, 1984
[11] Hattano, M., Kambara, S., Okamoto, S., *J. Polymer. Science* 1961, 51, 526.
[12] Faid, K., et al., *Macromolecules* 1995, 28, 284-287.
[13] HELL, M. G., et al., *Chem. Mat.* 1992, 4, 1106-1113.
[14] Guerrero, J. D.; Ren, X.; Ferraris, J. P.; *Chem. Mater.* 1994, 6, 1432-1443.
[15] Kaneto, K., Ura, S., Yashino, K., Jpn. *J. Appl. Phys.* 1984, 23, L189-L191.
[16] Pernaut, J. M.; Peres, R. C. D.; Juliano, V.F.; De-Paoli, M. A., *J. Electroanal. Chem.* 1989, 74, 225-233.
[17] Diaz. A. F.; et al.; J. Electrochemical Synthesis of Conducting Polymers. In: SKOTHEIN, T. A., Handbook of conducting polymers. 2. Ed. New York: Marcel Dekker. Cap.3, p.81-115, 1986.
[18] Roncali, J.; Garnier, F., *New Journal Chem.* 1986, 10, 237.
[19] Roncali, J.; *Chem. Rev.* 1992, 92, 711-738.
[20] Chen, S. A.; Tsai, C. C.; *Macromolecules* 1993, 26, 2234-2239.
[21] Andersson, M. R.; Selse, D.; Berggren, M.; Jarvinen, H.; Hjertberg, T.; Inganas, O.; Wennerstrom, O.; Osterholm, J. E.; *Macromolecules* 1994, 27, 6503-6506,.
[22] Bizzarri, P. C.; Andreani, F.; Casa, C. D.; Lanzi, M.; Salatelli, E. *Synth. Met.* 1995, 75, 141-147.
[23] Casa, D. C.; Fraleoni-Morgera, A.; Lanzi, M.; Costa-Bizzarri, P.; Paganin, L.; Bertinelli, F.; Schenetti, L.; Mucci, A.; Casalboni, M.; Sarcinelli, F.; Quatela, A. *Eur. Polym. J.* 2005, 41, 2360-2369.
[24] Hou, J.; Huo, L.; He, C.; Yang, C.; Li, Y.; *Macromolecules* 2006, 39, 594-603.
[25] Calado, H. D. R.; Matencio, T.; Donnici, C. L.; Cury, L. A.; Rieumont, J.; Pernaut, J. M..; *Synth. Met.* 2008, 158, 1037.

[26] Latonen, R. M.; Lonnqvist, J. E.; Jalander, L.; Ivaska, A.; Electrochim. *Acta* 2006, 51, 1244-1254.
[27] Pang, Y., Li, X., Ding, H., Shi, G.; Jin, L. *Electrochim. Acta* 2007, 52, 6172–6177.
[28] Heinze, J.; Frontana-Uribe, B. A.; Ludwigs, S.; *Chem. Rev.* 2010, 110, 4724–4771.
[29] Sotzing, G. A.; Reynolds, J. R.; Steel, P. J.; *Chem. Mat.* 1996, 8, 882-889.
[30] Sankaran, B.; Reynolds J. R.; *Macromolecules* 1997, 30, 2582-2588.
[31] Fouad, I.; Mechbal, Z.; Chane-Ching, K. I.; Adenier, A.; Maurel, F.; Aaron, J. J.; Vodicka, P.; Cernovska, K.; Kozmik, V.; Svoboda, J. *J. Mater. Chem.* 2004, 14, 1711.
[32] Tang, H.; Zhou, Z.; Zhong, Y.; Liao, H.; ZHU, L.; *Thin Solid Films* 2006, 515, 2447-2451.
[33] Blankespoor, R. L.; Miller, L. L.; *J. Chem. Soc., Chem. Commun.* 1985, 90.
[34] Pohjakallio, M.; Sundholm, G. *Synth. Met.* 1993, 55, 1590-1595.
[35] Mert, O.; Sahin, E.; Ertas, E.; Ozturk, T.; Aydin, E. A.; Toppare, L. *J. Electroanal. Chem.* 2006, 591, 53-58.
[36] Mortimer, R. J.; *Chem. Soc. Rev.* 1997, 26, 147-156.
[37] Granqvist, C. G. *J. Eur. Ceram. Soc.* 2005, 25, 2907-2912.
[38] Lee, H. J.; Park, S.; *J. Phys. Chem.* B 2004, 108, 1636.
[39] Monk, P. M. S.; Mortimer, R. J.; Rosseinsky, D. R.; Electrochromism: Fundamentals and Applications; John Wiley & Sons:New York, 1995
[40] Pang, Y.; Xu, H.; Li, X.; Ding, H.; Cheng, Y.; Shi, G.; Jin, L. *Electrochem. Commun.* 2006, 8, 1757–1763.
[41] Alves, M. R. A.; Calado, H. D. R.; Donnici, C. L; Matencio, T.; *Synth. Met.* 2010, 160, 22.
[42] Krishnamoorthy, K.; Kanungo, M.; Contractor, A. Q.; Kumar, A.; *Synth. Met.* 2001, 124, 471.
[43] Grätzel, M . *Nature* 2001, 414, 338.
[44] Winder, C.; Matt, H.; Hummelen, J. C.; Janssen, A.J.; Sariciftci, N.S.; Brabec, C. *J. Thin Solid Films* 2002, 403, 373
[45] Coakley, K. M.; McGehee, M. D.; *Chem. Mater.* 2004, 16, 4533-4542
[46] Kymakis, E.; Amaratunga, G. A. J.; *Appl. Phys. Lett.* 2002, 80, 112.

Chapter 7

POLYCRYSTALLINE SILICON THIN FILMS ON GLASS FOR PHOTOVOLTAIC APPLICATIONS

Nicolás Budini[1], Javier A. Schmidt[1,2], Fermín M. Ochoa[3], Pablo A. Rinaldi[1], Roberto D. Arce[1,2] and Román H. Buitrago[1,2]

[1]INTEC, CONICET-UNL, Güemes 3450, S3000GLN Santa Fe, Argentina
[2]Facultad de Ingeniería Química, UNL, Santiago del Estero 2829, S3000AOM Santa Fe, Argentina
[3]Instituto de Ciencias, BUAP, 14 Sur y San Claudio, Cdad. Universitaria, 72570, Puebla, México

ABSTRACT

In this chapter, we briefly present the state of the art in some deposition and processing techniques that can be used to obtain thin polycrystalline silicon (poly-Si) films on glass substrates. In particular, we focus on the ultimate advances to date related to the technique of metal induced crystallization of amorphous silicon, which allows to produce large crystalline grains (sizes over 100 µm) at low temperatures (lower than 600 °C) by means of thermally activated solid phase transformations. Specifically, we discuss in more detail the use of nickel as the metallic inductor of silicon crystallization, and its influence on grain nucleation and growth. We also analyze the relevant concept of the solid phase epitaxial crystallization of an amorphous silicon thin film, deposited onto a previously crystallized seed layer, to produce poly-Si. The effects of doping on crystallization are presented, together with the problems that it entails. Taking account of all the issues presented, we propose a suitable process, scalable to industrial levels, which will allow to produce a competitive large-grained poly-Si solar cell with high efficiency and at low costs.

1. INTRODUCTION

Solar energy, as well as other renewable energy sources, appears as a promising candidate to capture an important proportion of the world energetic matrix. Indeed, solar energy conversion and photovoltaic electricity consumption have been increasing in the last decades at high rates. At the same time, research in this field has also increased, giving rise to new technologies. One of the most powerful engines, or driving forces, for investigation in this subject is the overcoming of the current compromise existing between conversion efficiency and production cost of solar cells. The latter is hard to reduce while mantaining, or even without decreasing, the former. However, present research intends to force this restriction to the limit of developing low-cost materials and processes leading to higher conversion efficiencies. In Section 2, we present some aspects of the current state of photovoltaics technology in order to set the background of our research. Silicon based solar cells, the giants of the photovoltaic industry, are shortly addressed in Section 3, together with a descprition of one of the most successful approaches to commercial thin film solar cells production based on silicon. Some processes available for the crystallization of amorphous silicon thin films, in order to obtain polycrystalline silicon, are issued in Sections 4 and 5. Finally, one approach that we are currently investigating for the production of low-cost polycrystalline silicon solar cells, with large grain sizes, is presented in Section 6. The conclusion of the present chapter constitutes Section 7.

2. TRENDS IN PHOTOVOLTAICS

The need for development of alternative, renewable and/or ecological energy sources has turned photovoltaics (PV) one of the most promising technologies in this field. The main characteristic and great advantage of solar energy is the enormous and widely available power that our star offers to us. For practical purposes we can consider the Sun as being an "infinite" source of energy, since it will keep radiating for billions of years. Therefore, it is well worth taking advantage of that power to satisfy, at least partially, the requirements of our nowadays lifestyle, aiding and complementing the present and long-term worldwide energetic matrix.

Great efforts have been made in the last years in order to reduce the production costs and, at the same time, to improve the conversion efficiency of PV devices. This challenge has led to a classification of the different stages of technological innovations from the point of view of design of the solar cell and also of the materials involved. Actually, there are three well-established *generations* in the field of PV devices. The *first generation* makes reference mainly to silicon (Si) or gallium arsenide (GaAs) wafer-based, single-junction, solar cells. Typical efficiencies for Si solar cells are in the range between 17 and 18% at module level and go up to 24.7% at laboratory level [1]. GaAs cells have slightly higher efficiencies, reaching 26–29% at laboratoy level and averaging 20% at module level [2,3]. However, the production costs for both types are relatively high. The *second generation* is based on the thickness reduction and optimization of single-junction solar cells. It is also known as "thin film solar cell technology", which is thought to be suitable for the mid-term future of PV [4]. At the laboratory level, efficiencies as high as 20% have been reached so far [5,6,7] while theoretical considerations fix an upper limiting value of 31% under an illumination of one sun, extendable to a mere 41% under extreme sunlight concentration (46,200 suns) [8]. Amorphous silicon (a-Si), polycrystalline silicon (poly-Si), copper indium diselenide (CIS) and cadmium telluride (CdTe) solar cells are the principal contenders at this level. A *third generation* of PV devices is still emerging. Its distinctive feature is the utilization of novel polymers or dye-sensitized materials for a better capture of incident sunlight, together with multi-junction thin film structures which would allow to overcome the efficiency limit of 31–41% for single-junction solar cells (i.e. for first and second generation PV devices). The estimated enhanced range of conversion efficiencies for this kind of devices is 30–60%, keeping low production costs. Current research concerning third generation solar cells is being made over a wide variety of new approaches, such as tandem or quantum well cells, which make use of processes like multiple energy thresholds, multiple electron-hole pairs creation, hot carriers excitation, and so on [9]. In the rest of the chapter we will be particularly concerned with second generation solar cells.

3. SILICON-BASED SOLAR CELLS

Nowadays, first and second generation solar cells coexist at the industrial level with an incipient inclusion of third generation devices. In this scenario, crystalline silicon (c-Si, including single-crystalline and multi-crystalline) widely dominates the market, despite being an expensive material, with 80–90% of the worldwide solar cell production based on wafer technology [10,11]. This is due to the fact that silicon has plenty of advantages if compared to other materials. First of all, being a non-toxic element, its notable abundance is of great importance. Silicon composes a fraction of 30% of the Earth's crust, possessing excellent electronic, chemical and mechanical properties. Moreover, it has a bandgap of 1.1 eV, which is almost perfectly matched to the solar spectrum impinging on the terrestrial surface. Furthermore, it gives long-term stable solar cells and also offers the vast experience of processing and production acquired from the microelectronics industry. However, the need to reduce the costs of PV energy forces towards a reduction in the production costs. To achieve this goal, other approaches using less material and lowering the processing temperature are needed.

Several possibilities have emerged during the second generation stage as powerful substitutes, lowering drastically the cost/throughput relation. Most of these alternatives are based on silicon, anyway, but they take advantage of other methods of production and processing which are economically more convenient. Such is the case of amorphous (a-Si), nanocrystalline (nc-Si), microcrystalline (µc-Si) and polycrystalline (poly-Si) silicon. Extensive investigation has been made concerning these materials [12] since they can be deposited over different substrates by several methods and, as a remarkable characteristic from the point of view of cost reduction, in the form of thin films.

In particular, the use of glass substrates introduces an important cost reduction, being one of the advantages of thin film solar cells. An important requirement that the chosen substrate must fulfill is to tolerate the temperatures used during manufacturing, which are usually lower than 600 °C. Borosilicate, aluminosilicate and alkali-free glasses are among the most commonly used substrates. As we shall see later, in the next subsection and in Section 6, some particular processes (e.g. defect annealing or doping) may need higher temperatures

(~900 °C), which are almost at the limit or even higher than the softening point of commercially available and cost-effective glasses. However, there are possible ways to overcome this inconvenient. Such is the case of rapid thermal annealing (RTA), which allows to rise and fall the temperature of the sample at high rates without heating the glass substrate in excess.

3.1. CSG Solar Approach

One of the most successful processes for large-scale production of PV modules, using thin film silicon based technology, is the CSG (crystalline silicon on glass) approach. The project began at the University of New South Wales, Sydney, in the late 1980s, and evolved through the formation of the companies Pacific Solar in 1995 and CSG Solar in 2004 [13,14]. The basic idea is to produce devices having a $n^+/p^-/p^+$ structure, i.e. the same standard single-junction structure for a c-Si solar cell. As can be observed in Figure 1, this approach basically consists on the deposition of three a-Si doped layers ($n^+/p^-/p^+$, respectively) onto a glass substrate. The substrate is a borosilicate glass, previously textured and coated with a half-micron-thick layer of silica beads by a patented technique. Previous to the a-Si deposition, the glass is also covered with an amorphous silicon nitride (SiN) film, which acts as an antireflective and passivating coating. The whole structure is obtained in the same deposition chamber by plasma enhanced chemical vapor deposition (PECVD), being the active thickness of the device in the order of 2 μm. Finally, an upper capping layer of silicon dioxide (SiO_2) is deposited onto the amorphous $n^+/p^-/p^+$ structure. Subsequent processes include solid phase crystallization (SPC) of silicon at 600 °C in a batch oven, defect annealing at 900 °C by RTA, plasma hydrogenation by exposure of the material to atomic hydrogen at temperatures above 600 °C, patterning and contacting with a fault-tolerant metallization scheme [15]. After these treatments the poly-Si solar cell, which is just 2.2 μm thick, is able to provide a short circuit current density of up to 29.5 mA/cm^2 [16]. Very effective light-trapping, defect annealing and defect passivation are responsible for the high efficiency that is obtained and for the minimal use of materials achieved. The best efficiencies obtained by CSG Solar are in the order of 10.4% for 94 cm^2 mini-modules, and about 7% for 1.4 m^2 modules at factory-level [17].

The final material resulting from the CSG process, morphologically speaking, is a poly-Si thin film with crystalline grain sizes in the order of 1 μm. This is inherent to the annealing procedure at 600 °C used to achieve the SPC of a-Si [18]. It has been shown that hydrogen incorporated during the deposition stage plays an important role in the SPC process [19]. Two features emerge in relation with hydrogen presence in the start-up material. First, it limits the heating rates because violent dehydrogenations can produce severe damage to the films. Second, hydrogen concentration is related to the final grain size since it accelerates the nucleation rate. Despite the fact that hydrogen incorporation during PECVD deposition of silicon can be strongly reduced if the deposition temperature is increased, the disorder inside the amorphous material still prevents from obtaining a higher grain size.

The grain size is known to drastically limit the conversion efficiency [20,21]. Thus, it is of great interest to increase the sizes of crystalline domains. One approach that has been intensively investigated lately for this purpose is the induction of a-Si crystallization mediated by metals, which will be discussed in Section 5. First, in the following section, we will focus on SPC, being the most common and straightforward method for crystallization of a-Si.

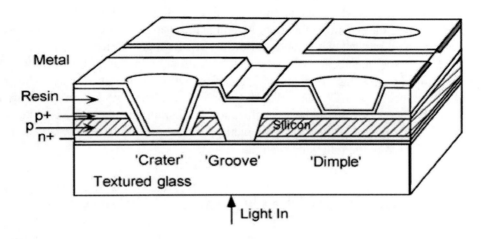

Figure 1. Scheme of the crystalline silicon on glass (CSG) solar cell. Reprinted with permission from Ref. [13].

4. SOLID PHASE CRYSTALLIZATION OF A-SI

After many years of research, the a-Si:H technology acquired a high degree of maturity, allowing the deposition of thin uniform films over large area substrates at elevated deposition rates. Despite considerable efforts, intrinsic stability problems still limit the performance and efficiency of a-Si:H-based devices, having worse electrical transport properties than those of c-Si [22,23]. However, renewed interest arose in using a-Si:H as a starting base material to produce thin poly-Si films on glass by means of a suitable crystallization process. This alternative to the direct deposition of poly-Si by chemical vapor deposition (CVD) has the advantage of requiring lower processing temperatures, making it compatible with the utilization of low-cost glass substrates. Some methods for crystallization of a-Si are the already mentioned SPC, zone melting recrystallization (ZMR) and metal induced crystallization (MIC) [2,24]. Each of these methods has its own advantages and disadvantages that we briefly enumerate in the following paragraphs.

SPC, the simplest and at the same time one of the first techniques explored to crystallize a-Si [25], is the direct annealing of the material for a sufficiently long time. The temperature used is generally in the range between 550 and 700 °C, and most commonly around 600 °C. This temperature is chosen as a compromise between high temperatures that produce faster crystallization but smaller grains, and low temperatures that produce larger grains but needing longer annealing times. The method is simple and straightforward, but the main drawbacks are the need for long annealing periods, generally tens of hours [26], and the relatively small grain size obtained, which is not larger than 1.5 μm [18]. In spite of these limitations, it is the basic process used to produce commercial poly-Si solar cells at the CSG Solar company with a great performance [13,14], as mentioned in the preceeding section.

The ZMR method consists in a hot scanner that creates a narrow molten zone which advances along the thin film. There are several choices for the heat source, which can be a laser, a halogen lamp, a resistor or an electron beam. When the melted material cools below the melting point, a heterogeneous nucleation starts. Due to the directional nature of the scan, grains with a particular crystal orientation are obtained, with dimensions that lay in the order of millimeters along the scanning direction. Although these are desirable grain characteristics, the method has the disadvantages of poor uniformity and high production costs [27,28].

By adding some metals to a-Si, the crystallization temperature can be reduced in comparison to SPC, leading to a larger grain size with a good uniformity. This is the innovation introduced by MIC. Some metals, like aluminum, indium, gold or antimony form eutectics with silicon, lowering the crystallization temperature. Other metals, like nickel, palladium or titanium, form silicides. This method has been intensively investigated lately and will be discussed in the following section.

5. METAL INDUCED CRYSTALLIZATION OF A-SI

The MIC method proved to be successful in terms of enlarging the grain size and lowering the annealing temperature to attain crystallization [29,30]. Both characteristics go also in the direction of costs reduction as is desired. By using this method, an increase in grain size of over one order of magnitude as compared to the grain size obtained by direct SPC of a-Si can be achieved. The method is based on the capability of some metals to form eutectics or silicides with silicon. Whether the former or the latter occurs depends on the metal used as the inductor. Several metals were studied so far, being Al and Ni the most interesting and, hence, those that have been more thoroughly investigated. Aluminum forms an eutectic with silicon at relatively low temperatures (< 600 °C), while nickel forms nickel disilicide ($NiSi_2$) at temperatures around 350 °C. Details on both methods, namely aluminum induced crystallization (AIC) and nickel induced crystallization (NIC), are given in the next subsections.

The main drawback of MIC in general is the metal contamination that may remain after crystallization. Due to the fact that the desired outcome of this method is a good quality poly-Si, metal impurities are thus detrimental and should be avoided. Taking account of MIC's advantages, concerning grain sizes and lower annealing temperatures, it is reasonable to develop suitable etching processes for applying them as impurity removals without affecting the overall quality of the poly-Si obtained. This aspect has been addressed and investigated so far [31]. Several alternatives are suitable to attack this problem and one of them is discussed later.

5.1. Aluminum Induced Crystallization

The AIC method of a-Si has been extensively studied for the production of thin film silicon solar cells [32,33]. Although large grains are obtained, the metal contamination problem is hard to overcome. Aluminum is a shallow acceptor in silicon and hence the layers result strongly p-doped, with aluminum contents as high as 1×10^{19} cm^{-3} [34]. Recombination processes at defect centers, introduced by metal impurities, reduce the carriers lifetime. Moreover, solar cells usually present a low parallel resistance due to the segregation of aluminum atoms in the grain boundaries. Therefore, layers prepared by AIC are commonly used as the back surface field (BSF) of the solar cell, but are not suitable to act as the absorber layer.

The physical process by which AIC acts is based in an exchange between an a-Si layer in contact with an Al layer during annealing (see Figure 2), resulting in a poly-Si film with an Al+Si capping layer. This process was named aluminum induced layer exchange (ALILE) [10]. The driving force during crystallization is the ability of aluminum atoms to diffuse into the a-Si structure producing a change in the Si–Si covalent bonds at the Si/Al interfaces due to the free electrons of the metallic phase [29]. Final grain sizes are generally in the order of 20–30 µm.

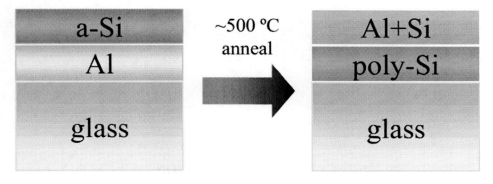

Figure 2. Aluminum induced layer exchange process at temperatures below the eutectic point.

5.2. Nickel Induced Crystallization

The NIC method has been thoroughly studied for the production of thin film transistors (TFTs) for crystal displays [35,36]. When crystallization advances in the lateral direction, starting from a Ni electrode, the process is named metal induced lateral crystallization (MILC) [37,38]. Nickel silicides can be formed at temperatures around 350 °C and, in this way, it is possible to obtain poly-Si at temperatures as low as 480 °C [39]. Most of the research concerning NIC and MILC for TFTs applications is done on thin samples, generally with film thicknesses that lay below 0.1 µm. Since we are concerned with solar cell applications, considerably thicker layers are needed.

The crystallization of a-Si by NIC is mediated by the formation of $NiSi_2$(111) octahedral precipitates, having a very similar lattice parameter (~99.6%) to that of c-Si(111) [30]. A simple diagram of the process can be seen in Figure 3. The detailed mechanism by which the crystallization is induced by Ni is still under investigation. However, it is thought that at a first stage of annealing Ni atoms diffuse over the a-Si surface until they form $NiSi_2$ nuclei. For this to accomplish, a minimum surface density of Ni atoms and a critical size of the silicide nuclei are required [40]. In a second stage, these $NiSi_2$ precipitates migrate through the a-Si volume leaving behind c-Si, until grain boundaries collide with those of neighbouring grains. Due to the nature of the growing process, the Ni concentration inside the grains is relatively low. It has been demonstrated by time-of-flight secondary ion mass spectroscopy (TOF-SIMS) analysis that after full crystallization the nickel concentration at grain boundaries is much higher than inside the grains. This proves that Ni atoms migrate during the grain growth accumulating at grain boundaries [41]. NIC has two important features. First, the required quantity of Ni atoms per unit of area is really small for crystallization to proceed and, second, the final grain size (depending upon Ni density) can be larger than that obtained by AIC. In a recent work it has been demonstrated that in a-Si samples with sputtered Ni surface densities lower than 1 $\times$ 10^{15} at./cm^2, disk-shaped grains with a diameter in excess of 100 µm can be obtained [42,43]. It is important to remark that those values of surface densities are equivalent to depositing less than a monolayer of Ni atoms.

There are several ways to deposit Ni atoms onto the a-Si layer. Such is the case of ion sputtering of a Ni target [30,42,43], ink-jet printing of an appropriate Ni-containing solution [44,45], among other possibilities as could be electroless deposition [46,47].

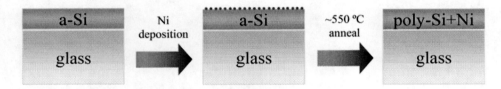

Figure 3. Nickel-induced crystallization of a-Si.

6. NICKEL INDUCED CRYSTALLIZATION AND EPITAXIAL GROWTH APPROACH (NICE) FOR SOLAR CELL DEVELOPMENT

Our current research consists on finding a suitable process, scalable to industrial levels, by which a low-cost solar cell with reasonable efficiency can be obtained.

6.1. n^+ Seed Layer Approach

We briefly described NIC of a-Si samples in Subsection 5.2, showing that it allows to produce large-grained poly-Si on glass by means of SPC in a conventional furnace. The a-Si crystallization does not seem to be altered if the glass is covered with a transparent conducting oxide (TCO), thus providing a front contact to the cell that also works as a surface passivator for the emitter. So, thinking on the design of a solar cell with a classical single-junction structure, glass/TCO/n^+/p^-/p^+/metal, it would be an advantageous process to start with a large-grained n^+ thin layer (emitter). This layer could serve as a crystalline seed for an epitaxial crystallization of the remaining amorphous p^-/p^+ layers (absorber/BSF), which complete the cell and can be deposited together by PECVD in the same process. However, it is known that high phosphorous concentrations affect the crystallization process [30]. Thus, the direct obtainment of a large-grained n^+ seed layer, starting from a strongly n-doped amorphous material, is neither simple nor straightforward.

A possible way to evade this problem is to combine NIC and epitaxial growth, leading to an approach that we call NICE. A potential complete process, summarized in Figure 4, is divided in two main stages denoted as *seed layer preparation* and *epitaxial process*. As can be observed, both stages consist of several sublevels. Those of the first stage, in particular the doping and etching steps, represent the most critical ones, as we shall see. We will briefly enumerate and describe each of them in the following items.

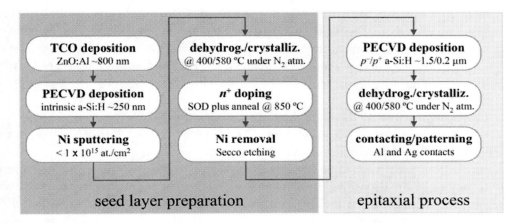

Figure 4. Tentative diagram of thin film poly-Si solar cell production by the NICE method.

TCO Deposition

As a first step, aluminum-doped zinc oxide (ZnO:Al) films are deposited onto the glass substrate by rf-sputtering from a ZnO/Al_2O_3 sintered disk with a respective weight proportion of 97.1/2.9%. The rf frequency is 13.56 MHz, and the generator is coupled to a 300 W amplifier which delivers this power to argon atoms inside a high-vacuum chamber. Ionization of argon atoms forms a plasma, providing the energetic ions required for bombardment of the target. By controlling the deposition parameters like substrate temperature, gas pressure and voltage polarization, the optical and electrical characteristics of the TCO can be optimized in order to obtain low-resistivity layers, having resistances lower than 10 Ω-square. The transparency and texture (also called Haze [48]) are both important factors as well and generally have to be in the order of 80% and 14%, respectively, for a better light-capture. The thickness of this layer is usually around 0.75 µm and the deposition rate is rather slow, say a few angstroms per second.

Seed Layer Deposition and Crystallization

An intrinsic ~200-nm-thick a-Si layer is deposited by PECVD onto the TCO-covered glass at a temperature of 200 °C. Silane (SiH_4) is used as the reactant gas, without hydrogen dilution, and it is excited with a relatively high rf-frequency of 50 MHz. This allows to obtain high deposition rates of aproximately 20 Å/s. The samples obtained in this way are slightly hydrogenated with hydrogen concentrations in the order of 10%, referring to atomic percentages.

Ion sputtering of a nickel target is used for the incorporation of metallic atoms in order to proceed with the thermal annealing. Argon is used as the ion source by exciting it in a vacuum chamber with a high dc-voltage (~800 VDC). The deposited atomic surface density can be controlled by changing the exposure times of the samples to the bombarded target. Due to the extremely low amounts of Ni atoms per surface unit required by NIC, the deposition only takes a few seconds.

To proceed with crystallization at 580 °C, a previous dehydrogenation step is needed to prevent abrupt hydrogen effusions which may severely damage the films. At temperatures slightly above 400 °C most of the bonded hydrogen atoms have enough energy to effuse from the sample, thus reducing the hydrogen concentration to an acceptable level. After this step, which takes generally a few hours, the temperature continues to raise up to 580 °C in order to allow for nucleation and migration of $NiSi_2$ nuclei. Crystalline grains grow until full crystallization is achieved after about 24 h, giving a large-grained intrinsic poly-Si film with grain sizes exceeding 100 μm.

Seed Layer Doping and Ni Removal

The poly-Si obtained in this way is intended to act as a crystalline seed for epitaxial crystallization of upper amorphous layers. Furthermore, it is required to act as the emitter of the solar cell for which it has to be strongly doped with phosphorous. Therefore, doping has to be performed before depositing the remaining p^-/p^+ structure in order to give the seed a n^+ character. The doping can be yielded by spin-on doping (SOD), which is a widely used method for junction formation by means of thermal atomic diffusion from some source containing *p*- or *n*-type dopants [49,50,51,52]. The dopant is generally dispersed in a viscous solution that is homogeneously spread onto the silicon film by means of a spinner. In the case of phosphorous, for *n*-type doping, its diffusion into the poly-Si film is promoted by temperatures in the order of 850 °C, easily attainable in a conventional or RTA furnace. The latter is more convenient from the point of view of the glass substrate softening point, which is generally below 900 °C for the kind of glasses used (borosilicate, aluminosilicate or alkali-free glasses). RTA allows rapid heating of silicon during short intervals by infrared radiation, which prevents the glass from reaching high temperatures since it is a poor infrared absorber (contrary to silicon, being a good one). The doping has to be optimized so as to produce sheet resistances lower than 10 Ω-square, which depends basically on the amount of dopant source used and the effectiveness of the doping process.

The high temperature of the doping stage produces the formation of an oxide on top of the silicon film that should be removed. As said before, it is also mandatory to remove the remaining nickel precipitated at grain boundaries during crystallization. An etchant that is able to perform both tasks at once is the so called "Secco" solution, generally composed of potassium dichromate ($K_2Cr_2O_7$), hydrofluoric acid (HF) and water. The HF removes the oxide grown during the doping process while potassium dichromate attacks preferentially the grain boundaries of the poly-Si, since it reacts with Si dangling bonds prevailing at defective zones. As a consequence, the Secco etching removes Ni atoms at grain boundaries in an indirect way, i.e. not reacting with Ni but instead removing the material zones where it is located. If Ni atoms are not successfuly removed from grain boundaries, they would induce crystallization to the upper layers during their future annealing stage, thus preventing an epitaxial crystallization as pointed out in Ref. [43].

It is important to note that both the doping and the etching processes have to be carefully performed and optimized since they interact directly with the main part of the solar cell, namely the semiconductor n^+/p^- junction. Moreover, the surface in question is the responsible of a good epitaxial process.

p^-/p^+ Structure Deposition and Expitaxial Crystallization

The next step consists on the deposition of the remaining amorphous p^-/p^+ layers by PECVD onto the already doped seed layer, i.e. the emitter of the solar cell. The deposition conditions in this case are the same as those used for the seed layer, except for the incorporation of diborane (B_2H_6) to the reactor, which provides the required boron source to give the layers a p^- and p^+ characters in the same deposition process. The total thickness of this structure has to be in the order of 1.5 µm, in an approximate proportion of 7:1, respectively.

For the epitaxial crystallization to proceed, a previous dehydrogenation step has to be fulfilled, at temperatures slightly above 400 °C, in order to eliminate the hydrogen incorporated during deposition. Subsequently, the required 580 °C thermal annealing is achieved. If a higher temperature is used, there exists the possibility of spontaneous nucleation and crystallization of a-Si, not giving place to an induction of epitaxial crystallization from the seed layer. This would result in a considerably smaller grain size as we already mentioned before, in Section 4, concerning SPC of a-Si. The dependence of epitaxial crystallization annealing times with the crystallizing thickness is still under research but, anyway, it is straightforward to see that the annealing time needed to achieve full epitaxial crystallization is directly proportional to the thickness of the crystallizing p^-/p^+ layer. Elapsed times ranging between 24 and 48 h are generally demanded, as we have experimentally obtained so far.

The last step to complete a solar cell following the NICE approach consists on the deposition of suitable rear and front contacts which would allow to characterize the electrical properties of the cell. By means of observing the electrical behavior of the cell it is possible to revise and optimize the different stages. At a laboratory level, it is enough to deposit simple contacts to have a first insight on the efficiency (η), the open-circuit voltage (V_{oc}), the short-circuit current (I_{sc}) and the fill factor (FF) of the I–V curve which are the main parameters to check together with the spectral response or external quantum efficiency, in order to analyze whether the solar cell is acceptable or not.

6.2. Alternative p^+/p^- Seed Layer Approach

One possible alternative to the n^+ seed layer approach is to start the solar cell backwards, maintaining the essence of the NICE method by combining NIC and epitaxial growth. This means depositing first a combined p^+/p^- amorphous thin layer on some metallic or conducting substrate (e.g. stainless-steel) and crystallize it to use as the seed layer. If the thickness of these structures is something like 80/320 nm, respectively, it has been shown that, when deposited on glass, they can be crystallized by NIC obtaining large grain sizes [43]. One could ask why not crystallizing a single p^+ layer to use it as a seed, but the answer is similar to the case of heavy phosphorous doping. Since boron is known to make Ni atoms to behave as more mobile interstitials, the nucleation of $NiSi_2$ is promoted and the final grain size decreases as shown in Refs. [30] and [43].

A subsequent deposition and epitaxial crystallization of another p^- layer, with a thickness of approximately 1 µm, would complete the absorber layer. Before this step it should be taken into account that the remaining Ni must be removed in order to obtain expitaxial crystallization [43], as mentioned before. The remaining and final step would be to deposit an amorphous n^+ thin layer (it suffices with a thickness of 80 nm) that does not need to be crystallized to act as the emitter.

This p^+/p^- seed layer approach is a readily-available alternative, although not so convenient from the point of view of cell design. First, it has the complication of incorporating a front glass for protection of the device and, second, the metallic substrate has detrimental effects on crystallization and final grain sizes. If the n^+ seed layer approach is used instead, the glass substrate is already the frontal protection of the device.

Up to date we are at the stage of investigating the epitaxial crystallization over the previously crystallized and externally doped seed layer which will act as the emitter. This led to a recently accepted article [53]. Thus, further investigation is still to be developed. The results obtained so far are promising for the achievement of a low-cost thin film poly-Si solar cell with large grains and efficiencies as high as 15%.

CONCLUSION

Summarizing, electricity generation by means of solar energy conversion is becoming an extremely important member inside the worldwide energetic matrix. Research is being intensively directed to cost reduction and several materials have appeared in scene, each of them having different benefits, characteristics and drawbacks. Copper indium gallium diselenide (CIGS) thin film solar cells reached the top-record efficiency of 20% for single-junction devices, pertaining to the *second* generation. The main problem of this technology is the utilization of contaminating and not-so-abundant materials. This turns poly-Si in the form of thin films to be the most promising material at mid- to long-term, with module efficiencies to date in the order of 10% (CSG Solar). However, *third generation* solar cells are emerging as good companions for the long-term perspective, elevating the efficiencies to above the c-Si limit of 31–41% and further reducing production costs.

Among the methods to produce poly-Si at low costs, NIC is the one that produces the largest crystalline grains, with the lowest amount of aggregated metal, after a SPC process. The metal contamination is thus not drastic in this case and can be easily removed by a suitable etching procedure. Grains larger than 100 μm have been reported, for nickel amounts below 1×10^{15} at./cm^2, making NIC a good candidate for solar cells applications due to the known fact that larger grains improve efficiency. The epitaxial thickening of a poly-Si seed layer is a flexible process that offers the possibility of decoupling the obtainment of the seed layer from the crystallization of a thicker layer. The main goal of the seed layer is to provide good crystallographic quality to the crystallizing layer, but it may also be tuned in terms of doping or contamination in order to serve itself as an active layer of the device. Therefore, the combination of NIC and solid phase epitaxial crystallization provides a viable way by which poly-Si thin film solar cells can be obtained, as proposed in this chapter. Direct epitaxial growth over a seed layer crystallized by NIC, and properly tuned in terms of doping, could also be obtained by PECVD or other related techniques such as hot-wire CVD (HWCVD), ion-assisted deposition or thermal CVD. The latter requires higher temperatures and, hence, substrates tolerating generally more than 1000 °C are needed. Taking into account the resulting relatively large grain size of the seed layer obtained by NIC, efficiencies above 10% are expected for solar cells developed by these methods, although this is still under research.

The key for success of the NICE method, particularly considering the n^+ seed layer approach described in Subsection 6.1, lays on the quality of the transference of crystalline and structural information from the seed layer to the crystallizing upper layers. This depends strictly on the state of the seed layer surface, which at the same time forms the solar cell junction. Thus, the etching procedure for Ni removal, previous to the deposition and epitaxial crystallization, has to be carefully controlled since it interacts with what will be the main part of the cell.

REFERENCES

[1] Green, M. A.; Jianhua, Z.; Wang, A.; Wenham, S. R.; Electron Devices, *IEEE Transactions on Electron Devices* 46 (10) (1999) 1940–1947.
[2] Bauhuis, G. J.; Mulder, P.; Haverkamp, E. J.; Huijben, J. C. C. M.; Schermer, J. J.; *Sol. Energy Mat. Sol. Cells* 93 (2009) 1488–1491.
[3] Green, M. A.; Emery, K.; Hishikawa, Y.; Warta, W.; *Prog. Photovolt.: Res. Appl.* 18 (2010) 346–352.
[4] Green, M. A.; *Sol. Energy* 76 (2004) 3–8.
[5] Ramanathan, K.; Teeter, G.; Keane, J. C.; Noufi, R.; *Thin Solid Films* 480–481 (2005) 499– 502.
[6] Jackson, P.; Würz, R.; Rau, U.; Mattheis, J.; Kurth, M.; Schlötzer, T.; Bilger, G.; Werner, J. H.; *Progr. Photovoltaics Res. Appl.* 15 (6) (2007) 507–519.
[7] Repins, I.; Contreras, M.; Romero, M.; Yan, Y.; Metzger, W.; Li, J.; Johnston, S.; Egaas, B.; DeHart, C.; Scharf, J.; McCandless, B. E.; Noufi, R.; *Photovoltaic Specialists Conference, 2008. PVSC '08. 33rd IEEE*, pp. 1–6, 11–16 May 2008.
[8] Shockley, W.; Queisser, H. J.; *J. Appl. Phys.* 32 (1961) 510.
[9] Green, M. A.; *"Third Generation Photovoltaics: Advanced solar energy conversion", Springer Series in Photonics (2005).*
[10] Gall, S.; Schneider, J.; Klein, J.; Hübener, K.; Muske, M.; Rau, B.; Conrad, E.; Sieber, I.; Petter, K.; Lips, K.; Stöger-Pollach, M.; Schattschneider, P.; Fuhs, W.; *Thin Solid Films* 511–512 (2006) 7–14.
[11] European Photovoltaic Industry Association (2010). Global Market Outlook for Photovoltaic until 2014. www.epia.org.
[12] Beaucarne, G.; *"Silicon Thin-Film Solar Cells", Advances in OptoElectronics, vol. 2007*, Article ID 36970 (2007) 12 pages.
[13] Green, M. A.; Basore, P. A.; Chang, N.; Clugston, D.; Egan, R.; Evans, R.; Hogg, D.; Jarnason, S.; Keevers, M.; Lasswell, P.; O'Sullivan, J.; Schubert, U.; Turner, A.; Wenham, S. R.; Young, T.; *Sol. Energy* 77 (2004) 857–863.
[14] Basore, P. A.; *In Conf. Record 2006 IEEE 4th World Conf. Photovoltaic Energy Conversion, IEEE, New York*, 2006, 2089–2093.
[15] Basore, P. A.; *In Proc. 19th European Photovoltaic Solar Energy Conf., WIP: Munich* (2004) 455–458.
[16] Keevers, M. J.; Young, T. L.; Schubert, U.; Green, M. A.; *In Proc. 22nd European Photovoltaic Solar Energy Conf., WIP: Munich* (2007) 1783–1790.
[17] This values were retrieved from the CSG Solar homepage: http://www.csgsolar.com.

[18] Song, D.; Inns, D.; Straub, A.; Terry, M. L.; Campbell, P.; Aberle, A. G.; *Thin Solid Films* 513 (2006) 356.
[19] Budini, N.; Rinaldi, P. A.; Schmidt, J. A.; Arce, R. D.; Buitrago, R. H.; *Thin Solid Films* 518 (18) (2010) 5349–5354.
[20] Focsa, A.; Gordon, I.; Beaucarne, G.; Tüzün, Ö.; Slaoui, A.; Poortmans, J.; *Thin Solid Films* 516 (20) (2008) 6896–6901.
[21] Carnel, L.; Gordon, I.; Van Gestel, D.; Beaucarne, G.; Poortmans, J.; *Thin Solid Films* 516 (2008) 6839–6843.
[22] Aberle, A. G.; *Thin Solid Films* 511–512 (2006) 26–34.
[23] Wronski, C. R.; Von Roedern, B.; Kolodziej, A.; *Vacuum* 82 (2008) 1145–1150.
[24] Catchpole, K. R.; McCann, M. J.; Weber, K. J.; Blakers, A. W.; *Sol. Energy Mat. Sol. Cells* 68 (2001) 173–215.
[25] Iverson, R. B.; Reif, R.; *J. Appl. Phys.* 62 (1987) 1675–1681.
[26] Matsuyama, T.; Terada, N.; Baba, T.; Sawada, T.; Tsuge, S.; Wakisaka, K.; Tsuda, S.; *J. Non-Cryst. Solids* 198–200 (1996) 940–944.
[27] Shibata, N.; Fukuda, K.; Ohtoshi, H.; Hanna, J.; Oda, S.; Shimizu, I.; *Jpn. J. Appl. Phys.* 26 (1986) L10–L13.
[28] Yoon, S. Y.; Park, S. J.; Kim, K. H.; Jang, J.; *Thin Solid Films* 383 (2001) 34–38.
[29] Dimova-Malinovska, D.; *J. Phys. Conf. Series* 223 (2010) 012013.
[30] Schmidt, J. A.; Budini, N.; Rinaldi, P. A.; Arce, R. D.; Buitrago, R. H.; *J. Cryst. Growth* 311 (2008) 54–58.
[31] Song, N. K.; Kim, M. S.; Kim, Y. S.; Han, S. H.; Joo, S. K.; *J. Korean Phys. Soc.* 51 (3) (2007) 1076–1079.
[32] Nast, O.; Puzzer, T.; Koschier, L. M.; Sproul, A. B.; Wenham, S. R.; *Appl. Phys. Lett.* 73 (1998) 3214-3216.
[33] Gall, S.; Becker, C.; Lee, K. Y.; Sontheimer, T.; Rech, B.; *J. Cryst. Growth* 312 (2010) 1277–1281.
[34] Nast, O.; Brehme, S.; Neuhaus, D. H.; Wenham, S. R.; *IEEE Trans. Electron. Dev.* 46 (1999) 2062–2068.
[35] Zhao, S.; Meng, Z.; Wu, C.; Xiong, S.; Wong, M.; Kwok, H. S.; *J. Mater. Sci.: Mater. Electron.* 18 (2007) S117–S121.
[36] Zhang, B.; Meng, Z.; Zhao, S.; Wong, M.; Kwok, H. S.; *IEEE Trans. Electr. Dev.* 54 (2007) 1244–1248.
[37] Lee, S. W.; Joo, S. K.; *IEEE Electron Device Lett.* 17 (1996) 160–162.
[38] Ma, T.; Wong, M.; *J. Appl. Phys.* 91 (2002) 1236–1241.
[39] Miyasaka, M.; Makihira, K.; Asano, T.; Polychroniadis, E.; Stoemenos, J.; *Appl. Phys. Lett.* 80 (2002) 944–946.
[40] Hayzelden, C.; Batstone, J. L.; *J. Appl. Phys.* 73 (1993) 8279.
[41] Kim, K. H.; Oh, J. H.; Kim, E. H.; Jang, J.; *J. Vac. Sci. Technol.* A 22 (2004) 2469.
[42] Schmidt, J. A.; Budini, N.; Rinaldi, P. A.; Arce, R. D.; Buitrago, R. H.; *J. Phys. Conf. Ser.* 167 (2009) 012046.
[43] Schmidt, J. A.; Budini, N.; Arce, R. D.; Buitrago, R. H.; *Phys. Status Solidi* C 7 (3-4) (2010) 600–603.
[44] Ishida, Y.; Nakagawa, G.; Asano, T.; *Jpn. J. Appl. Phys.* 46 (2007) 6437.
[45] Lee, J. S.; Kim, M. S.; Kim, D.; Kim, Y. M.; Moon, J.; Joo, S. K.; *Appl. Phys. Lett.* 94 (2009) 122105.

[46] Jeske, M.; Schultze, J. W.; Thönissen, M.; Münder, H.; *Thin Solid Films* 255 (1995) 63–66.
[47] Liu, Y. M.; Sung, Y.; Pu, N. W.; Chou, Y. H.; Yeh, K. C.; Ger, M. D.; *Thin Solid Films* 517 (2008) 727–730.
[48] Granqvist, C. G.; Sol. *Energy Mat. Sol. Cells* 91 (2007) 1529–1598.
[49] Bourdais, S.; Beaucarne, G.; Slaoui, A.; Poortmans, J.; Semmache, B.; Dubois, C.; *Sol. Energy Mat. Sol. Cells* 65 (2001) 487.
[50] Liu, Z.; Takato, H.; Togashi, C.; Sakata, I.; *Proceedings 24th European Photovoltaic Solar Energy Conference* (2009) 1971.
[51] Bentzen, A.; Svensson, B. G.; Marstein, E. S.; Holt, A.; *Sol. Energy Mat. Sol. Cells* 90 (2006) 3193.
[52] Tüzün, Ö.; Slaoui, A.; Roques, S.; Focsa, A.; Jomard, F.; Ballutaud, D.; *Thin Solid Films* 517 (2009) 6358.
[53] Budini, N.; Buitrago, R. H.; Schmidt, J. A.; Risso, G.; Rinaldi, P. A.; Arce, R. D.; *Proceedings 25th European Photovoltaic Solar Energy Conference* (2010) *In-Press*.

In: Advances in Materials Science Research, Volume 7
Editor: Maryann C. Wythers
ISBN 978-1-61209-821-0
© 2012 Nova Science Publishers, Inc.

Chapter 8

CATALYTIC EFFECT OF TRANSITION METAL OXIDES NANOPARTICLES IN COMPOSITE PROPELLANTS

José Luis de la Fuente[*]
Instituto Nacional de Técnica Aeroespacial "Esteban Terradas", INTA,
Ctra. de Ajalvir, Torrejón de Ardoz. Madrid. Spain

ABSTRACT

The use of powder additives to modify the burning rate of solid propellant and other energetic materials has been the topic of much study for several decades. This paper is a brief review to describe the effects on the performances of ammonium perchlorate (AP) based composite propellants by mean of employing nano-powder of transition metal oxides (TMOs), as for example iron (III) oxide (Fe_2O_3), cupric oxide (CuO) and copper chromite ($CuCr_2O_4$), as burning rate catalysts in comparison to traditional micro-fillers. The solid composite propellants with microparticles as catalyst seem to be less stable due to oversensitivity to pressure variations, but the nanostructured composite propellants yield high stable burning rates over a broad pressure range. In addition, the incorporation of these nanoparticles in the formulations of these energetic materials can also improve their combustion and thermal properties, according to the characterization obtained by thermal analysis, with techniques such as differential scanning calorimetry (DSC) and thermogravimetry analysis (TGA). These results indicate the excellent benefits found in using these nanoparticles as an additive for solid rocket propulsion applications.

Keywords: nanoparticles, nanocomposites, AP based composite propellants, combustion catalysis, transition metal oxides

---

[*] Instituto Nacional de Técnica Aeroespacial "Esteban Terradas", INTA, Ctra. de Ajalvir, Km 4. 28850 Torrejón de Ardoz. Madrid. Spain. Fax: 0034 91 5206611; E-mail Address: fuentegj@inta.es.

1. INTRODUCTION

Nanostuctured materials are distinguished from conventional ones by the size of structural units that compose them, and they often exhibit properties that are drastically different from those of conventional materials [1-2]. Important areas of relevance for nanoparticles and nanotechnology are advanced materials, electronics, biotechnology, pharmaceutics, and sensors. One of the potential applications for materials with such fine grain sizes is in improving the combustion characteristics of energetic materials [3], which are substances that store energy chemically, and are typically categorized as explosives, pyrotechnics and propellants [4-10].

As it is well known, composite propellants are highly particle-filled elastomers used in solid rocket motors (SRMs) technology, both in launch vehicle and missile applications. Mission requirements often place a demand for very high thrust on SRMs. This is provided for, either by increasing the burning rate surface, or by convoluting the regressing surface or by increasing the burning rate of the propellant. If the burning surface area is increased, the volumetric loading of the propellant decreases. Hence, whenever possible, it would be judicious to increase the burning rate of the propellant. High burning rate composite propellants are needed for future programmes; particularly in surface to air missions to reduce the operational time of the missiles.

The composite propellants are made by embedding a finely divided solid oxidizing agent, generally micrometric ammonium perchlorate (AP) in a plastic, resinous, or elastomeric matrix. The matrix material usually provides the fuel for the combustion reaction, although solid reducing agents are frequently included in the compositions, such as aluminum. Composite solid propellants based on hydroxyl-terminated polybutadiene (HTPB) have become the workhorse propellants in the present-day SRMs world-wide. The urethane network obtained by curing HTPB with a suitable diisocyanate (curative) provides a matrix (binder) for inorganic oxidizer and metallic fuel which are dispersed in the propellant grain. Others ingredients (additives) are plasticizers, bonding agents, stabilizers and burn rate modifiers. Additives are normally added at a small percent of the propellant binder. Their function is to modify the propellant properties in order to improve them, except specific impulsive, which is often decreased (not more than 1%-wt, in weight) due to their low energy content [11, 12].

From the expressed above, it is important to note that the burning rate is a relevant parameter regarding the performance of composite solid propellants and the ballistics of the rocket design [13, 14]. A number of ways are suggested in the literature to modify the burning rate of aluminized composite propellants. These can be broadly classified into the following types:

1) Reducing the AP particles size or increasing the fine to coarse ratio.
2) Adding ultra fine aluminum particles.
3) Adding a burn rate modifier.

In the first method, the burn rate is increased by decreasing the NH_4ClO_4 particles size making the propellant combustion more premixed, and consequently having a higher burning rate pressure index, and one can find a good review in Kubota [4]. A well-known way to increase the reactivity of powders is to decrease their particle size.

The addition of aluminum (normally its content will vary between 5 and 25%-wt with 15%-wt being common, and the typical diameter of particles used in propellants is in the order of ~ 30 µm) generally decreases the burn rate of composite propellants. However, significant increases in propellant burning rates, shorter ignition delays and shorter agglomerates burning times were obtained for composite solid propellant formulation containing ultra-fine Al, and a lot of work in this area has already been reported [15-27]. With the use of ultrafine aluminum in composite propellants, the burn rates were observed to increased by around 100% relative to the propellants prepared with micron sized aluminum, but poor mechanical properties of the propellant and a higher content of condensed Al_2O_3 in the combustion residues, which is responsible for the specific impulse reduction, at present restrict the use of nanosized Al propellants and suggest further investigation.

In the last approach, additives for altering burning rate of propellant (the so-called burning rate catalyst) are always of great interest in order to improve the ballistics of SRMs. The best additives are the ones which can be employed in small quantities (~ 3%-wt), in order to increase/decrease the burning rate, to decrease the sensitivity of burning rate to pressure or temperature in a controlled way. Burning rate catalysts are normally added in a small amount, and are expected to modify only the propellant burning rate, with minor effect on processing, mechanical properties and energetic performances. Transition metals oxides (TMOs) and their mixture are most effective, and are commonly used as burning rate catalyst in practical applications of composite solid propellants. These oxides have great influence on thermal decomposition behaviour of oxidizers. The burning rate modifiers catalyze the decomposition of AP and its smaller particle size enhances the catalysis in the gaseous phase of the combustion. Much work has been done to study the catalyst effect of these inorganic salts on the thermal decomposition of the NH_4ClO_4 and also on the combustion of the AP based propellants [28-32].

The most frequently employed TMOs as burning rate catalysts are Fe_2O_3, CuO, Cr_2O_3, Co_2O_3, etc, and also mixed metal oxides, as for example copper chromite ($CuO \cdot Cr_2O_3$) and the much less known and recently described as copper ferrite ($CuFe_2O_4$) [33] and/or perovskite-type oxides $LaMO_3$ (M = Fe, Co, Ni) [34]. A comprehensive survey of the literature on burning rate catalysts suggests that their effect depends on catalyst concentration, surface area of added catalyst, particle size and state of aggregation [28, 29]. The optimum concentration of the catalyst produces effective results. High burning rate composite propellants are required, particularly in those missions where a reduced operational time is needed, as it was mentioned before. However, besides a high burning rate insensitive to pressure variations is preferable. Typical propellant formulations that achieved a high burning rate by increasing the value of the pressure exponent are unstable, due to their oversensitivity to pressure variations, which can lead to catastrophic failure. The latter is traditionally difficult to achieve over a wide pressure range, and to date, the ability of certain additives to modify the burning rate of propellants has met with mixed success [15].

One of the most important investigation areas in the composite propellants is centered in the use of nano-fillers in their formulations, and more specifically the incorporation of metal nanoparticules, such as aluminum, as previously commented. However, the works where nano-additives, as for example burning rate catalysts, are used are scarce [30-43]. Therefore, the main objective of this work is to describe the use of nano-TMOs as combustion additives.

The literature was reviewed in order to discover the current state of knowledge of these nano-sized energetic materials of application in SRMs technology. The effect of these nanoparticles on the burning rate behaviour of composite propellant will be analyzed, showing the recent advances in this exciting and rapidly growing area of research interest.

2. NANO-TMOs AS CATALYST BURN RATE

Metal oxides particles ranging in size from nanometers to micrometers and possessing unique electronic, and optical properties are of great interest for a variety of applications. In particular, their performance in catalysis and photocatalysis, which is primarily determined by their physical and chemical characteristics such as composition, crystallinity, morphology, and surface area, has received much attention in the last few decades. [44]

Logically, all these characteristics of the TMOs powders, which are usually studied by methods of scanning and transmission electro microscopy (SEM and TEM), X-ray photoelectron spectroscopy, X-ray diffraction analysis and Brunauer-Emmett-Teller (BET) technique, will be conditioned by the synthetic method employed in their preparation. Thus there are several methods for the synthesis of ordinary nanosized particles, including physical vapour deposition, chemical vapour deposition, reactive precipitation, micro-emulsion, supercritical chemical processing, hydrothermal synthesis, phase shift reaction, sol-gel method, electrochemistry synthesis, etc. Among these methods, perhaps reactive precipitation is the most popular, and this is attributed to its ease of operation, low cost, and massive process.

Specifically, the properties of nano-TMOs as burning rate catalysts in composite propellants are a very less explored research area. In addition, and in many occasions these works are more concentrated on the study of their catalytic activity in the thermal decomposition of AP, performed through thermal analysis (with analytical techniques such as differential scanning calorimetry (DSC) and thermogravimetry analysis (TGA)), than their real catalytic effect on the burning rate of AP based composite propellants. This section will focus principally on those nano-TMOs more commonly used in the application under study, Fe_2O_3, CuO and $CuCr_2O_4$. Also, it is important to note that though TiO_2 has not habitually been used as burning rate catalyst on this energetic material (previous studies using traditional micron powders found that titanium dioxide had practically no effect on AP thermal decomposition), the recent relevant results published concerning the strong catalytic activity of its nanoparticles towards this reaction has motivated the inclusion of this novel nano-additive.

2.1. Nano Fe_2O_3

Perhaps the most commonly used TMO as burning rate modifier in practical applications of composite propellants is Fe_2O_3. This may be because iron (III) oxide is known to be structurally simple, highly stable, easy to synthesize and inert to side reactions.

Joshi's group described for the first time the effect of nano ferric oxide particles on the thermal decomposition behaviour of an AP/HTPB composite propellant (85% of AP and

15%-wt of binder) through a comparative study in the presence and absence of this common burning rate additive [30]. The synthesis of α-Fe$_2$O$_3$ nanoparticles was carried out by an electrochemical route, and the propellant formulation was made using a 2%-wt of the synthesized 3.5 nm sized nano iron (III) oxide, and compared with the same propellant without a catalyst. The experimental results found by mean of DSC measurements (at different heating rates in a dynamic nitrogen atmosphere) demonstrated a lowering of the high temperature decomposition by 49 °C. A higher heat release up to 40% was also observed in the presence of nano α-Fe$_2$O$_3$. The kinetic combustion parameters were evaluated using the well-known Kissinger method, and the increase of the rate constant in the catalyzed propellant confirmed the enhancement of the catalytic activity of AP. The scanning electron micrographs of the nano Fe$_2$O$_3$ incorporated in HTPB revealed a well-separated characteristic necklace like structure of the α-Fe$_2$O$_3$ particles at high magnification.

Similar conclusions were achieved in different works where the particles size of the ferric oxides used in these studies varied from nanometer size to submicron size [34, 35]. Thus, for example Liu et al. [34] described the effect of the iron (III) oxide nanoparticles with four different morphologies such as spherical, cubic, spindle and acicular shape, prepared by the sol-gel methods, hydrothermal reaction and forced hydrolysis, respectively, on the thermal decomposition of AP. The spindle and acicular shape Fe$_2$O$_3$ nanoparticles presented bigger specific surface area, so they had better catalytic effect than those of the cubic and spherical shaped ones. The acicular shape Fe$_2$O$_3$ nanoparticles with maximum specific surface area could decrease the higher thermal decomposition of AP by 67 °C, and in fact increased the exothermic heat by 785 J/g, showing good catalytic effect.

2.2. Nano CuO and Cu-Cr-O Nanocomposites

Among the different TMOs, the CuO has been studied as a unique and attractive monoxide material, due to its both fundamental investigations and practical applications [45, 46]. CuO is a p-type metal oxide semiconductor with narrow band-gap, and exhibiting versatile range of applications. On the other hand, the Cu-Cr-O composite oxides have also long been recognized as versatile functional materials due to their wide commercial applications as catalysts for several chemical reactions between them as a burning rate catalyst for solid propellants.

Joshi and co-worker have also focussed their researches on the effect of nano copper (II) oxide and cooper chromite on the thermal decomposition reaction of AP [31]. Their DSC study revealed the nanosized p-type CuO (5.2 nm) and CuCr$_2$O$_4$ (4.0 nm), synthesized by an electrochemical method, have the best catalytic effect on the high temperature exothermic thermal decomposition peak of AP, compared with commercial microscale CuO and CuCr$_2$O$_4$. AP was mixed with the synthesized nano oxides in the ratio of 2 and 1%-wt. The nano copper chromite showed the best catalytic effects as compared to nanometric cupric oxide in lowering the high temperature decomposition by 118 °C at 2wt-%. High heat releases of 5.430 and 3.921 kJ/g were observed in the presence of nano CuO and CuCr$_2$O$_4$, respectively. The kinetic parameters were evaluated using the Kissinger method. The decrease in the activation energy and the increase in the rate were constant for both oxides and confirmed the enhancement in catalytic activity of AP. The following mechanism based on an electron transfer has also been proposed for AP in the presence of these p-type semiconducting oxides like CuO, according with the reports of Freeman and Anderson [47], based on electron transfer from the perchlorate ion to a positive hole (formed at high temperature) of the oxide along with the abstraction of atomic oxygen from the perchlorate ion.

$$e^-_{oxide} + ClO_4^- \rightarrow O_{oxide} + ClO_3^- \rightarrow \tfrac{1}{2} O_2 + ClO_3^- + e^-_{oxide}$$

where e^-_{oxide} represents a positive hole in the valence band of the oxide and O_{oxide} is an oxygen atom abstracted from the oxide. These positive holes and the corresponding active sites in CuO and $CuCr_2O_4$ are highly associated with the adsorption of gaseous products and electron exchange.

Other different synthetic methodologies were applied by Heng et al. [36] and Xu et al. [37] in order to prepare CuO nanopowders with distinct morphologies. However, the results obtained in both works based on AP thermal decomposition behaviour, by means of DSC analysis, were very similar to those commented on previously.

Li and Cheng presented the synthesis of Cu-Cr-O nanocomposites via a citric acid complexing approach, and their evaluation as additives for the catalytic combustion of a typical AP based composite propellant, formulated with a 5%-wt of catalyst [38]. The propellant's burning rates and the pressure exponents were measured and compared for different samples prepared. The burning rate was determined by performing burning test in a Crawford bomb between 7 and 18 MPa. The pressure dependence of the burning rate is normally of great interest for describing the combustion performance of these solid energetic materials. This function, in accordance with the so-called Vieille's law, can be expressed as follows:

$$r_b = a P^n \tag{1}$$

where r_b, a, P, and n are denoted as burning rate, pre-exponential factor, pressure, and pressure exponent, respectively. For a typical composite propellant, the exponent in the range 0.25 - 0.60 is usual. In particular, a high value of the burning rate is favorable for a high power, and a small value (propellant possessing a zero "plateau burning") of the pressure exponent is aimed for a stable burning at high pressures. The authors reported these important ballistic parameters for the composite propellants containing the different Cu-Cr-O nano-catalysts prepared. For comparison, results of the propellants without using any catalyst were also included. The pressure exponents were deduced from the slope of the $\ln(r_b)$-$\ln(P)$ plots (as that showed in Figure 2) and some of the results found are summarized in Table 1.

In brief, addition of the nano-additives prepared in this work obviously enhances the burning rate as well as lowering the pressure exponent of the AP based composite propellants. Noticeably, the catalyst with Cu/Cr molar ratio of 0.7 exhibits the most promising catalytic activity, high burning rate and low pressure exponent (0.43) at all pressures under study.

Similar conclusions were obtained in the work of Liu et al. where they described the preparation of a novel nanostructured copper-chromium oxide, which consisted of copper-chromium oxide nanoparticles and an inert component, and its catalytic effect on RDX/AP/Al/HTPB propellants [39]. Thus, the burning rate at 6 MPa increased from 6.31 mm/s (without catalyst) to 8.82 mm/s with this novel nano-additive (0.5%-wt). However, in this case the pressure exponent increased from 0.35 (without catalyst) to 0.38 in the presence of this catalyst in the pressure range 4-10 MPa.

Table 1. Burning rate parameters of the AP based composite propellant with different burning rate catalysis, Cu-Cr-O composite oxides [38], CuO [40] and TiO$_2$ [41]

Catalyst/Sample	Grain size (nm)	Fitting burning equation $\ln(r_b /\text{mm·s}^{-1}) = \ln a + n \ln (P/\text{MPa})$	
		ln a	n
Cu-Cr-O/(Cu/Cr = 0.3)	100	3.00	0.53
Cu-Cr-O/(Cu/Cr = 0.5)	100	3.12	0.55
Cu-Cr-O/(Cu/Cr = 0.7)	100	3.44	0.43
Cu-Cr-O/(Cu/Cr = 1.0)	100	2.66	0.60
Cu-Cr-O/(Cu/Cr = 0.5)	> 10 µm	2.34	0.75
Without catalyst	-	1.91	0.85
CuO/1	42	2.31	0.37
CuO/2	34 µm	1.37	0.51
CuO/3	11 µm	1.28	0.55
TiO$_2$/amorphous	-	1.48	0.31
TiO$_2$/anatase	15	0.69 (pre plateau)[a] 2.11 (plateau)[b]	0.80 (pre plateau)[a] 0.0052 (plateau)[b]
TiO$_2$/rutile	50	1.57	0.27
Without catalyst	-	1.84	0.10

[a] From 3-6 MPa.
[b] From 6-20 MPa.

One recent work performed in our laboratory describes the use of commercial nano CuO particles as an effective burn rate enhancer of aluminized composite propellants based on HTPB [40]. Experiments were also carried out using a strand burner at pressures ranging from 1.5 to 12 MPa. With the addition of nanometric CuO as catalyst, the burning rate of the propellant greatly enhances at all pressures under study. Therefore, for example for a pressure of 7 MPa, the burning rate of the nano-catalyzed energetic composite materials is double the burning rate found in those propellants which incorporate conventional micro CuO catalysts, which different morphologies can be seen by mean of the SEM micrographs shown in Figure 1.

At the same time the pressure exponent is reduced, as can be observed from Figure 2, which shows the combustion pressure dependence of the burring rate for the composite propellant containing (2%-wt), the nano-additive (designed as 1), and two different micro-additives (designed as 2 and 3). A n value of 0.37 is obtained for the nanocomposite, while that an exponent of around 0.53 is exhibited by the microcomposite, so a decrement of 30% is found in this relevant combustion parameter (see Table 1). This produces a more stable combustion in this nano energetic material at the pressure range under analysis. It is important to note that typical fast-burning propellants are habitually unstable due to oversensitivity to pressure variations, but this nano CuO yields a propellant with high, yet stable burning rates over a broad pressure range. On the other hand, it also indicates that the composite solid propellants including CuO particles in the micrometric range do not exhibit appreciable difference between them.

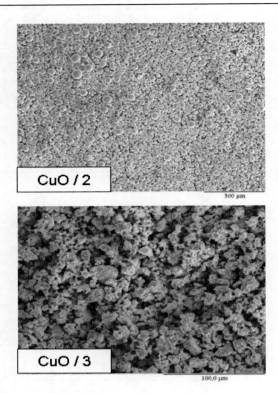

Figure 1. SEM micrographs of the different micrometric CuO powders employed in the preparation of the AP based composite propellants [40].

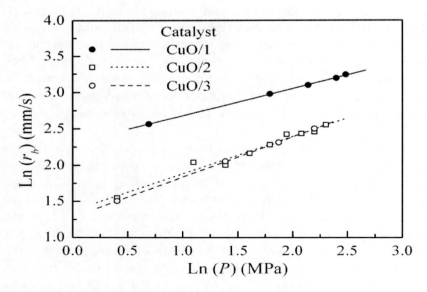

Figure 2. Burning rate versus pressure for the AP based composite solid propellants with incorporating different CuO powders [40].

The thermal behavior of these energetic composites was also investigated using different techniques such as TGA and DSC. From the TGA curves it is observed that for all the samples, the weight loss takes place in one-step and is 100% in the temperature range of 300-325 °C (as shown in Figure 5a) in reference [40] and the initial temperature decomposition and maximum rate temperature of weight loss (T_{max}) were determined from these curves. The decomposition of all these energetic materials is produced in a narrow single step, as illustrated in Figure 5b) in reference [40], where the differential termogravimetric (DTG) curves are represented, obtained from the original TGA. As can be seen in its DTG curves, T_{max}s increases in the same manner that the particle size of the CuO incorporated in the formulations of the composite materials, whose degradation develops homogeneously. It is important to note that the sample with the nano-catalyst gives logically a lower T_{max} (around 305 °C) with a narrower thermal degradation peak. The activation energy (E_a) of the thermal degradation for the different energetic composites was calculated from the thermal degradation curves, using the Horowitz and Metzger method [48]. The E_as for the thermal decomposition of these materials increase as the particle size of the CuO employed increases. This fact indicates again that the introduction of a nano-catalyst in the structure of these materials gives a higher combustion performance. The results extracted from the thermogravimetric analysis were similar to those achieved across the DSC measurements, finding an intense exothermic peak, corresponding to the thermal decomposition of the composite samples under studied, given an average enthalpy value of - 5.4 kJ/g.

2.3. Nano TiO$_2$

In 1972, Fujishima and Honda demonstrated the splitting of water on a TiO$_2$ electrode [49], and in the years since, titania, a material traditionally remarkable for its use in pigments, has become one of the most intensely studied materials for catalytic applications. Focus has shifted more recently to nanostructured titania, which exhibits remarkable properties, due to the combined phenomena of high surface area and quantum size effects [50]. While much of the current literature concentrates on its photoinduced activity, nanoscale titania is finding use in other new areas as well, for example, as an additive for improving the burning characteristics of solid rocket propellants [15].

A pioneer and complete work about the use of nanoparticles of this TMO as a burning rate catalyst is that of Seal and co-workers [41]. These authors investigated the effect of anatase, rutile, and amorphous TiO$_2$ nanoparticles on the combustion of nonmetallized propellants (AP/HTPB). Amorphous titanium oxide nanoparticles were prepared by a typical sol-gel method using acetylacetone as a stabilizing agent. Subsequent annealing at 250, 400 and 800 °C produced amorphous, anatase, and rutile powders, respectively. Each additive increased the burning rate of propellant strands by 30%, but the anatase structure yielded propellants with high, yet stable burning rate over a broad pressure range. However, it is important to note that the propellants containing anatase exhibited two distinct regions: from 3-6 MPa, the burning rates increased exponentially with pressure (pre plateau region), but from 6-20 MPa, the burning rate was constant (plateau region). The combustion parameters calculated from the combustion kinetic measurements are collected in Table 1. The plateau observed with the anatase additive may indicate that a surface reaction becomes the rate-limiting step at pressure above 6 MPa, while at lower pressures; the reaction is diffusion-limited. This simplified mechanism agrees with the DSC/TGA results showing the catalytic oxidation/reduction of gaseous AP decomposition products on the anatase nanoparticles.

However, amorphous and rutile additives, increasing the propellant burning rates, but did not catalyze AP thermal decomposition. From this result, it is easy to think that other mechanisms are certainly involved in the complex reactions of solid propellant combustion, and at the same time emphasizes the importance of control over nanostructure in the new generations of nano-additives for tailoring energetic reactions.

3. CONCLUSION

In the wake of a great expansion of interest in the design and synthesis of materials with tailored chemical and physical properties, materials scientists have focused much attention on controlled composite materials, with the goal being to create systems in which there is functional cooperation between the overall nanoscale architecture of material and its physical properties. The nanocomposite propellants, used for solid rocket propulsion applications, here described are a clear example in this sense.

From the contents of this review, it is also clear that the area of these energetic nanomaterials has expanded in the last several years, and that the real-work applications of them may not be too far off. In addition, this review demonstrated the feasibility of TMOs nano-powder as combustion rate catalyst to enhance the performances of AP based composite propellants. The ballistic characterization of these materials with differently size TMOs powders, as for example CuO and $CuCr_2O_4$, in terms of steady burning rates showed how the performances of using nano-additives are better compared to traditional ones, yielding higher burning rates in a broad pressure range and a lower pressure exponent. Also, the thermal analysis studies indicated that the incorporation of nano-TMOs particles reduces the decomposition (ignition) temperature of these composite materials giving a more homogeneous, stable and consequently also safer combustion. The use of nano-additives provides a new approach for improving the thermal and combustion characteristics of this highly energetic material, establishing a better comprehension of the complex interactions among the nano-TMOs powders and all other composite solid propellant ingredients.

Since this area of nanocomposite propellants is still in its infancy, it is clear that we are only witnessing the beginning of what has already developed into one of the most active an exciting area of energetic materials science.

REFERENCES

[1] Suryanarayana, C. *Adv. Eng. Mater.* 2005, 7, 983.
[2] Gleiter H. *Acta Mater.* 2000, 48, 1.
[3] Iyer S.; Slagg, N. *Adv. Mater.* 1990, 2, 174.
[4] Kubota N. *Propellant and Explosives, Thermochemical Aspects of Combustion*; Wiley-VCH GmbH: Weinheim, GE, 2002.
[5] Teipel, U.; Ed.; *Energetic Materials, Particle Processing and Characterization,* Wiley-VCH, Weinheim, GE, 2005.
[6] Son S. F.; Yetter R. A.; Yang V. *J. Propul. Power* 2007, 23, 643.

[7] Armstrong, R. W.; Thadhani, N. N.; Wilson, W. H.; Gilman, J.; Simpson, R.; Eds.; *Synthesis, Characterization and Properties of Energetic/Reactive Nanomaterials*, Mater. Res. Soc. Sym. Proc. 800, Warrendale, PA, 2003.
[8] Thadhani, N. N.; Armstrong, R. W.; Gash, A. E.; Wilson, W. H.; Eds.; *Multifunctional Energetic Materials*, Mater. Res. Soc. Sym. Proc. 896, Warrendale, PA, 2006.
[9] Dlott, D. D. *Mater. Sci. Technol.* 2006, 22, 463.
[10] Tillotson, T. M.; Gash, A. E.; Simpson, R. L.; Hrubesh, L. W.; Satcher, J. H.; Poco, J. F. *J. Non-Cryst. Solids* 2001, 285, 338.
[11] Boyars, C.; Klager, K.; Eds.; *Propellants, Manufacture, Hazards, and Testing*, America Chemical Society, Washington DC, 1969.
[12] Kuo, K. K.; Eds.; *Challenges in Propellants and Combustion 100 Years After Nobel*, Begell House, NY, 1997.
[13] Davenas, A.; Ed.; *Solid Rocket Propulsion Technology*, Pergamon Press, Oxford, UK, 1993.
[14] Sutton, G. P.; Biblarz, O. *Rocket Propulsion Elements*, Wiley and Sons: NY, 2001.
[15] Brill, T. B.; B. Budenz, B. T. *Solid Propellant Chemistry, Combustion, and Motor Interior Ballistics* AIAA: Reston, VA, 2000.
[16] Diaz, E. ; Brousseau, P. ; Ampleman, G. ; Emery Prud'homme, R. *Propellants Explos. Pyrotech.* 2003, 28, 210.
[17] Jayaraman, K.; Anand, K. V.; Chakravarthy, S. R.; Sarathi, R. *AIAA Paper* 2007- 1430.
[18] Gokalp, I.; Bocanegra, P. E.; Chauveau, C. *Aerosp. Sci. Technol.* 2007, 11, 33.
[19] Galfetti, L.; De Luca, L. T.; Severini, F.; Colombo, G.; Meda, L.; Marra, G. *Aerosp. Sci. Technol.* 2007, 11, 26.
[20] Baschung, B. in: *Multifunctional Energetic Materials*, Thadhani, N. N.; Armstrong, R. W.; Gash, A. E.; Wilson, W. H.; Eds.; Mater. Res. Soc. Sym. Proc. 896, Warrendale, PA, 2006, p 111.
[21] Trunov, M A; Umbrajkar, S.; Schoenitz, M.; Mang J. T.; Dreizin, E. L. in: *Multifunctional Energetic Materials*, Thadhani, N. N.; Armstrong, R. W.; Gash, A. E.; Wilson, W. H.; Eds.; Mater. Res. Soc. Sym. Proc. 896, Warrendale, PA, 2006, p 171. .
[22] Galfetti, L.; De Luca L. T.; Severini, F.; Meda, L.; Marra, G.; Marchetti, M.; Regi M.; Bellucci, S. *J. Phys.-Condens. Mater.* 2006, 18, S1991.
[23] Risha G. A.; Boyer, E.; Evans, B.; Kuo, K. K.; Malek, R. in: *Synthesis, Characterization and Properties of Energetic/Reactive Nanomaterials*, Armstrong, R. W.; Thadhani, N. N.; Wilson, W. H.; Gilman, J.; Simpson, R.; Eds.; Mater. Res. Soc. Sym. Proc. 800, Warrendale, PA, 2003, p. 243.
[24] Meda, L.; Marra, G.; Galfetti, L.; Severini F.; De Luca, L. *Mater. Sci. Eng. C-Biomimetic Supramol. Syst.* 2007, 27, 1393.
[25] De Luca, L. T.; Galfetti, L.; Severini, F.; Meda, L.; Marra, G.; Vorozhtsov, A. B.; Sedoi, V. S.; Babuk, V. A. *Combust. Explos.* 2005, 41, 681.
[26] Meda, L.; Marra, G.; Galfetti, L.; Inchingalo, S.; Severini, F.; De Luca, L. *Compos. Sci. Technol.* 2005, 65, 769.
[27] Armstrong, R. W.; Baschung, B.; Booth D. W.; Samirant, M. *Nano Lett.* 2003, 3, 253.
[28] Ma, Z.; Li, F.; Bai, H. *Propellants Explos. Pyrotech.* 2006, 31, 447.
[29] Kawamoto, A. M.; Pardini, L. C.; Rezende, L. C. *Aerosp. Sci. Technol.* 2004, 8, 591.
[30] Patil, P. R.; Krishnamurthy, V. N.; Joshi, S. S. *Propellants Explos. Pyrotech.* 2006, 31, 442 and references herein.

[31] Patil, P. R.; Krishnamurthy, V. N.; Joshi S. S. *Propellants Explos. Pyrotech.* 2008, 33, 266.
[32] Liu, T.; Wang, L. Yang, P.; Hu, B. *Mater. Lett.* 2008, 62, 4056.
[33] Wang, Y. ; Yang, X.; Lu, L.; Wang, X. *Thermochim. Acta* 2006, 443, 225.
[34] Liu, J. X.; Li, F. S.; Chen, A. S.; Yang, Y. L. ; Ma, Z. Y. *J. Propul. Technol.* 2006, 27, 381.
[35] Fujimura, K. ; Miyake, A. *Sci. Technol. Energ. Mater.* 2008, 69, 149.
[36] Heng, Q. L.; Xiao, F.; Luo, J. M.; Sun, Q. J.; Wang, J. D.; Su, X. T. *Chin. J. Inor. Chem.* 2009, 25, 359.
[37] Xu, Y.; Chen, D.; Jiao, X.; Xue, K. *Mater. Res. Bull.* 2007, 42, 1723.
[38] Li, W; Cheng, H. *Solid State Sci.* 2007, 9, 750.
[39] Liu, J. M.; Zhou, W. L.; Gong, L.; Xu, F. M.; Zhu, L. X.; Wang, Z. Y.; Zhao, J. *J. Propul. Technol.* 2005, 26, 462.
[40] De la Fuente J. L.; Mosquera, G.; París, R. *J. Nanosci. Nanotechnol.* 2009, 9, 6851.
[41] Reid, D. L.; Russo, A. E.; Carro, R. V.; Stephens, M. A.; LePage, A. R.; Spalding, T. C.; Petersen, E. L.; Seal, S. *Nano Lett.* 2007, 7, 2157.
[42] Stephens, M.; Sammet, T.; Carro, R.; LePage, A.; Reid, D.; Seal S.; Petersen E. *AIAA Paper* 2006-4948.
[43] Kappor, I. P. S.; Srivastava, P.; Singh, G. *Propellants Explos. Pyrotech.* 2009, 34, 351.
[44] Fierro, J. L. G.; Ed.; *Metal Oxides: Chemistry and Applications*; Chemical Industries Series Vol. 108. CRC Press and Francis Group; Boca Raton, FL, 2006.
[45] Cao, M. H.; Hu, C. W.; Wang, Y. H.; Guo, Y. H.; Guo, C. X.; Wang, E. B. *Chem. Commun.* 2003, 15, 1884.
[46] Vaseem, M.; Umar, A.; Kim, S. H.; Hahn, Y. B. *J. Phys. Chem.* 2008, C 112, 5729.
[47] Freeman, E. S.; Anderson, D. A. *Nature* 1965, 206, 378
[48] Horowitz, H. H.; Metzger, G. *Anal. Chem.* 1963, 35, 1464.
[49] Fujishima, A; Honda, K. *Nature* 1972, 283, 37.
[50] Abrams, B. L.; Wilcoson; J. P. *Crit. Rev. Solid State Mater. Sci.* 2005, 30, 153.

Chapter 9

A SIMPLE MODEL FOR SURFACE RELIEF IN THE OCEAN CRUST

A. L. Volynskii[*1] and S. L. Bazhenov[2†]

[1]Faculty of Chemistry, Moscow State University, Vorob'evy gory,
Moscow, Russia

[2]Institute of Synthetic Polymeric Materials, Russian Academy of Sciences,
Moscow, Russia

ABSTRACT

The method for modeling of the development of the surface relief near the mid-ocean ridges is developed. Structural and mechanical behavior of polymer films with a thin rigid coating is analyzed. The behavior of such systems under applied tensile and compressive stress is accompanied by the formation of a regular wavy surface relief and by regular fragmentation of the coating. The above phenomena are universal. Both phenomena (stress-induced development of a regular wavy surface relief and regular fragmentation of the coating) are provided by the specific features of mechanical stress transfer from a compliant soft support to a rigid thin coating. Structural and mechanical properties of the Earth are modeled by a soft polymer substrate coated with a thin rigid film. This method models the development of a system of folds and transform faults near the mid-ocean ridges was modeled. This approach allows estimate the effective thickness of the ocean crust.

INTRODUCTION

Recent years are characterized by an ever-growing interest in the systems composed of a thin rigid coating on a soft polymer substrate [1–6]. A special attention is focused on studying

---

[*] Faculty of Chemistry, Moscow State University, Vorob'evy gory, Moscow, 119899 Russia. Fax number: +7 (095) 9390174; e-mail: volynskii@mail.ru.
[†] Institute of Synthetic Polymeric Materials, Russian Academy of Sciences, Profsoyuznaya ul. 70, Moscow, 117393 Russia; e-mail: Bazhenov@ispm.ru.

both the structural and mechanical aspects of their deformation. Tensile drawing of such two-layer systems is accompanied by two general phenomena. The first is development of a regular wavy microrelief on the initially smooth surface [4–10]. The second is regular fragmentation (disintegration) of a rigid coating on the bands [1–4].

The above systems are called "a rigid coating on a soft substrate". Theoretical expressions describing the period of regular wavy microrelief and average width of fractured fragments were derived. The above relationships look to be universal. Hence, one may assume that they may be applied for description of various "rigid coating on a soft substrate" systems. Therefore, a polymer coated with a thin rigid layer may be used as an accessible and simple model for the description of various natural phenomena. In this work, the above approach is used for modeling of the natural phenomena in the Earth crust.

PHYSICAL BASIS OF THE MODELING METHOD

Deformation of polymer films with thin coating is accompanied by a surface relief formation. Figure 1 shows the SEM images of the coated polymer samples after stretching. Dark and light bands are perpendicular to the direction of stretching. Dark bands correspond to cracks (grooves) in the coating, whereas light bands are the fragments of the fractured coating.

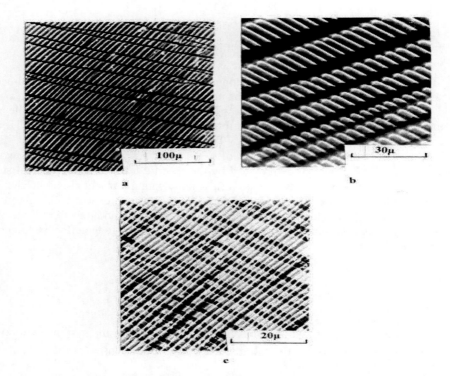

Figure 1. SEM image of the natural rubber coated with thin (15 nm) gold film after stretching at room temperature to strain 50% (a); PVC coated with a thin (21 nm) platinum film after stretching at 90°C to a strain of 20% (b) and PET coated with a thin (15 nm) platinum film after its stretching at 90°C to a strain of 100% (c).

Two regular structures are formed on the surface of the deformed samples. Regular wavy relief and fragments of a thin rigid coating are observed. In Figure 1, dark and light fragments have a regular wavy pattern. Both grooves and rises are oriented along the direction of tensile drawing. The most striking feature is the regularity of the spontaneously formed relief and its alignment along the tensile axis.

The formation of the above structures has a general character and is independent of the nature of the substrate and the coating materials. The surface pattern is practically identical if natural rubber (an elastic polymer), PVC (ductile polymer) or PET (amorphouse polymer) were used as a substrate. The requirements for a formation of the relief are: 1) the coating must be much thinner than the substrate; 2) the Young's modulus (rigidity) of the coating should be much higher than that of the substrate. Such systems are called a rigid coating on a soft substrate.

The formation of the relief presented in Figure 1 is caused by contraction of the polymer substrate in the direction lateral to the stretching axis. The volume of polymers at stretching does not change, and elongation leads to contraction of the substrate in the lateral direction. The elongation of the polymer causes multiple fragmentation of the coating, while its contraction in the transverse direction causes appearance of a regular wavy relief.

MECHANISM OF APPEARANCE OF REGULAR PERIODIC STRUCTURES IN COATED POLYMERS

The mechanisms of appearance of regular surface wave and multiple fragmentation of the coating are different. Fragmentation of coating is caused by tension, while the appearance of regular wave is related with its compression in the lateral direction. On this reason these phenomena are considered separately.

A. Mechanism of Appearance of Regular Surface Wave

Formation of regular wave is explained by buckling of elastic coating at compression. According to Euler, an elastic rod under compression buckles at a critical load as shown in Figure 2. If the elastic plate is bonded to a soft elastic substrate (Figure 2c), the buckled plate (coating) takes the sine shape. The detailed theoretical analysis of this phenomenon gives the following equation for wavelength λ [9]:

$$\lambda = 2\pi \sqrt[6]{\frac{1+\upsilon}{18}} h \sqrt[3]{\frac{E_1}{E}} \tag{1}$$

where υ is the Poisson ratio of polymer substrate, h is coating thickness, E_1 and E are moduli of the coating and the substrate respectively. Equation (1) was checked experimentally, and a satisfactory agreement between the theory and the experimental data was observed [9].

In the following experiments the rubber was stretched and coated with the Pt in the elongated state. After that the stretched sample was unloaded so that it contracted to its initial length. Figure 3 compares surface relieves of the coated rubber samples appearing at stretching (a) and shrinkage (b). The surface relieves are similar. Similarity of the considered surface structures is caused by similar mechanism of surface relief formation. However, there is an important difference between Figures 3a and 3b. The wave at tention is oriented perpendicularly to the deformation axis.

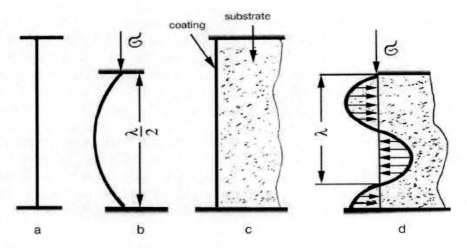

Figure 2. Buckling instability of an isolated elastic beam (a, b) and elastic beam on a soft elastic substrate (c, d).

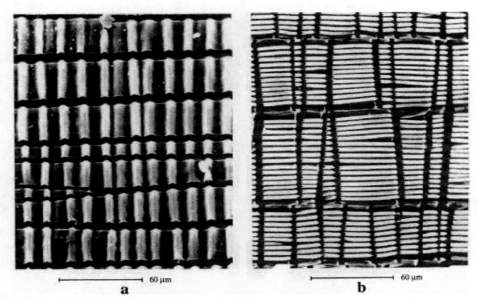

Figure 3. SEM images of rubber samples coated with a thin (15 nm thick) platinum film. (a) – stretching of a coated sample to 50%; (b) – shrinkage of rubber coated with Pt layer. Axis of extension is vertical.

The microrelief at shrinkage is oriented at 90° to that at stretching. This is caused by different direction of the coating compression at shrinkage and stretching. In both cases coating compression is caused by compressive deformation of the polymer substrate. At stretching, the rubber contracts in the lateral direction. As a result, the coating is compressed perpendicularly to the stretching direction. At substrate shrinkage, the direction of compression coincides to the direction of the initial stretching. In other words, directions of the coating compression at tension and at shrinkage are perpendicular. This explains the perpendicular orientation of the waves in Figures 3a and 3b. However, in both cases the regular microrelief appear on the surface of a coated sample.

B. Multiple Fragmentation of Coating at Stretching

The second surface phenomenon is regular multiple fragmentation of a coating. Tensile failure of solids is caused by defects. At tension, sample usually breaks on two parts. However, failure of coating in Figure 1 is quite different. The coating is fragmented on a number of bands of about the same width. To explain the cause of regular fragmentation of the coating at stretching, this process is considered below.

Initially at stretching the width of coating bands is different, and this is explained by an occasional defects in the coating. However, at some moment the mechanism of fragmentation changes. The stress in the coating in a crack is equal to zero. The stress in the fragment increases from its edges to the center (Figure 4). The maximum stress in the center of each fragment increases with stretching. At some moment the coating strength is reached, and the fragment breaks in its center. As a result, the fragment splits into two almost equal halves, and the process repeats with both fragments.

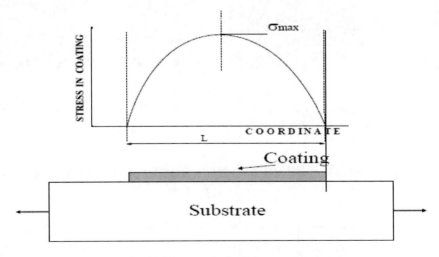

Figure 4. Schematic drawing of a coating fragment adhered to a polymer substrate.

This failure mechanism was observed in experiments. Figure 5 shows two polyethyleneterephtalate (PET) samples coated with platinum and stretched by 100%. The first sample (Figure 5a) was stretched at 100°C by 100%. The coating has broken into several fragments with the width of approximately 5 μm. The second sample (Figure 5b) was stretched at 100°C by 50%. Then the temperature was lowered to 80°C and the sample was stretched again by 50%. The decrease in temperature resulted in an increase in the substrate stress causing further fragmentation of the coating. At 100°C the coating was fragmented on much wider bands. The reduction of temperature to 80°C and further sample stretching caused fragmentation of the initial bands on two halves (b). The fragment width, L, is given by:

$$L = 4h\sigma^*/\sigma_0, \qquad (2)$$

where h is the thickness of the coating, σ^* is its strength, and σ_0 is the stress in the substrate. Thus, the surface wave length and the width of the coating fragments are determined by the properties of the coating and the substrate.

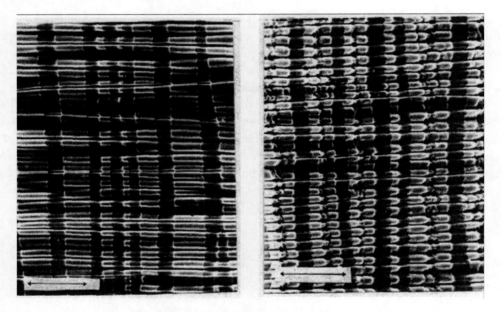

Figure 5. (a) - SEM micrograph of Pt/PET, elongated to 100% at temperature 100°C. (b) Pt/PET elongated to 50% at temperature 100°C, and after that elongated to total strain of 100% at 85°C. The secondary cracks are in the middle of the initial fragment bands.

MODELLING OF GEODYNAMIC PROCESSES BY TWO-LAYER STRUCTURES

We may suppose that appearance of a wave and coating fragmentation are general for any two-layer systems consisting of a soft substrate and thin rigid coating. Hence, we may try to apply the above approach to the Earth. The Earth consists of thick soft mantle (the substrate) and the rigid crust (coating). The movement of the crust is explained by circulated convection currents in the underlaying Earth lower mantle. This causes the continental drift and ocean floor spreading. The continental drift is caused by compression of the ocean crust due to volcanic process in the mid-ocean ridges.

Figure 6 shows the ocean floor relief of the northern part of the Eastern Pacific Rising. Transform faults, i.e. a system of almost parallel, many thousand kilometer long cracks in the Earth crust are observed. The faults are almost parallel and the distances between them are comparatively close. Thus, the ocean floor reminds a regular system of parallel fragments.

The Earth crust between the faults has a number of folds with the period of 70 km. The folds are perpendicular to the direction of transform faults. These structures are characteristic for many oceans (the Arctic Ocean is the exception), taking up more than one third of the entire ocean floor area. These structures are typical for the mid-ocean ridges.

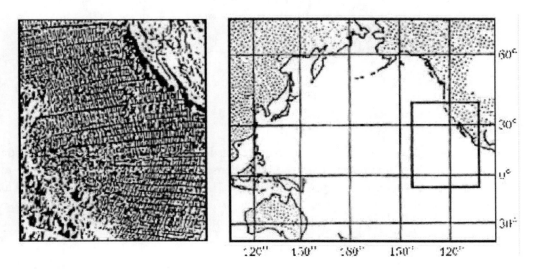

Figure 6. Fragment of ocean map of the Middle Ocean Rise.

Figure 1a shows an SEM micrograph of the natural rubber sample with thin (15 nm) gold coating after its stretching at 50% at room temperature. Morhological similarity of the natural (Figure 6a) and the model (Figure 1a) structures is evident. In our opinion, this similarity is caused by similar structure ("rigid coating on a soft substrate").

The above approach may be used for analysis of the ocean floor relief. Below the following questions are analysed: 1) what is the cause of regular folding of the ocean floor? 2) what is the cause of close distance between the transform faults? 3) why the faults and the folds are always perpendicular? 4) why both structures always coexist (are superimposed) on the same areas of the ocean floor?

ANALYSIS OF THE OCEAN FLOOR FOLDING

Regular folds are typical for the ocean floor and rocks. Buckling instability of the layered rocks was studied in the beginning of the XX-th century by Smoluchowskii [13,14]. Later this problem was considered by Biot [15-17] and Ramberg [18-20].

Figure 7 illustrates the experimental method used in [20]. The load is applied to the elastic layer (rubber or gelatin film) floating on a sufficiently dense liquid (mercury or saturated KI solution). This system models folding of the rigid layer and does not model the coating cracking. To our opinion, application of the compressive force directly to the rigid coating has essential disadvantages. The only driving force of the tectonic relief formation is the movement of the plastic upper Earth mantle. This movement transfers the stress to the Earth crust laying on it.

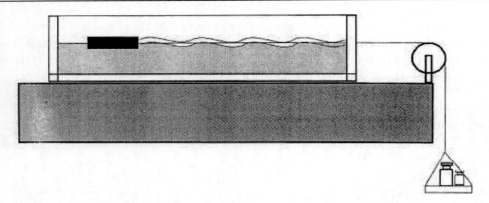

Figure 7. Schematic of folding of elastic gelatin film floating on a liquid mercury [20].

To analyze buckling of the ocean crust, an elastic plate floating on a liquid mantle material with density ρ and covered with water is considered similarly to [9]. In contrast to [9], the ocean crust under a water layer is considered. The liquid mantle density is assumed to be equal to that of the rigid crust. The displacement of an element of the crust plate from an equilibrium position leads to emergence of a returning force determined by hydrostatic pressure (Figure 8):

$$F = (\rho - \rho_w) gx \qquad (3)$$

where ρ is the density of the crust and liquid mantle, ρ_w is the water density, g is gravitational constant and x is displacement of the crust from equilibrium position.

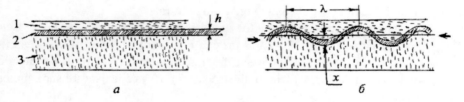

Figure 8. Schematic of the ocean floor before (a) and after buckling under compression (b). 1 – water, 2 – elastic beam (ocean crust), 3 – soft substrate.

Taking into account equation (3), the Earth crust shape is described by equation [21]:

$$EI\frac{d^4x}{dy^4} + N\frac{d^2x}{dy^4} + (\rho - \rho_w)gx = 0 \qquad (4)$$

where E is the Young modulus of the ocean crust, N is compressive force, $I = h^3/12$ is the second momentum of area of the crust layer cross-section, h is the crust thickness. Solution of equation (4) is a periodic function

$$x = C \sin\frac{2\pi y}{\lambda} \qquad (5)$$

where C is the wave amplitude and λ is the buckling wavelength. Substituting equation (5) in (4), the compressive stress is found:

$$N = EI\left(\frac{4\pi^2}{\lambda^2} + \frac{(\rho - \rho_w)g\lambda^2}{4\pi^2 EI}\right) \tag{6}$$

Function $N(\lambda) > 0$ at any λ and has a minimum N^* at some wavelength λ^*. The ocean crust is stable and flat if the load in the crust $N < N^*$. In contrast, if the load $N > N^*$, the crust is unstable and folds appear. In the minimum, the derivative $dN/d\lambda$ is equal to zero. Considering that the Young modulus E and the speed of sound c of a rigid material are related as $E = \rho c^2$, the buckling wavelength is given by

$$\lambda^* = \pi \sqrt[4]{\frac{\rho c^2 h^3}{3(\rho - \rho_w)g}} \tag{7}$$

where c is the speed of sound in the crust. Equation (7) allows to estimate the thickness of the buckled crust layer from the buckling wavelength value in Figure 9. Using equation (7) with the typical buckling wavelength value of $\lambda^*=70$ km, $c = 6{,}5$ km/s, basalt density $\rho = 2900$ kg/m^3 and water density $\rho_w = 1000$ kg/m^3 the crust thickness is estimated as ≈ 5 km. This estimate gives a reasonable ocean crust thickness, close to the ocean crust thickness value obtained by studying propagation of seismic waves created by earthquakes. The obtained result confirms plausibility of assumptions made at derivation of equation (7).

SIMULATION OF FAULTING IN THE OCEAN CRUST

Let us apply the above approach to dynamic processes in the Earth crust. Cracks in the coating (Figure 1a) are similar to the transform faults in the ocean crust. The transform faults have two features. First, they are oriented almost perpendicularly to the mid-ocean ridges and are parallel to each other. Second, the distance between the transform faults is rather close. We may suppose that the transform faults and crust foldings are related.

The morphological similarity of the surface relief of deformed coated polymer films and the ocean floor relief [11] can be explained by similarity of their structure. The thickness of the Earth's mantle is estimated as 2900 km, which is much higher than the thickness of the ocean crust (~5 km). The tectonic stresses are transferred from a viscous upper Earth's mantle to the crust. Stresses in the crust lead to the drift of continents and, in addition, to development of the surface relief (wavy relief) and the crust cracking (transform faults).

Uniaxial compression of the polymer substrate leads to lateral tension of the coating. Uniaxial tension of the coated polymer models causes lateral compression [9]. As a result, both at stretching and compression surface morphologies of coating are similar.

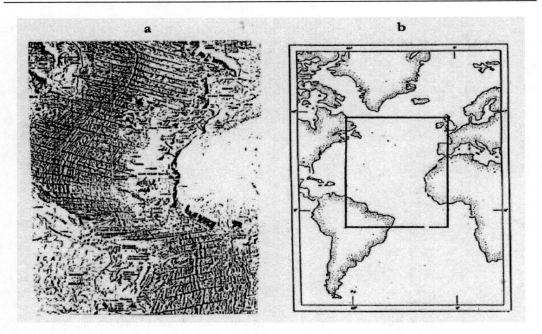

Figure 9. Fragment of the ocean floor relief in the region of the Middle Atlantic Ridge (a) and its location on the map of Atlantic Ocean (b).

We may apply the above analysis to the ocean crust relief near the Atlantic mid-ocean ridge (Figure 9) [9–11]. Figure 10 presents the lithosphere plates in this region. Similar structure is typical for other mid-ocean ridges. Spreading of the ocean crust is caused by the extrusion of the material of the upper Earth mantle through the mid-ocean ridge [22]. The Atlantic Ocean broadens, and the American plate is drifted to the west with a speed of several centimeters per year. At the same time, the Pacific plate moves in the opposite direction under the American continent with the same speed. As a result of this movement, the American plate is compressed. The compressive tectonic stress in the American plate is directed along the direction of spreading (perpendicularly to the axis of the mid-ocean ridge). Comparison of Figures. 3b and 9a allows assume that the ocean crust is compressed in the direction perpendicular to the direction of the Atlantic mid-ocean ridge. As a result, it is extended in the lateral direction, parallel to the axis of the Atlantic mid-ocean ridge. If a lithospheric plate behaves as a rigid coating on a soft substrate, its axial compression causes the lateral tension of the crust. In other words, compressive stress is perpendicular to the axis of the Atlantic mid-ocean ridge. Hence, the mid-ocean ridge is under tension in the transverse direction, i.e. along the mid-ocean ridge. This may explain why transform faults are always perpendicular to the axis of spreading.

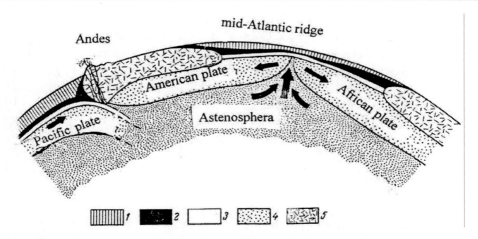

Figure 10. Middle Atlantic Ridge in terms of the tectonics of lithospheric plates. 1 - water; 2 - sediment; 3 - basaltic ocean crust; 4 - upper mantle; 5 - continental crust.

Any crack propagates perpendicularly to the tension axis. This is illustrated by the map of the ocean floor relief near the Atlantic mid-ocean ridge (Figure 9). The tensile stress acts along the axis of the mid-ocean ridge. Equation (2) determines the distance between cracks in the coating. If we apply equation (2) to the transform faults, L corresponds to the distance between them. L value may be estimated from the ocean floor map. For application of equation (2), the stress σ_0 in the upper Earth's mantle should be known. The force of continental spreading is provided by a viscous convective flow of the material in the upper Earth's mantle (Figure 8). Viscous flow of a mantle may be estimated with equation

$$\sigma_0 = \eta \, d\varepsilon/dt, \tag{8}$$

where η is the viscosity and $d\varepsilon/dt$ is the strain rate. η value is estimated as $\sim 10^{21}$ Pa s [23]. Let us assume that the rate of the drift of continents is equal to the rate of viscous displacement of the Earth's mantle under lithosphere. The rate of flow of the material in the upper mantle is assumed to be equal to rate of spreading (the drift rate of the continents). The spreading rate is

$$d\varepsilon/dt = (\Delta l/l_0)/\Delta t$$

where l is the initial distance between the continents, which is assumed to be equal to ≈ 5000 km; Δl is the continent displacement per year ($\Delta l = 0.1$ m); Δt is the time (1 year) of spreading.

Substituting the above values in equation (8), $\sigma_0 = 0.6$ MPa is obtained. This value estimates the effective stress in the mantle. This value is unexpectedly low. However, this stress acts over rather long periods of time.

The above estimation allows to estimation of the strength of the Earth ocean crust with equation (2). Hence, if the thickness of the Earth crust in the vicinity of the mid-ocean ridge h = 5 km, the mean distance between the transform faults L = 200 km, and the tectonic stress σ_0 = 0.6 MPa, the strength of the Earth ocean crust in the regions of transform faults is estimated as approximately 6 MPa. This value is amazingly low.

Below we try to explain so low strength of the Earth ocean crust. According to Griffith [24], the strength of a solid body decreases with the length of the crack in that body. In addition, the strength decreases with time of loading. Although the strength of solids is listed in many handbooks, it depends on temperature and loading velocity [25]. Figure 11 presents the strength plotted against the time of loading for the two different materials: rock salt (I) and semicrystalline aluminum (II) [25]. In both cases the strength decreases with the time of application of stress and temperature.

The thermofluctuation fracture of a solid may be described in terms of the Zhurkov equation:

$$\tau = \tau_0 \exp[(U_0 - \gamma\sigma)/RT], \qquad (9)$$

where τ is the time to fracture; U_0 is the activation energy of the fracture; γ is the so-called activation volume where an elementary fracture event takes place; τ_0 is the frequency of thermal vibrations of atoms; σ is the applied constant stress; R is the gas constant, and T is temperature.

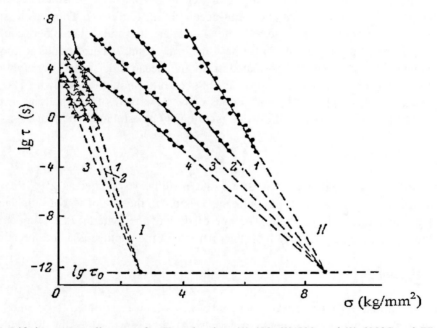

Figure 11. Lifetime vs. tensile stress for (I) rock salt at (1) 400, (2) 500, and (3) 600°C and (II) semicrystalline aluminum at (1) 18, (2) 100, (3) 200, and (4) 300°C [18].

Figure 11 may be used to estimate the parameters of equation (9). To estimate the Earth's crust strength, we consider the Earth's crust as an integral solid. It is characterized by a spherical shape, variable chemical composition and thickness, temperature and density gradient, defects and other parameters. Nevertheless, it is considered as an integral solid laying on a soft substrate. This approach models the Earth as an integral solid, which is able to transfer the mechanical stresses over rather long distances (within oceans). Evidently, the strength of basalt, which may be estimated in experiments, is quite different from the strength of the Earth's crust as a whole. Obviously, this is inaccurate estimation. However, the strength of the Earth crust can not be estimated by any other method.

Table. 1 Parameters of the Zhurkov equation [25]

	τ_0, s	U_0, kCal/mol	γ, (kCal·mm^2) mol/kg
Metals	$10^{-12} \sim 10^{-13}$	25 – 170	0.7 – 9.6
Ionic crystals	$10^{-12} \sim 10^{-13}$	31 – 74	14 – 60
Covalent bonds	$10^{-12} \sim 10^{-13}$	91 – 113	-
Glasses	$10^{-12} \sim 10^{-13}$	45 – 90	-
Polymers	$10^{-12} \sim 10^{-13}$	25 – 53	0.14 – 0.9

Table 1 lists the experimental estimates of the parameters of the equation (9) for different solids. The parameter τ_0 remains almost unchanged for various solids. This is quite expected because the frequency of thermal oscillations in a solid is almost independent of its nature. The parameter U_0 is equal to several dozens of kcal/mol, and this corresponds to the energy of rupture of various intermolecular bonds. The physical meaning of the parameters given in Table 1 for such a complex subject as the Earth's crust is still vague. Therefore, let us invoke the mean values of the parameters in the equation (9). For rough estimation of the lifetime of the Earth's crust, let us assume $\tau_0 = 10^{-12}$ s, $U_0 = 50$ kcal/mol, $\gamma = 1$ (kcal mm^2/mol kg), R = 0.002 kcal/mol K, T = 400 K, σ = 6 MPa. By substituting the above-estimated values to equation (9), we get the time to fracture of the Earth's crust of 30 millions years. This estimate is very approximate.

The goal of this work is to demonstrate the use of two-layer models for geodynamic processes. This approach allows to estimate the average strength of the Earth crust and the period of development of the various morphological forms of the Earth surface.

ACKNOWLEDGMENT

We would like to thank Profs. V.E. Khain, N.V. Koronovsky, M.A. Goncharov, Yu.A.Taran for stimulating discussions. We acknowledge the financial support from the Russian Foundation for Basic Research (project #05-03-32538).

REFERENCES

[1] Wheeler, D.R.; Osaki, H. *ACS Symp. Ser.* 1990, vol. 440 - 500.
[2] Horniq, T.; Sokolov, I.M.; Blumen, A. *Phys. Rev. E* 1996, vol.54-4293.
[3] Leterrier, Y.; Boogh, L.; Andersons, J.; Manson, J.-A.E. *J. Polym. Sci. Part B: Polym. Phys.* 1997, vol.35-1449.
[4] Bazhenov, S.L.; Chernov, I.V.; Volynskii, A.L.; Bakeev, N.F. *Dokl. Acad. Nauk* 1997, vol.356-199.

[5] Bowden, N.; Brittain, S.; Evans, A.G.; Hutchinson, J.W.; Whitesides, G.M. *Nature*, 1998, vol.393-146.
[6] Volynskii, A.L.; Bazhenov, S.; Lebedeva, O.V.; Ozerin, A.N.; Bakeev, N.F. *J. Appl. Polym. Sci.* 1999, vol.72-1267.
[7] Bowden, N.; Huck, W.T.S.; Paul, K.E.; Whiteside, G.M. *Appl. Phys. Lett.* 1999, vol.75-2557.
[8] Huck, W.T.S.; Bowden, N.; Onck, P.; Pardoen, T.; Hutchinson, J.W.; Whiteside, G.M. Langmuir 2000, vol.16-3497.
[9] Volynskii, A.L.; Bazhenov, S.L.; Lebedeva, O.V.; Bakeev, N.F. *J. Mater. Sci.* 2000, vol. 35-547.
[10] Bazhenov, S.L.; Volynskii, A.L.; Alexandrov, V.M.; Bakeev, N.F. *J. Polym. Sci. Part B: Polym. Phys.* 2002, vol.40-10.
[11] Volynskii, A.L.; Bazhenov, S.L. *Geofis. Int.* 2001, vol.40-87.
[12] Heezen, B.C.; Tharp, M. World Ocean Floor, Office of Naval Research, US Navy, 1977.
[13] Smoluchowskii, M. Abh. Akad. *Wiss. Krakau, Math.* 1909, vol.K1-3.
[14] Smoluchowskii, M. Abh. Akad. *Wiss. Krakau, Math.* 1910, vol.K1-727.
[15] Biot, M.A.; *J. Appl. Phys.* 1954, vol.25-2133.
[16] Biot, M.A.; Q. *Appl. Math.* 1959, vol. 17-722.
[17] Biot, M.A. *Mechanics of Incremental Deformations*, Wiley, New York, 1965.
[18] Ramberg, H. *Bull. Am. Assoc. Petrol. Geologists* 1963, vol.47-484.
[19] Ramberg, H. *Tectonophysics* 1964, vol.9-307.
[20] Ramberg, H.; O. *Stephansson Tectonophysics* 1964, vol.1-101.
[21] Landau, L.D.; Lifshitz, E.M. Elasticity theory, *Nauka*: Moscow, 1965.
[22] Shaw, P.R. *Nature* 1992, v.358-490.
[23] Cathes, L.M. The Viscosity of the Earth Mantle, Princeton University Press: NY, 1975.
[24] Griffith A.A. Theory of Rupture, in Proceedings of the 1st International Conference on Applied Mechanics, Delft, Holland, 1924 pp. 55.
[25] Regel, V.R.; Slutsker, A.I.; Tomashevskii, E.E. Kinetic Nature of Strength in Solids, *Nauka*: Moscow, 1974.

INDEX

#

20th century, 29

A

abstraction, 290
acetic acid, 181, 182, 204
acetone, 151, 181, 182, 209, 213
acid, 115, 170, 171, 173, 174, 175, 176, 177, 178, 179, 183, 184, 186, 187, 188, 189, 190, 192, 193, 196, 202, 203, 204, 207, 290
acidity, 179, 186
acidulated phosphate, viii, 167, 170, 192, 193
activation energy, ix, 202, 207, 210, 211, 221, 230, 289, 293, 308
active site, 146, 260, 290
additives, x, 78, 216, 222, 285, 286, 287, 290, 291, 294
adhesion, 124, 148, 169, 174, 179, 180, 193, 195, 197, 199
adhesive strength, 197
adhesives, 180, 199
adjustment, 179
adsorption, 114, 290
advancement, 121
aerogels, 2, 36, 53
AFM, 124
aggregation, 148, 287
alcohols, 179, 202
all-ceramic restorations, 169, 174, 188, 194
aluminium, viii, 167, 170, 171, 172, 174, 177, 178, 183, 184, 185, 188, 189, 190
aluminum oxide, 193
amino, 179, 187
ammonia, 98, 115

ammonium, x, 204, 225, 237, 285, 286
ammonium perchlorate (AP) based composite propellants, x, 285
amorphous polymers, 91
amorphous silicon thin film, x, 269, 270
anatase, 291, 293
anisotropy, 7, 9, 73, 78, 81, 92, 93
annealing, viii, 78, 113, 122, 123, 124, 125, 126, 128, 129, 130, 133, 135, 136, 138, 139, 140, 141, 142, 143, 144, 151, 155, 158, 159, 271, 272, 273, 274, 275, 277, 278, 279, 293
antimony, 274
aqueous solutions, 151
ARC, 211, 218
argon, 131, 152, 211, 277
arithmetic, 7, 10, 11, 12, 28, 31, 40, 55, 56, 57, 90
aromatic hydrocarbons, 202
assessment, viii, 2, 4, 88, 113, 115, 206, 264
asymmetry, 5, 135
atmosphere, 77, 123, 136, 151, 152, 153, 211, 213, 227, 246, 260, 289
atmospheric pressure, 124, 223, 227
atomic force, 124
atomic force microscope, 124
atoms, 48, 76, 78, 84, 88, 116, 125, 137, 139, 142, 149, 155, 168, 176, 274, 275, 276, 277, 278, 279, 308

B

backscattering, 131, 152
band gap, 122, 137, 138, 139, 140, 266, 271
barium, 68, 74
base, ix, 116, 124, 145, 184, 186, 189, 190, 197, 199, 201, 204, 209, 214, 215, 219, 221, 222, 224, 226, 236, 237, 242, 244, 248, 273
basic raw materials, 71

batteries, 144, 255
bending, 168
benefits, x, 280, 285
benzene, 209, 216
beryllium, 90, 184
bias, 4, 149, 156
biocompatibility, 168
biological processes, 218
biotechnology, 286
Boltzmann constant, 43, 51, 52, 77
bonding, vii, viii, 167, 169, 170, 171, 175, 176, 178, 180, 183, 184, 185, 186, 189, 190, 191, 192, 194, 195, 196, 197, 198, 199, 286
bonds, viii, 76, 91, 168, 180, 195, 201, 202, 243, 256, 278, 309
bosons, 48
bounds, vii, 1, 3, 4, 7, 8, 9, 10, 11, 12, 13, 14, 15, 16, 19, 20, 21, 23, 24, 26, 27, 31, 37, 38, 40, 41, 43, 47, 88, 92, 93
breakdown, 122
building blocks, 151, 157, 159
bulk materials, 89
burn, 175, 286, 287, 291
by-products, ix, 202, 210, 214, 215, 216, 218, 220

C

C.F. Schönbein, viii, 201
CAD, 174
cadmium, 79, 271
calcium, 204
calcium carbonate, 204
calibration, 206
CAM, 174
carbides, vii, 1, 79, 81, 85, 89
carbon, 2, 78, 81, 87, 89, 92, 112, 116, 121, 177, 204, 212, 266
carbon dioxide, 177
carbon monoxide, 116, 121
carbon nanotubes, 78, 81, 89, 92, 112
carbon tetrachloride, 204
carbonyl groups, 208, 209
casting, 196, 197, 199
catalysis, 285, 287, 288, 291
catalyst, viii, x, 113, 135, 245, 285, 287, 288, 289, 290, 291, 293, 294
catalytic activity, 288, 289, 290
catalytic effect, 288, 289, 290
catastrophic failure, 287
category a, 225
cation, 114, 150, 258, 259, 260
C-C, 91, 243
cellulose, viii, 201, 202, 203, 204, 205, 207, 209

ceramic materials, 53, 68, 88, 89
ceramic restorations, viii, 167, 168, 169, 174, 183, 188, 190, 191, 194, 196
chain scission, 207
challenges, viii, 113, 159
charge density, 263
chemical characteristics, 288
chemical etching, viii, 167, 183, 190, 196
chemical properties, ix, 201, 225, 241, 255
chemical reactions, 289
chemical stability, 122
chemical vapor deposition, 130, 141, 272, 273
chemical vapour deposition, 122, 135, 288
chemicals, 178
chemiluminescence, 216
chloroform, 204
chromatograms, 206
chromium, 184, 186, 197, 290
CIS, 271
classes, vii, 1, 3, 4, 5, 87, 88, 179, 241, 242
classification, 77, 169, 170, 270
clinical application, viii, 167, 169, 184, 189, 195
clustering, 30
clusters, 26, 118, 121, 148, 154, 158
C-N, 243
CO2, 52, 98, 114, 177, 212, 213, 214, 215, 216, 244
coal, 81, 204
coatings, 81, 129, 143
cobalt, 56, 63, 184
collisions, 48, 49
color, 53, 185, 223, 236, 255, 262, 263, 264
combustion, ix, x, 208, 221, 222, 223, 237, 239, 241, 242, 244, 246, 248, 285, 286, 287, 289, 290, 291, 293, 294
commercial, 80, 134, 135, 145, 178, 203, 270, 273, 289, 291
community, ix, 5, 135, 201, 254
compensation, 257
complex interactions, 294
complexity, 217, 260
complications, 191
composite resin, 192, 193, 194, 195, 196, 197, 198
composites, vii, 1, 2, 3, 4, 5, 6, 9, 10, 11, 12, 13, 14, 15, 16, 22, 23, 25, 29, 30, 31, 34, 39, 40, 41, 42, 82, 88, 100, 106, 176, 177, 178, 179, 188, 192, 199, 292
composition, vii, ix, 1, 2, 4, 39, 48, 67, 71, 75, 77, 78, 88, 121, 122, 123, 135, 136, 151, 153, 173, 187, 196, 201, 204, 241, 260, 288, 308
compounds, 78, 79, 81, 85, 89, 90, 179, 203, 204, 208, 213, 215, 216, 222, 248, 266
comprehension, 294
compression, 168, 209, 299, 300, 302, 304, 305, 306

Index

computer, 16, 174, 231
computer-aided design, 174
condensation, 179, 180
conditioning, 191, 192, 198
conductance, 114, 117, 122, 148, 149, 155, 156, 168, 255
conducting polymer composites, 78
conduction, vii, 1, 3, 4, 54, 74, 111, 114, 139, 256, 257, 265
conductor, 262
configuration, 17, 120, 131, 145, 151, 152, 214, 250, 261, 266
conjugated polymers, x, 249, 254, 256, 262, 263, 265, 266
conjugation, 260
connectivity, 20
conservation, 218
constant rate, 262
constituents, 10, 71, 150
construction, 2
consumption, 270
contamination, 115, 123, 124, 136, 145, 274, 280
contradiction, 8
controversial, 189
conviction, 3
cooling, 120, 158
cooperation, 294
copolymers, 261
copper, x, 55, 90, 189, 198, 208, 271, 285, 287, 289, 290
copyright, 208, 210, 212, 214, 215, 217
correlation, 10, 11, 13, 16, 30, 177, 231
correlation coefficient, 231
correlation function, 10, 11, 13, 16, 30
corrosion, 204
cosmetic, 191
cost, viii, 114, 122, 135, 167, 169, 189, 190, 265, 270, 271, 273, 276, 280, 288
cotton, viii, 134, 184, 185, 201, 202, 203, 206, 207
coughing, 189
Coulomb energy, 139
covalent bond, 91, 179, 275
covering, 39, 174
cracks, 19, 37, 51, 168, 217, 298, 302, 307
criminals, 205
crown, 74, 185, 196
crowns, 167, 194, 198
crust, xi, 71, 271, 297, 298, 302, 303, 304, 305, 306, 307, 308, 309
crystal growth, 134, 151
crystal structure, 9, 151, 152, 174
crystalline, x, 39, 45, 46, 47, 48, 49, 50, 55, 58, 59, 71, 73, 78, 88, 90, 91, 116, 118, 123, 125, 134, 146, 147, 149, 156, 158, 159, 169, 174, 188, 269, 271, 272, 273, 276, 278, 280, 281
crystalline solids, 48, 49, 50, 59, 88
crystallinity, 117, 125, 127, 128, 140, 141, 142, 158, 288
crystallisation, 129
crystallites, 6, 8, 9, 10, 12, 17, 39, 78, 132
crystallization, x, 117, 121, 125, 177, 269, 270, 272, 273, 274, 275, 276, 278, 279, 280, 281
crystals, 5, 6, 7, 9, 61, 62, 82, 83, 88, 89, 90, 117, 125, 130, 141, 146, 150, 173, 174, 175, 176, 309
CSCE, 219
cubic boron nitride, 89
current ratio, 149, 155, 156
CVD, 79, 80, 102, 273, 280

D

damages, 122, 136
data collection, 4
decomposition, ix, 202, 209, 210, 211, 212, 213, 215, 216, 217, 218, 219, 221, 222, 227, 228, 230, 231, 232, 235, 237, 243, 244, 246, 287, 288, 289, 293, 294
decomposition temperature, ix, 202, 211, 212
decoupling, 280
defective zones, 278
defects, viii, 48, 50, 53, 57, 60, 88, 89, 113, 114, 122, 129, 130, 132, 133, 134, 135, 137, 142, 143, 154, 158, 168, 254, 261, 301, 308
deficiencies, 24
deficiency, 114, 150
deformation, 298, 300
degenerate, 12, 14, 20
degradation, 207, 209, 211, 212, 217, 218, 260, 264, 293
degradation process, 207, 217, 218
degree of crystallinity, 91, 127
denitrifying, 216
dense polycrystalline materials, vii, 1, 4
dental caries, 169
dental ceramics, 168, 169, 174, 177, 182, 188
dental resins, 193
dentin, 168
dentist, 190
dentures, 167, 169, 184, 197
depolarization, 23, 27, 28, 29, 94, 95, 96
deposition, viii, x, 113, 115, 116, 122, 123, 124, 130, 135, 136, 137, 141, 145, 146, 148, 150, 152, 157, 158, 162, 254, 269, 272, 273, 277, 279, 280, 281, 288
deposition rate, 273, 277
deposits, 148, 176

derivatives, 260, 261, 263, 266
destruction, 175
detection, 117, 205, 206, 207, 239
detection system, 205, 207
deviation, 141, 231, 260
diamonds, 60, 66, 85
dielectric constant, 22, 139, 149, 156
dielectrics, 53, 55, 87, 89, 104
Differential Scanning Calorimetry (DSC), ix, x, 202, 208, 209, 210, 211, 212, 218, 223, 227, 228, 229, 230, 231, 232, 235, 285, 288, 289, 290, 292, 293
Differential Thermal Analysis (DTA), ix, 202, 210, 211, 218
diffraction, 117, 125, 127, 128, 129, 131, 146, 152
diffusion, 151, 251, 253, 254, 260, 266, 278, 293
dimensionality, 150
dimethacrylate, 182
discontinuity, 56, 63
discs, viii, 115, 167, 183
disorder, 42, 48, 49, 70, 272
dispersion, 26, 169, 208
displacement, 304, 307
distilled water, 151, 182, 185, 192
distribution, ix, 14, 16, 17, 18, 26, 29, 119, 127, 128, 131, 142, 153, 205, 209, 221, 223, 239, 240, 241, 246
diversity, 123, 130, 141
dopants, 58, 278
doping, x, 123, 136, 254, 256, 257, 258, 259, 260, 262, 269, 271, 277, 278, 279, 280
double bonds, 254
drawing, 298, 299, 301
drying, 115, 206
DSC, ix, x, 202, 208, 209, 210, 211, 212, 218, 223, 227, 228, 229, 230, 231, 232, 235, 285, 288, 289, 290, 292, 293
DTA curve, 210, 211
dualism, 48
durability, 174, 190, 194
dye-based solar cells, viii, 113

E

earthquakes, 305
elastomers, 286
election, 191
electric conductivity, 78, 79, 81
electric field, 264, 265, 266
electrical conductivity, 94, 122, 141, 256
electrical properties, viii, 113, 114, 141, 145, 150, 151, 152, 156, 159, 254, 279
electricity, 184, 266, 270, 280
electrochemistry, x, 249, 288

electrochromic films, 263
electrode surface, 253, 261
electrodes, viii, 113, 122, 131, 135, 141, 144, 145, 150, 152, 158, 265
electroless deposition, 276
electroluminescence, 265
electroluminescent devices, x, 249
electrolyte, 250, 255, 260, 261, 262
electromagnetic, 48, 52, 106, 262
electromagnetic waves, 48
electron diffraction, 145, 151
electron microscopy, 48, 117, 130, 145, 151, 152
electronic structure, x, 249
elongation, 299
elucidation, 250
emission, 215, 265
enamel, 168, 189, 197, 199
endothermic, 227
energetic materials, x, 225, 226, 243, 285, 286, 288, 290, 293, 294
energy, ix, 48, 52, 115, 120, 123, 124, 132, 135, 136, 137, 138, 139, 148, 151, 152, 153, 158, 178, 199, 210, 211, 221, 222, 223, 224, 225, 226, 243, 246, 255, 256, 257, 259, 260, 262, 264, 265, 266, 270, 271, 278, 280, 281, 286, 309
energy density, 222
engineering, 49, 53, 55, 73, 78, 105
enlargement, 227
environment, 114, 142, 155, 211, 212, 213
EPA, 216
epitaxial growth, 276, 279, 280
equality, 8
equilibrium, 123, 130, 141, 304
equipment, 2, 4, 61, 109, 184, 247
erbium, 177
erosion, 169
ester, viii, 184, 188, 201, 202
etching, viii, 167, 170, 171, 173, 174, 175, 176, 177, 178, 183, 184, 188, 189, 190, 192, 193, 196, 274, 277, 278, 280, 281
ethanol, 115, 145, 152, 181
etherification, 204
ethylene, 98
evidence, 115, 157
evolution, 115
excitation, 131, 152, 265, 271
experimental condition, 131, 260
explosives, viii, 201, 204, 205, 207, 209, 210, 213, 218, 286
exposure, 175, 272, 277
extraction, 260, 262
extrusion, 81, 222, 224, 306

F

fabrication, 114, 115, 124, 135, 136, 140, 145, 168, 188
ferrite, 287
fiber, 7, 12, 53, 77, 81, 93, 217
fibers, 12, 15, 17, 31, 77, 99, 204, 210, 211
filament, 99, 213
fillers, x, 285, 287
film thickness, 215, 254, 260, 275
films, viii, x, 79, 80, 81, 113, 114, 115, 117, 118, 121, 122, 123, 124, 125, 126, 127, 128, 129, 130, 131, 132, 133, 134, 135, 136, 137, 139, 140, 141, 142, 143, 150, 158, 159, 204, 210, 213, 217, 260, 262, 264, 269, 272, 273, 277, 278
financial support, 100, 309
first generation, 270
flame, ix, 216, 221, 223, 236, 237, 239, 242, 246
flaws, 168, 169, 181
flexibility, 123, 130, 141, 254, 263
flight, 275
flour, 204
fluctuant, 239
fluorescence, 264
fluoride gel, viii, 167, 192, 193
fluorine, 151
foams, 26, 81
force, 124, 270, 275, 303, 304, 307
formation, x, 48, 115, 119, 120, 121, 126, 142, 148, 155, 157, 158, 159, 218, 222, 253, 256, 257, 261, 264, 266, 272, 275, 278, 297, 298, 299, 300, 303
formula, viii, 22, 23, 24, 26, 43, 68, 117, 125, 137, 139, 156, 179, 181, 201, 202, 225, 242, 243
fractal cluster, 115, 116, 118, 119, 120, 121, 157
fractal dimension, 115, 116, 118, 119, 121, 157
fractal growth, 119
fractal structure, 115, 119, 120, 121, 158
fractal theory, 115, 116, 121
fracture modes, 197
fractures, 169, 170, 189
fragments, 298, 299, 301, 302
free volume, 15
fruits, 92
FTIR, 208, 209, 211, 212, 213, 216, 217, 218
FTIR spectroscopy, 213, 216
fullerene, 266
fusion, 2

G

gallium, 79, 89, 90, 177, 270, 280

gas sensors, viii, 113, 114, 117, 119, 122, 131, 135, 136, 140, 144, 151
gel, viii, 73, 167, 170, 171, 175, 176, 177, 188, 189, 190, 192, 193
geometry, 18, 39, 114, 116, 124, 131, 134, 145, 151
glasses, vii, 1, 3, 4, 6, 43, 48, 49, 50, 55, 68, 69, 71, 73, 74, 75, 76, 77, 87, 89, 91, 271, 278
glycol, 151
grain boundaries, 38, 48, 50, 53, 114, 274, 275, 278
grain size, vii, 1, 2, 3, 4, 12, 38, 39, 42, 43, 44, 45, 46, 47, 53, 60, 67, 69, 85, 88, 114, 115, 117, 124, 125, 131, 135, 136, 140, 158, 270, 272, 273, 274, 275, 276, 278, 279, 280, 286
grants, 100
graph, 56, 64, 139, 142, 143
graphene sheet, 81, 89, 92
graphite, 2, 7, 8, 9, 78, 81, 86, 87, 92, 93, 107
gravitational constant, 304
grids, 153
growth, x, 115, 119, 123, 124, 130, 134, 137, 141, 144, 145, 146, 148, 150, 153, 154, 155, 158, 159, 235, 246, 269, 276
growth mechanism, 130, 141, 148, 159
growth rate, 115, 123, 124
guidance, viii, 113, 136
gunpowder, ix, 201, 203

H

hafnium, 78, 85
half-life, 210
halogen, 273
Hartree-Fock, 243
hazardous materials, 216
hazards, 175
HDPE, 91
heat conductivity, 5, 6
heat release, 121, 213, 289
heat transfer, vii, 1, 3, 4, 48, 50, 51, 52, 53, 54, 55, 58, 64, 68, 73, 74, 75, 77, 88, 89, 90
heat transfer mechanisms, vii, 1, 4, 53
heating rate, 210, 212, 223, 230, 231, 272, 289
height, 49, 88, 114, 122, 227, 252
helium, 211
high strength, 174
history, 191
HNIW-CMDB propellant, ix, 221, 222, 223, 224, 230, 235, 236, 237, 239, 240, 241, 242, 243, 244, 245, 246
holmium, 177
host, 25, 81
HRTEM, 130, 131, 132, 133, 134, 146, 147, 152, 153, 154, 158

hybrid, 78, 198
hydrocarbons, 98, 213, 216, 241, 244
hydrofluoric acid, viii, 167, 175, 192, 278
hydrogen, 77, 180, 207, 215, 272, 277, 278, 279
hydrogen atoms, 278
hydrogen bonds, 207
hydrogen cyanide, 215
hydrogenation, 272
hydrolysis, 83, 179, 207, 289
hydrothermal synthesis, 288
hydroxyapatite, 177
hydroxyl, viii, 180, 201, 202, 204, 216, 286
hydroxyl groups, viii, 180, 201, 202, 204, 216
hypothesis, 244

I

illumination, 270
image analysis, 16, 48
immersion, 196
impurities, 48, 50, 60, 89, 117, 125, 130, 141, 146, 152, 274
incidence, 264
incomplete combustion, 216
independence, 89, 211
indium, 271, 274, 280
induction, 213, 272, 279
induction period, 213
induction time, 213
inductor, x, 269, 274
industry, 2, 71, 179, 270, 271
infrared spectroscopy, 216
ingredients, 286, 294
initiation, 213, 250, 260
inoculum, 216
inorganic glasses, vii, 1, 71, 91
insertion, 262
insulation, vii, 1, 47, 51, 77
insulators, 55, 71, 79, 83, 255
integrated circuits, 83
integration, 253
integrity, 197
interface, 38, 120, 122, 136, 150, 186, 195, 197, 250, 266
intermetallic compounds, 88
intrinsic viscosity, 206, 207
inversion, 17, 24
ionization, 213, 216, 256, 260
ions, 43, 57, 73, 77, 217, 254, 257, 261, 262, 277
iron, x, 67, 77, 204, 285, 288, 289
irradiation, viii, 167, 170, 171, 177, 193
ISC, 264
isolation, 175, 189

isotope, 88, 90
isotropic media, 13
issues, x, 135, 150, 169, 269

K

ketones, 202
kinetic model, 230, 231
kinetic parameters, 230, 231, 246, 289
kinetics, ix, 202, 207, 209, 210, 211, 221, 222, 231, 247
Kissinger method, 289

L

laminar, 51, 54
lanthanum, 68
laser ablation, 116, 121, 124
laser irradiation, viii, 167, 170, 171, 177, 193
lasers, 177, 193
lattice parameters, 116, 124, 147, 153
lattices, 20, 130, 134
lead, 36, 52, 56, 60, 73, 74, 76, 99, 119, 138, 204, 205, 262, 287, 305
leakage, 168
lens, 115, 124
light, 74, 77, 123, 129, 136, 143, 205, 206, 214, 215, 217, 237, 255, 262, 264, 265, 272, 277, 298, 299
light emitting diode, 255, 265
light scattering, 129, 143
limestone, 71, 73
linear dependence, 33, 259
liquid phase, 47
liquids, vii, 1, 3, 48, 49, 51, 52, 87, 91
lithium, 144, 174, 177
lithography, 145, 152
localization, 139
low temperatures, x, 50, 55, 60, 85, 87, 89, 90, 91, 269, 273, 274
luminescence, 264, 265
Luo, 101, 247, 296

M

magnitude, 2, 40, 46, 55, 60, 62, 64, 67, 72, 73, 74, 87, 88, 89, 90, 212, 254, 274
majority, 2, 82, 84, 89
manganese, 63, 90
manipulation, 207, 209
mantle, 302, 303, 304, 305, 306, 307
manufacturing, 174, 179, 211, 224, 271

Marx, 191
mass, ix, 43, 48, 88, 138, 201, 202, 205, 207, 211, 212, 214, 215, 217, 218, 223, 224, 226, 227, 228, 244, 252, 275
mass loss, 211, 212, 227, 228
mass spectrometry, ix, 202, 214
materials science, viii, 3, 114, 159
matrix, 6, 10, 14, 15, 17, 20, 22, 23, 24, 25, 31, 32, 34, 36, 41, 94, 127, 173, 174, 176, 179, 180, 205, 270, 280, 286
matter, 52, 123, 260
measurement, 4, 16, 43, 61, 80, 92, 118, 121, 125, 158, 207
measurements, ix, 43, 56, 60, 62, 64, 67, 73, 82, 102, 123, 124, 136, 142, 149, 151, 156, 157, 158, 159, 201, 206, 211, 213, 263, 265, 289, 293
meat, 92
mechanical degradation, 209
mechanical properties, x, xi, 39, 78, 88, 169, 209, 249, 271, 287, 297
mechanical stress, xi, 168, 297, 308
media, 4, 9, 18, 21, 24, 78, 105, 106, 252
melon, 244
melt, 47, 75, 155
melting, 74, 174, 177, 212, 273, 274
melting temperature, 74, 212
melts, 51, 74, 174
memory, 263, 264
mercury, 51, 52, 90, 99, 303, 304
mesoporous materials, 2, 51
metal ions, 142
metal oxides, 144, 155, 180, 287
metal surface treatment, vii, viii, 167, 169, 183, 186, 187, 190, 199
metals, vii, viii, 1, 2, 3, 4, 49, 53, 55, 76, 79, 81, 87, 89, 90, 91, 92, 99, 167, 179, 198, 254, 272, 274, 287
methanol, 115, 124, 145, 184, 216
methodology, 115, 127, 158, 261
microcrystalline, 271
microelectronics, 271
micro-emulsion method, viii, 113, 157
micromechanical bounds, vii, 1, 3, 4, 8, 37, 93
micrometer, 60, 150
microscope, 116, 124, 131, 146, 152, 171, 217
microscopy, 130, 144, 288
microstructure(s), viii, 2, 4, 7, 8, 9, 10, 11, 12, 13, 14, 15, 16, 17, 18, 19, 23, 24, 30, 32, 36, 38, 39, 40, 41, 60, 62, 71, 81, 88, 94, 113, 115, 122, 123, 129, 134, 135, 141, 158, 159, 174
migration, 278
military, ix, 201, 205
miniature, 223, 246

miniaturization, 150
Ministry of Education, 100
missions, 286, 287
mixing, 204, 213, 266
MMA, 182, 197
model microstructures, 17
models, vii, xi, 1, 2, 3, 12, 16, 24, 26, 34, 39, 41, 44, 45, 46, 47, 55, 93, 207, 211, 297, 303, 305, 308, 309
modifications, 2, 78, 84, 89, 208
modules, 77, 81, 272
modulus, 21, 43, 45, 194, 299, 304, 305
moisture, 206
moisture content, 206
mold, 174
mole, 242
molecular mass, ix, 221, 223, 225, 246
molecular weight, 260
molecular weight distribution, 260
molecules, 48, 114, 122, 148, 179, 180, 181, 205, 207, 212
molybdenum, 78, 81
momentum, 304
monolayer, 266, 276
monomers, viii, 180, 184, 186, 191, 193, 197, 201, 260, 261
morphology, 110, 121, 123, 131, 144, 146, 153, 196, 288
motivation, viii, 113, 159
multiphase materials, 2, 4, 6, 11

N

NaCl, 90, 151, 155
nanobelts, 130, 141, 144, 150
nanocomposites, 100, 266, 285, 290
nanocrystalline ceramics, vii, 1, 3, 46
nanocrystals, 119, 120, 121, 139, 158
nanoelectronics, 135, 151, 157, 159
nanomaterials, 39, 135, 144, 294
nanometer(s), 126, 145, 146, 150, 151, 153, 159, 288, 289
nanoparticles, x, 117, 123, 125, 130, 134, 135, 136, 141, 144, 150, 158, 159, 285, 286, 288, 289, 293
nanoribbons, 144
nanorods, viii, 113, 130, 141, 144, 150, 151, 152, 153, 154, 155, 156, 157, 159
nanostructured materials, viii, 113, 151, 159
nanostructures, 117, 144, 148, 150, 159
nanotechnology, 286
nanotube, 266
nanowires, viii, 113, 130, 141, 144, 145, 146, 147, 148, 149, 150, 157, 159

National Bureau of Standards, 101
negative effects, 136
neodymium, 177
neutral, 151, 256, 257, 258, 259, 262
next generation, 121, 222
NH2, 179
nickel, x, 63, 184, 186, 197, 269, 274, 275, 277, 278, 280
nitration of cellulose, viii, 201
nitrides, vii, 1, 79, 81, 85, 89
Nitrocellulose, v, viii, ix, 201, 202, 203, 204, 205, 207, 208, 209, 210, 216, 218
nitrogen, ix, 202, 203, 204, 205, 206, 207, 208, 209, 210, 211, 212, 213, 214, 216, 217, 218, 223, 289
nitrogen gas, 223
N-N, 222, 243
noble metals, 199
non-polar, 186
non-radiative transition, 264
novel materials, 150
novel polymers, 271
nucleation, x, 119, 120, 150, 155, 192, 235, 246, 269, 272, 274, 278, 279
nuclei, 119, 121, 275, 278

O

oceans, 303, 308
oligomers, 179, 180
one dimension, 244
one-dimension gas phase reaction theory, ix, 221, 242, 246
optical parameters, 136, 158
optical properties, 50, 71, 73, 76, 122, 136, 158, 162, 174, 254, 255, 262, 263, 288
optimization, 270
orbit, 243
ores, 53
organic compounds, 215, 254
organic polymers, 87, 255
organic solvents, 204
overlap, 29, 243
oxidation, 124, 152, 209, 211, 252, 253, 254, 255, 262, 263, 293
oxide nanoparticles, 289, 290, 293
oxygen, ix, 70, 83, 114, 116, 122, 123, 132, 136, 141, 142, 145, 148, 150, 151, 152, 153, 155, 176, 221, 246, 290

P

paints, viii, 201, 204

palladium, 186, 189, 198, 274
parallel, 6, 12, 33, 37, 54, 73, 87, 145, 147, 152, 153, 274, 302, 305, 306
passivation, 272
PCR, 67
perchlorate, x, 237, 285, 286, 290
percolation, 19, 20, 21, 24, 26, 30, 36, 41
percolation theory, 36
permission, 176, 208, 210, 212, 214, 215, 217, 273
PET, 298, 299, 301, 302
pH, 115, 175, 179, 181, 182, 204
pharmaceutics, 286
phase boundaries, 10, 12
phase mixture models, vii, 1, 39
phase transitions, 56
phonons, 4, 48, 49, 50, 55, 57, 87, 88, 90, 264
phosphate(s), viii, 68, 69, 167, 170, 182, 184, 186, 187, 188, 192, 193, 197
phosphorous, 276, 278, 279
photocatalysis, 288
photoelectron spectroscopy, 288
photographs, 208, 212, 223
photolithography, 144
photoluminescence, 265
photonics, 150
photons, 3, 48, 50, 55, 77
photovoltaic cells, 122, 135, 140, 255, 266, 267
photovoltaic devices, vii, 266
photovoltaics, x, 249, 270
physical properties, 150, 168, 294
physics, viii, 3, 5, 48, 61, 113, 159
plasticizer, 205
platelets, 15, 17
platinum, 298, 300, 301
plausibility, 3, 37, 305
PM3, 243
PMMA, 91
point defects, 88
Poisson ratio, 21, 299
polar, 186
polarizability, 26, 94
polarization, 26, 27, 28, 94, 95, 97, 277
pollutants, 119, 121
pollution, 114
polybutadiene, 286
polydispersity, 26
polymer, viii, ix, x, 91, 201, 202, 203, 204, 205, 206, 207, 209, 212, 218, 222, 241, 249, 254, 255, 256, 257, 258, 260, 261, 262, 263, 265, 266, 267, 297, 298, 299, 300, 301, 305
polymer chain, 91, 207, 256, 257, 266
polymer films, x, 297, 298, 305
polymer matrix, 257, 265, 266

polymer structure, 212, 254
polymeric materials, 264
polymerization, 180, 202, 260, 261
polymers, vii, x, 1, 3, 4, 87, 89, 91, 206, 249, 254, 255, 256, 261, 262, 263, 265, 266, 267, 299
polymethylmethacrylate, 91, 206
polymorphism, 222
polypropylene, 91
polysaccharide, 202
polystyrene, 206
polyurethane, 91
polyvinyl alcohol, 223
polyvinylchloride, 91
population, 49, 243
pore openings, 51
porosity, vii, 1, 2, 3, 4, 12, 14, 16, 17, 20, 29, 32, 33, 35, 36, 39, 41, 42, 46, 47, 51, 53, 54, 55, 56, 57, 59, 60, 61, 62, 63, 64, 65, 66, 71, 72, 81, 85, 87, 88, 108
porous materials, 10, 11, 13, 14, 15, 16, 18, 19, 20, 25, 26, 30, 32, 33, 34, 36, 37, 41, 43, 51, 53, 54
porous media, 3, 4, 9, 10, 14, 23, 30
potassium, 174, 204, 278
precipitation, 176, 288
preparation, iv, 39, 115, 150, 151, 276, 288, 290, 292
priming, 199
principles, 10, 13, 267
probability, 13, 16, 20, 49, 119
probability density function, 20
probe, 154, 184
production costs, 270, 271, 274, 280
project, 100, 105, 218, 272, 309
propagation, 57, 168, 260, 305
propane, 98
proportionality, 94, 263
propylene, 179
prostheses, 191, 197, 198, 199
prosthesis, 196
protection, 109, 152, 189, 280
protons, 174
pulp, 203, 207, 211
pulsed laser deposition techniques, viii, 113, 123, 136, 141, 145, 150, 157
purity, 4, 61, 90, 115, 124, 130, 145, 152, 223
PVC, 91, 298, 299
pyrolysis, ix, 122, 135, 202, 210, 213, 214, 215, 216
pyrolytic graphite, 81, 89, 92

Q

quality control, 204
quanta, 48
quantum confinement, 138
quantum mechanics, 48
quantum theory, 3
quantum well, 271
quartz, 19, 55, 69, 70, 71, 73, 93, 99, 124, 180

R

radial distribution, 29
radiation, vii, 1, 3, 33, 48, 52, 53, 54, 55, 58, 62, 68, 73, 74, 77, 88, 89, 113, 116, 124, 145, 151, 152, 262, 278
radius, 14, 19, 124, 131, 138, 149, 156
Raman scattering measurements, 131, 152
Raman spectra, 130, 134, 135
Raman spectroscopy, 130, 135, 141
rash, 175, 189
raw materials, 204
reactant, 277
reaction mechanism, 231, 235, 246
reaction medium, 204
reactions, 12, 215, 243, 250, 251, 252, 253, 257, 288, 294
reactive groups, 179
reactivity, 178, 198, 287
reality, 19, 24, 44
recognition, 130, 141
recrystallization, 78, 273
reflectance spectra, 136, 138
refraction index, 206
refractive index, ix, 50, 77, 136, 139, 143, 144, 158, 202, 205, 207, 218
refractive index variation, 143
regression line, 57
rehabilitation, 197
relaxation, 119, 208, 264
relevance, 115, 264, 286
reliability, 4, 64
relief, x, 169, 197, 297, 298, 299, 300, 302, 303, 305, 306, 307
renewable energy, 270
renormalization, 138
repair, viii, 167, 169, 178, 190, 191, 192, 193, 194, 196, 197, 198
repulsion, 256
requirements, 60, 169, 270, 286, 299
researchers, 266
residues, 228, 244, 287
resins, 196, 197
resistance, 2, 39, 114, 122, 168, 169, 260, 264, 274
resolution, 48, 124, 130, 131, 145, 146, 151, 152, 169, 212, 214
response, 114, 195, 253, 254, 263, 279
response time, 263

restoration, viii, 167, 169, 174, 178, 184, 185, 190, 196
restorative dentistry, 181
restorative materials, 168, 194
rheology, 35, 93
rings, 127, 128, 129, 131
room temperature, 2, 35, 49, 51, 53, 55, 57, 58, 59, 61, 62, 64, 65, 67, 68, 69, 71, 72, 73, 74, 75, 77, 78, 79, 80, 81, 82, 83, 84, 85, 87, 89, 90, 91, 92, 93, 98, 121, 124, 126, 129, 142, 145, 148, 151, 155, 185, 204, 298, 303
root(s), 123, 129, 141, 143, 158, 192
root-mean-square, 123, 129, 141, 143, 158
roughness, 53, 123, 124, 126, 129, 141, 143, 158, 159, 168, 174, 186, 191
rubber, 91, 175, 189, 298, 299, 300, 303
rules, 10
rutile, 43, 68, 69, 70, 116, 124, 145, 146, 147, 151, 152, 153, 155, 159, 291, 293

S

safety, 195, 209
salts, 151, 287
sawdust, 204
scaling law, 21
scandium, 177
scanning electron microscopy, 115, 118, 123
scatter, 53, 55, 85
scattering, 10, 19, 49, 57, 88, 155, 205, 206
science, vii, 55, 114, 190, 205, 294
scope, 7, 16, 21, 51, 55, 78, 81, 88
second generation, 270, 271, 280
security, 10, 20
sediment, 307
seed, x, 269, 276, 278, 279, 280, 281
segregation, 274
selected area electron diffraction, 123, 127, 131, 146, 153
self-consistency, 23, 243
self-similarity, 24
SEM micrographs, 291, 292
semiconductor, 114, 122, 135, 136, 138, 141, 148, 155, 254, 256, 266, 278, 289
semiconductors, 49, 55, 79, 81, 90, 91, 151, 157, 159, 255, 265
sensing, 114, 115, 116, 117, 121, 130, 135, 151, 157, 158, 159, 161
sensitivity, 114, 115, 121, 122, 134, 135, 158, 185, 211, 287
sensors, 114, 117, 144, 151, 158, 286

shape, 27, 28, 33, 36, 37, 44, 45, 46, 51, 57, 60, 95, 129, 131, 134, 135, 146, 153, 174, 227, 289, 299, 304, 308
sharp notch, 168
shear, 21, 171, 172, 173, 174, 176, 188, 189, 193, 194, 195, 196, 198, 199
shelf life, 180
shock, 2
shortness of breath, 189
showing, x, 115, 213, 246, 249, 276, 288, 289, 293
Si3N4, 43, 78, 81, 83, 84, 85, 86
side chain, x, 249
signals, 208, 211, 213, 214, 216, 217, 250
signal-to-noise ratio, 135
silane, viii, 167, 170, 177, 178, 179, 180, 181, 182, 183, 185, 186, 189, 190, 194, 195, 198
silane treatment, viii, 167, 170, 179, 185
silanol groups, 179, 180
silica, viii, 50, 55, 69, 71, 73, 74, 75, 76, 99, 115, 124, 167, 170, 173, 174, 176, 177, 178, 180, 183, 185, 186, 188, 189, 190, 191, 194, 196, 199, 272
silica coated aluminium oxide, viii, 167, 170, 178, 183, 185, 189, 190
silicon, vii, x, 46, 78, 81, 83, 84, 89, 90, 145, 149, 152, 156, 173, 178, 186, 189, 269, 270, 271, 272, 273, 274, 278
silver, 55, 90, 198
simulation, 16, 43
single crystals, vii, 1, 4, 5, 6, 7, 53, 55, 60, 62, 64, 67, 70, 77, 79, 80, 82, 83, 89, 92, 93
single-base gunpowders, ix, 201, 204
sintering, 37, 47, 57, 78, 83, 84
SiO2, 55, 69, 70, 71, 73, 76, 89, 93, 99, 145, 152, 173, 174, 180, 272
SiO2 films, 152
Size Exclusion Chromatography (SEC), ix, 202, 205, 206, 207, 218
skeleton, 6, 32, 51, 54, 243
slavery, 3
sludge, 216
smoothness, 128, 129, 143
sodium, 175
solar cells, viii, 113, 141, 150, 265, 266, 270, 271, 273, 274, 280
sol-gel, 115, 130, 141, 150, 288, 289, 293
solid matrix, 12, 19, 54
solid phase, x, 2, 10, 14, 32, 50, 51, 54, 56, 269, 272, 280
solid solutions, 88, 90
solid state, 3, 161, 210
solid surfaces, 51
solitons, 256
solubility, 12, 202, 207

solution, 7, 8, 23, 30, 33, 94, 115, 146, 151, 179, 180, 181, 186, 192, 203, 206, 207, 217, 250, 253, 260, 261, 276, 278, 303
solvation, 260
solvents, 204
species, 114, 115, 117, 122, 123, 125, 129, 151, 216, 250, 253, 254, 256, 257, 259, 260
specific heat, 48, 49, 50, 56
specific surface, 114, 289
spectroscopy, 132, 151, 152, 208, 213, 222, 275
speculation, 256
speed of response, 114
spin, 256, 278
spindle, 289
stability, 122, 135, 179, 209, 210, 211, 212, 217, 218, 260, 264, 273
stabilization, 115, 204
stabilizers, 286
standard deviation, 71
steel, 89, 99, 197, 204, 279
stimulus, 263, 264
stoichiometry, 141, 150, 156
storage, 135, 180, 195, 204, 207, 209, 213, 262
stress, x, 141, 169, 195, 217, 297, 301, 302, 303, 305, 306, 307, 308
stretching, 91, 298, 299, 300, 301, 303, 305
strong interaction, 78
strontium, 68
structural characteristics, 115, 255
substitutes, 271
substitution, 202, 203
substrate, xi, 78, 115, 116, 117, 118, 119, 120, 121, 123, 124, 135, 145, 148, 157, 179, 180, 262, 271, 272, 277, 278, 279, 280, 297, 298, 299, 300, 301, 302, 303, 304, 305, 306, 308
substrates, x, 83, 123, 124, 129, 131, 136, 141, 145, 152, 158, 194, 262, 269, 271, 273, 280
sulfate, 212
sulfur, 186, 204
sulfuric acid, 184, 203, 204
Sun, 39, 43, 68, 103, 110, 164, 187, 194, 270, 296
surface area, 122, 171, 286, 287, 288, 289, 293
surface chemistry, 123, 130, 141
surface energy, 37, 120
surface hardness, 168, 171, 186
surface layer, 132, 142
surface properties, 129
surface region, 132, 135, 142
surface structure, 300
surface tension, 186
surface treatment, vii, viii, 167, 169, 170, 171, 173, 176, 177, 182, 183, 184, 185, 186, 187, 188, 189, 190, 192, 193, 194, 195, 196, 198, 199

surfactant, 151, 155, 159
suspensions, 36
swelling, 175
symmetry, 5, 6, 8, 11, 17, 24, 45, 95, 111
synthesis, viii, x, 113, 130, 135, 141, 144, 148, 150, 151, 157, 203, 222, 249, 254, 255, 261, 288, 289, 290, 294

T

tantalum, 81
target, 115, 120, 123, 124, 136, 145, 148, 276, 277
techniques, viii, ix, x, 113, 115, 122, 123, 130, 135, 136, 141, 144, 145, 150, 151, 157, 158, 159, 162, 169, 174, 184, 185, 188, 191, 196, 202, 208, 218, 249, 261, 262, 263, 269, 273, 280, 285, 288, 292
technology(ies), 2, 3, 4, 16, 39, 43, 100, 114, 174, 222, 223, 236, 265, 270, 271, 272, 273, 280, 286, 288
teeth, 191, 194
teflon, 91
TEM, 40, 48, 124, 125, 127, 128, 130, 131, 141, 146, 147, 148, 152, 153, 154, 288
temperature dependence, vii, 1, 2, 49, 50, 51, 53, 55, 56, 57, 58, 59, 67, 72, 76, 77, 79, 82, 83, 84, 85, 88, 91, 117, 125, 138
tension, 174, 299, 301, 305, 306, 307
testing, 223, 242, 243
textbook, 2, 3
textbooks, 2, 3, 80
texture, viii, 73, 201, 202, 277
TGA, x, 285, 288, 292, 293
thermal analysis, x, 285, 288, 294
Thermal conductivity, vii, 1, 2, 57, 60, 61, 62, 63, 64, 65, 66, 67, 72, 74, 75, 79, 80, 82, 83, 84, 85, 86, 91, 92, 98
thermal decomposition, ix, 202, 210, 211, 212, 218, 221, 227, 228, 230, 231, 237, 243, 244, 287, 288, 289, 290, 293, 294
thermal degradation, ix, 202, 210, 211, 212, 213, 217, 293
thermal evaporation, 122, 130, 135, 141
thermal expansion, 168
thermal insulation applications, vii, 1
thermal properties, vii, ix, x, 39, 202, 205, 218, 266, 285
thermal resistance, 13, 38, 53
thermal stability, 112, 210, 211, 212, 264
thermal treatment, 207
thermodynamics, 5
thermogravimetric analysis, 293
ThermoGravimetry (TG), ix, x, 202, 210, 211, 212, 218, 223, 227, 228, 229, 230, 285, 288

thermogravimetry analysis (TGA), x, 285, 288, 292, 293
thermolysis, 207
thin films, vii, viii, 113, 114, 115, 116, 117, 118, 119, 121, 122, 123, 124, 125, 126, 127, 128, 129, 130, 131, 132, 133, 134, 135, 136, 137, 138, 139, 140, 141, 142, 143, 144, 150, 157, 158, 159, 162, 254, 262, 270, 271, 280
thin polycrystalline silicon (poly-Si), x, 269, 271, 272, 273, 274, 275, 276, 277, 278, 280
thin rigid film, xi, 297
three-dimensional space, 10
tin, vii, viii, 113, 116, 122, 123, 125, 132, 135, 136, 144, 146, 148, 150, 152, 155, 167, 183, 185, 186, 189, 190, 196, 197, 199
tin oxide, viii, 113, 117, 122, 123, 125, 135, 136, 146, 153, 185, 189, 196
tissue, 92, 175, 189
titania, 43, 57, 58, 68, 69, 293
titanium, 78, 81, 85, 90, 197, 262, 265, 274, 288, 293
toluene, 216
tooth, 171, 194
topology, 21, 32
tourmaline, 73
toxic gases, 114
transference, 231, 281
transistor, 145, 148, 149, 151, 152, 155, 156, 157, 159
transition metal oxides (TMOs), x, 285, 287, 288, 289, 294
transmission, 123, 124, 130, 131, 141, 145, 146, 151, 152, 154, 288
transmission electron microscopy, 123, 130, 141, 145, 146, 151, 152, 154
transmittance spectra, 136, 137
transparency, 73, 122, 135, 277
transport, 77, 112, 121, 130, 134, 141, 150, 204, 207, 252, 257, 260, 273
treatment, viii, 3, 5, 6, 16, 38, 78, 81, 151, 167, 169, 170, 171, 173, 176, 177, 179, 183, 185, 186, 187, 188, 190, 195, 198, 199, 203, 211, 213, 215, 216, 218, 254
treatment methods, 183
trial, 10
tungsten, 78, 81, 262
tungsten carbide, 78, 81
tunneling, 121, 144
twins, 130

U

uniform, 10, 23, 54, 94, 134, 145, 146, 151, 153, 159, 184, 273

uranium, 81
urethane, 286

V

vacancies, 48, 60, 132, 141, 142, 148, 151, 155
vacuum, 51, 52, 88, 115, 123, 124, 148, 149, 156, 277
valence, 139, 142, 255, 256, 257, 265, 290
vapor, 51, 98, 117, 144, 146, 148, 203, 213
vapor-liquid-solid, 144, 146, 148
variations, x, 129, 143, 203, 285, 287, 291
vector, 5, 94, 95
velocity, ix, 48, 49, 221, 223, 246, 308
vibration, 48, 189
viscosity, ix, 36, 51, 94, 155, 202, 203, 205, 206, 207, 218, 307
VLS, 144, 148
volatile organic compounds, 215
volatilization, 228, 246
vulnerability, 205

W

waste, 216
water, 51, 71, 92, 115, 174, 177, 179, 181, 182, 185, 195, 204, 209, 211, 278, 293, 304, 305, 307
water vapor, 71
wavelengths, 53
wear, 168, 197
weight loss, 293
weight ratio, 151
wettability, 180, 184
wetting, 169, 186
wide band gap, 135
windows, 51, 223, 263
wires, 144
wood, 92, 203, 206, 207
workers, 171, 173, 176, 177, 179, 184, 186, 188, 189, 293
worldwide, 270, 271, 280

X

X-ray diffraction (XRD), 115, 116, 117, 123, 124, 125, 130, 136, 141, 142, 145, 146, 147, 151, 152, 158

Y

yield, x, 25, 47, 73, 260, 285

ytterbium, 90
yttrium, 67, 68, 174, 177

Z

zinc, 79, 277

zinc oxide (ZnO), 57, 59, 70, 76, 150, 277
zirconia, 41, 43, 45, 46, 47, 50, 56, 57, 58, 60, 66, 68, 69, 70, 89, 108, 110, 169, 177, 178, 182, 188, 189, 190, 194, 196, 198
zirconium, 78, 81, 174